THE
HANDY
SCIENCE
ANSWER
BOOK

THE HANDY SCIENCE ANSWER BOOK

**Compiled by the Science and Technology Department
of the Carnegie Library of Pittsburgh**

VISIBLE INK PRESS

DETROIT   WASHINGTON D.C.   LONDON

THE HANDY SCIENCE ANSWER BOOK

Copyright © 1994 by The Carnegie Library of Pittsburgh

Published by Visible Ink Press™
a division of Gale Research Inc.
835 Penobscot Building
Detroit, MI 48226-4094

Visible Ink Press is a trademark of Gale Research Inc.

Most Visible Ink Press books are available at special quantity discounts when purchased in bulk by corporations, organizations, or groups. Customized printings, special imprints, messages, and excerpts can be produced to meet your needs. For more information, contact Special Markets Manager, Gale Research Inc., 835 Penobscot Bldg., Detroit, MI 48226. Or call 1-800-877-4253, extension 1033.

ISBN 0-8103-9451-0

Contents

PHYSICS AND CHEMISTRY

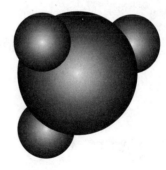

SPACE

EARTH

CLIMATE AND WEATHER

MINERALS AND OTHER MATERIALS

ENERGY

ENVIRONMENT

BIOLOGY

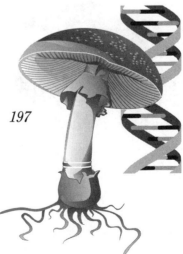

PLANT WORLD

ANIMAL WORLD

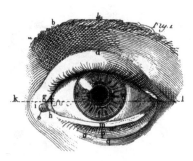

HUMAN BODY

HEALTH AND MEDICINE

WEIGHTS, MEASURES, TIME, TOOLS, AND WEAPONS

BUILDINGS, BRIDGES, AND OTHER STRUCTURES

BOATS, TRAINS, CARS, AND PLANES

COMMUNICATIONS

GENERAL SCIENCE AND TECHNOLOGY

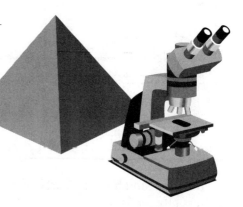

Introduction

We live in an age of science and technology, talking on cellular phones, cooking in microwave ovens, and watching our cable television. From the trivial to the complex, gizmos and whatchamacallits dominate modern life. Information whizzes by faster than a speeding bullet. Discoveries are made daily, hourly, by the minute. The horizons of knowledge are becoming mighty distant for many of us. That quizzical staring into space and idle scratching of the head are symptoms of the condition known as *information overload*. Though not life-threatening, it certainly is frustrating.

Wouldn't it be great if someone would just collect the answers to some of the mysteries of daily life? Like, just what is a syzygy? When will the sun die? Why is the sky blue? How deep is the ocean? What are PCBs? What is genetic engineering? When did the Ice Age begin? When did it end? Does rock music kill plants? Why do cats like catnip? How long do specific animals live? What is a male lobster called? Why do people snore? How much blood is in the average human body? How many miles of veins do I have? What is the purpose of goose-bumps? How often does the human eye blink? What is dead reckoning? Why do AM radio stations have a wider broadcast range at night? What is high-definition television? What is a computer "virus" and how is it spread?

The Handy Science Answer Book is just such a collection. Each year, the staff of the Science and Technology Department at The Carnegie Library of Pittsburgh receives nearly 100,000 reference questions and inquiries by the telephone, through the mail, and from personal in-house contact. Even though armed with 390,000 books, 425,000 bound periodicals, and thousands of government documents, technical reports, and microfiche, the Department finds that it requires a special file to answer these questions quickly and reliably. Some of this wealth of cumulative information has been in existence for at least 42 years—as far back as any of the current Carnegie staff can remember. It may even date back to 1905, the year when The Carnegie Library of Pittsburgh became the first major public library in the United States to establish a separate Science and Technology Department. The gems from this ready reference file are shared with you in *The Handy Science Answer Book*.

Handy Science provides answers to more than 1,200 unusual, interesting, or frequently asked questions in the areas of science, pseudo science, and technology. More than

100 illustrations and many tables augment the text. In some ways, science touches so much of our lives—whether it be our environment, our homes, our workplaces, right down to our physical bodies themselves—that it can become difficult to categorize what actually constitutes science. *Handy Science* makes no particular effort to restrict the questions to pure science, but focuses on those questions that have achieved noteworthiness either through their popularity, the time-consuming nature of their research, or their uniqueness.

The Carnegie staff has verified figures and dates to the best of their ability. Keep in mind that even in science, figures can seem to be in conflict; many times such discrepancies may be attributable to the authority perspective, or more commonly, to the results of simple mathematical rounding of figures. Occasionally, the figure or date listed is a consensus of the consulted sources; other times the discrepancy is noted and alternatives given. *Handy Science* rounds off figures whenever it seems that such precision is unnecessary.

The answers are written in non-technical language and provide either a succinct response or a more elaborate explanation, depending on the nature of the question. Definitions to scientific terminology are given within the answer itself, and both metric and U.S. customary measurements are listed. Following the main Q&A section are suggestions for further reading (most of which were used to answer various questions) and the index.

So, when you're stuck wondering why the clothes you tried on in the store under florescent light look so much different in the light of day, reach for *Handy Science*. Or if you suddenly need to know the lifespan of a chimpanzee (51 years maximum) or the number of horses in the world (65,292,000) or how many muscles it takes to produce a smile (17) or a frown (43), reach for *Handy Science*. You'll find it downright handy.

Acknowledgments

Many people have made significant contributions to *The Handy Science Answer Book* from its initial stage of conception to its realization as a reference work. Margery Peffer kept the work going day after day for almost seven months, as well as answering hundreds of questions, checking leads, verifying sources, assisting other staff members, and generally being one of the finest science reference librarians that I have ever worked with and have known.

The staff librarians in the Science and Technology Department of The Carnegie Library of Pittsburgh know how much I have appreciated both their individual and collective efforts in gathering, reviewing, answering, revising, and verifying many more questions than the number included in this volume. I want to thank Joan Anderson, Naomi Balaban, Jan Comfort, Diane Eldridge, Susan Horvath, Dorothy Melamed, Kristine Mielcarek, David Murdock, Gregory Pomrenke, Gertrude Ross, and Donna Strawbridge for their sustained endeavors. These librarians were remarkable at balancing the never-ending needs of our patrons with the frequent deadlines required for submitting batches of questions. Among those who need to be thanked are Marilyn Macevic, who typed most of the questions in the first draft; Helen Yee, who formatted every question and typed the final copy; Mary Shields, who read and proofed most of the questions, verified citations, and solved numerous problems relating to the citations.

Staff members throughout The Carnegie Library of Pittsburgh system submitted numerous questions for possible inclusion in this book. I want to thank the following for their response to my request: Ernestine Audin, Judy Beczak, Elaine Bisiada, Dallas Clautice, Lisa Dennis, Clare DiDominicis, Cheryl Engel, Phoebe Ferguson, Constance J. Galbraith, Mary Ellen Gildroy, Mildred Glenn, Esther Hamiel, Kathy Herrin, Thomas Holmes, Meeghan Humphrey, Sheila Jackson, Patricia James, Nancy Jessup, Andrea Jones, Amy Korman, Rebecca Kosanovich, Pamela Kuchta, Kathryn Logan, Mary Long, James Lutton, Stephanie Messina, Charmaine Spaniel Mozlack, David Murray, Anne New, Julianna Posch, Barbara Preusser, Barbara Rogers, Barbara Rush, Marsha Schafer, Deborah Silverman, Christine Stein, Viktoria Strod, Helene Tremaine, Pamela Van Eman, and Gloria Zuckerman.

I thank Robert B. Croneberger, Director of The Carnegie Library of Pittsburgh, for initially approving this project and signing the contract and Joseph Falgione, Gladys Shapera, and Loretta O'Brien, senior library administrators at The Carnegie Library of Pittsburgh, who offered encouragement as well as the submission of various questions.

Finally, thanks to my wife, Sandi, and sons, Andrew and Michael, for their patience and understanding as well as for their interesting discussions of the various questions and answers that were "hot off the press."

James E. Bobick
Head, Science and Technology Department
The Carnegie Library of Pittsburgh

THE
HANDY
SCIENCE
ANSWER
BOOK

PHYSICS AND CHEMISTRY

ENERGY, MOTION, FORCE, AND HEAT

See also: Energy

How is "absolute zero" defined?

Absolute zero is the theoretical temperature at which all substances have zero thermal energy. Originally conceived as the temperature at which an ideal gas at constant pressure would contract to zero volume, absolute zero is of great significance in thermodynamics and is used as the fixed point for absolute temperature scales. Absolute zero is equivalent to 0°K, -459.69°F, or -273.16°C.

The velocity of molecules of substances determines their temperature; the faster they move, the more volume they require, and the higher the temperature becomes. The lowest actual temperature ever reached was two-billionth of a degree above absolute zero (2 x 10-9 Kelvin) by a team at the Low Temperature Laboratory in the Helsinki University of Technology, Finland, in October, 1989.

Does hot water freeze faster than cold?

A bucket of hot water will not freeze faster than a bucket of cold water. However, a bucket of water that has been heated or boiled, then allowed to cool to the same temperature as the bucket of cold water, may freeze faster. Heating or boiling drives out some of the air bubbles in water; because air bubbles cut down thermal conductivity, they can inhibit freezing. For the same reason, previously heated water forms denser ice than unheated water, which is why hot-water pipes tend to burst before cold-water pipes.

1

What is **superconductivity**?

Superconductivity is a condition occurring in many metals, alloys, etc., usually at low temperatures, involving zero electrical resistance and perfect diamagnetism. In such a material an electric current will persist indefinitely without any driving voltage, and applied magnetic fields are exactly cancelled out by the magnetization they produce. Superconductivity was discovered by Heike Kamerlingh Onnes (1853–1926) in 1911.

What is **inertia**?

Inertia is a tendency of all objects and matter in the universe to stay still, or if moving, to continue moving in the same direction, unless acted on by some outside force. This forms the first law of motion formulated by Isaac Newton (1642–1727). To move a body at rest, enough external force must be used to overcome the object's inertia; the larger the object is, the more force is required to move it. In his Philosophae Naturalis Principia Mathematica, published in 1687, Newton sets forth all three laws of motion. Newton's second law is that the force to move a body is equal to its mass times its acceleration ($F = MA$), and the third law states that for every action there is an equal and opposite reaction.

Why do golf balls have dimples?

The dimples minimize the drag (a force that makes a body lose energy as it moves through a fluid), allowing the ball to travel further than a smooth ball would. In a dimpled ball the air, as it passes a ball, tends to cling to the ball longer, reducing the eddies or wake effect that drain the ball's energy. Hit with the same force, the dimpled ball will travel up to 300 yards (275 meters) but the smooth ball only goes 70 yards (65 meters). A ball can have 300 to 500 dimples, each of which can be 0.01 inch (0.25 millimeters) deep. Another effect to get distance is to give the ball a backspin. With a backspin there is less air pressure on the top of the ball, so the ball stays aloft longer (much like an airplane).

Who is the father of the science of **magnetism**?

The English scientist William Gilbert (1544–1603) regarded the Earth as a giant magnet, and investigated its field in magnetism terms of dip and variation. He explored many other magnetic and electrostatic phenomena. The Gilbert (symbol Gb), a unit of magnetism, is named for him.

Why is **Van Vleck** considered to be one of the fathers of modern magnetic theory?

Awarded the Nobel Prize in 1977, John H. Van Vleck (1899–1980), an American physicist, made significant contributions in the field of magnetism. He explained the magnetic, electrical, and optical properties of many elements and compounds with the ligand field theory, demonstrated the effect of temperature on paramagnetic materials (called Van Vleck paramagneticism), and developed a theory on the magnetic properties of atoms and their components.

What is **Maxwell's demon**?

An imaginary creature who, by opening and shutting a tiny door between two volumes of gases, could, in principle, concentrate slower molecules in one (making it colder) and faster molecules in the other (making it hotter), thus breaking the second law of thermodynamics. This law states that heat cannot by its own accord flow from a colder to a hotter body. The hypothesis was formulated in 1871 by James C. Maxwell (1831–1879), who is considered to be the greatest theoretical physicist of the 19th century. Maxwell discovered that light consists of electro-magnetic waves, proved the nature of Saturn's rings, set forth the principles of color vision and established the kinetic theory of gases, wherein the nature of heat resides in the motion of the molecules.

When was **spontaneous combustion** first recognized?

Spontaneous combustion is the ignition of materials stored in bulk. This is due to internal heat build-up caused by oxidation (generally a reaction in which electrons are lost, specifically when oxygen is combined with a substance, or when hydrogen is removed from a compound). Because this oxidation heat cannot be dissipated into the surrounding air, the temperature of the material rises until the material reaches its ignition point and bursts into flame.

A Chinese text written before 290 C.E. recognized this phenomenon in a description of the ignition of stored oiled cloth. The first Western recognition of spontaneous combustion was by J.P.F. Duhamel in 1757, when he discussed the gigantic conflagration of a stack of oil-soaked canvas sails drying in the July sun. Before spontaneous combustion was recognized, such events were usually blamed on arsonists.

What is **phlogiston**?

Phlogiston was a name used in the 18th century to identify a supposed substance given off during the process of combustion. The phlogiston theory was developed in the early 1700s by the German chemist and physicist, Georg Ernst Stahl (1660–1734).

In essence, Stahl held that combustible material such as coal or wood was rich in a material substance called "phlogiston." What remained after combustion was without phlogiston and could no longer burn. The rusting of metals also involved a transfer of phlogiston. This accepted theory explained a great deal previously unknown to chemists. For instance, metal smelting was consistent with the phlogiston theory. Charcoal in burning lost weight. Thus the loss of phlogiston either decreased or increased weight.

The French chemist Antoine Lavoisier (1743–1794) demonstrated that the gain of weight when a metal turned to a calx was just equal to the loss of weight of the air in the vessel. Lavoisier also showed that part of the air (oxygen) was indispensable to combustion, and that no material would burn in the absence of oxygen. The transition from Stahl's phlogiston theory to Lavoisier's oxygen theory marks the birth of modern chemistry at the end of the 18th century.

Does **water running down a drain** rotate in a different direction in the Northern and Southern Hemispheres?

If water runs out from a perfectly symmetrical bathtub, basin, or toilet bowl, in the Northern Hemisphere it would swirl counterclockwise; in the Southern Hemisphere, the water would run out clockwise. This is due to the Coriolis effect (the Earth's rotation influencing any moving body of air or water). However, some scientists think that the effect does not work on small bodies of water. Exactly on the equator, the water would run straight down.

What is a **Leyden jar**?

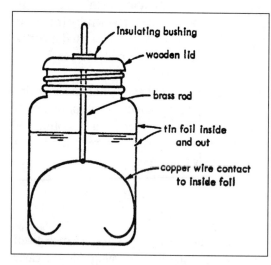

Insulating bushing
wooden lid
brass rod
tin foil inside and out
copper wire contact to inside foil

A Leyden jar, the earliest form of capacitor, is a device for storing an electrical charge. First described in 1745 by E. Georg van Kleist (c.1700–1748), it was also used by Pieter van Musschenbroek (1692–1761), a professor of physics at the University of Leyden. The device came to be known as a Leyden jar and was the first device that could store large amounts of electric charge. The jars contained an inner wire electrode in contact with water, mercury, or wire. The outer electrode was a human hand holding the jar. An improved version coated the jar inside and outside with separate metal foils with the inner foil connected to a conducting rod and terminated in a conducting sphere. This eliminated the need for the liquid electrolyte. In use, the jar was normally charged from an electrostatic generator. The Leyden jar is still used for classroom demonstrations of static electricity.

LIGHT, SOUND, AND OTHER WAVES

What is the **speed of light**?

In round numbers, the speed of light in a vacuum is 186,000 miles (300,000 kilometers) per second. The exact figure is 186,282 miles (299,792.458 kilometers) per second.

What are the **primary colors** in light?

Color is determined by the wavelength of its light (the distance between one crest of the light wave and the next). Those colors that blend to form "white light" are from shortest wave length to longest: red, orange, yellow, green, blue, indigo, and violet. All these monochromatic colors, except indigo, occupy large areas of the spectrum (entire range of wavelengths produced when a beam of electromagnetic radiation is broken up). These colors can be seen when a light beam is refracted through a prism. Some consider the primary colors to be six monochromatic colors that occupy large areas of the spectrum: red, orange, yellow, green, blue, and violet. Many physicists recognize three primary colors: red, yellow, and blue; or red, green, and blue; or red, green, and violet. All other colors can be made from these by adding two primary colors in various proportions. Within the spectrum, scientists have discovered 55 distinct hues. Infrared and infra-violet rays at each end of the spectrum are invisible to the human eye.

Why does the color of clothing appear different in direct light than it does in a store under **fluorescent light**?

White light is a blend of all the colors, and each color has a different wavelength. Although sunlight and fluorescent light both appear as "white light," they each contain slightly different mixtures of these varying wavelengths. The mixture of wavelengths determines the color perceived. When sunlight and fluorescent light (white light) are absorbed by the piece of clothing, only some of the wavelengths (composing white light) reflect from the clothing. When the retina of the eye perceives the "color" of the clothing, it is really perceiving these reflected wavelengths.

What were **Anders Angstrom's** contributions to the development of spectroscopy?

Swedish physicist and astronomer, Anders Jonas Angstrom (1814–1874), was one of the founders of spectroscopy. His early work provided the foundation for spectrum analysis (analysis of the ranges of electromagnetic radiation emitted or absorbed). He investigated the sun spectra as well as that of the Aurora Borealis. In 1868, he established measurements for wavelengths of greater than 100 Frauenhofer. In 1907, the angstrom (Å, equal to 10^{-10m}), a unit of wavelength measurement, was officially adopted.

Why was the **Michelson-Morley** experiment important?

This experiment on light waves, first carried out in 1881 by physicists Albert A. Michelson (1852–1931) and E.W. Morley (1838–1923) in the United States, is one of the historically significant experiments in physics and led to the development of Einsteins's theory of relativity. The original experiment, using the Michelson interferometer, attempted to detect the velocity of the Earth with respect to the hypothetical "luminiferous ether," a medium in space proposed to carry light waves. The procedure measured the speed of light in the direction of the Earth and the speed of light at right angles to the Earth's motion. No difference was found. This result discredited the ether theory and ultimately led to the proposal by Albert Einstein (1879–1955) that the speed of light is a universal constant.

Why does a **double sonic boom** occur when the space shuttle enters the atmosphere?

As long as an airborne object, such as a plane, is moving below the speed of sound (called Mach 1), the disturbed air remains well in front of the craft. But as the craft passes Mach 1 and is flying at supersonic speeds, a sharp air pressure rise occurs in front of the craft. In a sense the air molecules are crowded together and collectively impact. What is heard is a claplike thunder called a sonic boom or a supersonic bang. There are many shocks coming from a supersonic aircraft but these shocks usually combine to form two main shocks, one coming from the nose of the aircraft and the other from the aft. Each of the shocks moves at different velocities. If the time difference between the two shock waves is greater than 0.10 seconds apart, two sonic booms will be heard. This usually occurs when the aircraft ascends quickly. If the aircraft ascends more slowly, the two booms will sound like only one boom to the observer.

What causes the **sounds heard in a seashell**?

When a seashell is held to an an ear, the sounds heard are ambient, soft sounds that have been resonated and thereby amplified by the seashell's cavity. The extreme sensitivity of the human ear to sound is illustrated by the seashell resonance effect.

What is the **Doppler effect**?

The Austrian physicist, Christian Doppler (1803–1853) in 1842 explained the phenomenon of the apparent change in wavelength of radiation—such as sound or light—emitted either by a moving body (source) or by the moving receiver. The frequency of the wave-lengths increases and the wavelength becomes shorter as the moving source approaches, producing high-pitched sounds and bluish light (called blue shift). Likewise as the source recedes from the receiver the frequency of the wavelengths decreases, the sound is pitched lower and light appears reddish (called red shift). This Doppler effect is commonly demonstrated by the train whistle of an approaching train or jet aircraft.

There are three differences between acoustical (sound) and optical (light) Doppler effects: The optical frequency change is not dependent on which is moving— the source or observer—nor is it affected by the medium through which the waves are moving, but acoustical frequency is affected by such conditions. Optical frequency changes are affected if the source or observer moves at right angles to the line connecting the source and observer. Observed acoustical changes are not affected in such a situation. Applications of the Doppler phenomenon include the Doppler radar and the measurement by astronomers of the motion and direction of celestial bodies.

What is the sound frequency of the **musical scale**?

EQUAL TEMPERED SCALE

Note	Frequency	Note	Frequency
C♭	261.63	G	392.00
C#	277.18	G#	415.31
D	293.67	A	440.00
D#	311.13	A#	466.16
E	329.63	B	493.88
F	349.23	C n	523.25
F#	369.99		

Notes: ♭ indicates flat; # indicates sharp; n indicates return to natural.

The lowest frequency distinguishable as a note is about 30 hertz. The highest audible frequency is about 40,000 hertz. A hertz (symbol Hz) is a unit of frequency that measures the number of the wave cycles per second frequency of a periodic phenomenon whose periodic time is one second (cycles per second).

What is the **speed of sound**?

The speed of sound is not a constant; it varies depending on the medium in which it travels. The measurement of sound velocity in the medium of air must take into account many factors, including air temperature, pressure, and purity. At sea level and 32°F (0°C), scientists do not agree on a standard figure; estimates range from 740 to 741.5 miles (1191.6 to 1193.22 kilometers) per hour. As air temperature rises, sound velocity increases. Sound travels faster in water than in air and even faster in iron and steel. Sounds traveling a mile in 5 seconds in air, will travel the same distance in 1 second underwater and travel ⅓ of a second in steel.

What are the characteristics of **alpha, beta, and gamma radiation?**

Radiation is a term that describes all the ways energy is emitted by the atom as x-rays, gamma rays, neutrons, or as charged particles. Most atoms, being stable, are nonradioactive; but some are unstable and give off either particles or gamma radiation. Substances bombarded by radioactive particles can become radioactive and yield alpha particles, beta particles, and gamma rays.

Alpha particles, first identified by Antoine Henri Becquerel (1852–1908), have a positive electrical charge and consist of two protons and two neutrons. Because of their great mass, alpha particles can travel only a short distance, around 5 centimeters in air, and can be stopped by a sheet of paper.

Beta particles, identified by Ernest Rutherford (1871–1937), are extremely high-speed electrons or protons that move at the speed of light. They can travel far in air and can pass through solid matter several millimeters thick.

Gamma rays, identified by Marie (1867–1934) and Pierre Curie (1859–1906), are similar to x-rays, but usually have a shorter wave length. These rays, which are bursts of protons, or very short–wave electromagnetic radiation, travel at the speed of light. They are much more penetrating than either the alpha or beta particles and can go through 7 inches of lead.

MATTER

What is the **fourth state of matter?**

Plasma, a mixture of free electrons and ions or atomic nuclei, is sometimes referred to as a "fourth state of matter." Plasmas occur in thermonuclear reactions as in the sun, in fluorescent lights, and in stars. When gas temperature is raised high enough the collision of atoms become so violent that electrons are knocked loose from their nuclei. The result of a gas having loose negatively charged electrons and heavier positively charged nuclei is called a plasma.

All matter is made up of atoms. Animals and plants are organic matter; minerals and water are inorganic matter. Whether matter appears as a solid, liquid, or gas depends on how the molecules are held together in their chemical bonds. Solids have a rigid structure in the atoms of the molecules; in liquids the molecules are close together but not packed; in a gas, the molecules are widely spaced and move around, occasionally colliding but usually not interacting. These states—solid, liquid, and gas—are the first three states of matter.

Who is generally regarded as the discoverer of the **electron**?

The British physicist, Sir Joseph John Thomson (1856–1940), in 1897 researched electrical conduction in gases, which led to the important discovery that cathode rays consisted of negatively charged particles called electrons. The discovery of the electron inaugurated the electrical theory of the atom, and this with other work entitled Thomson to be regarded as the founder of modern atomic physics.

How did the **quark** get its name?

This mathematical particle, considered to be the fundamental unit of matter, was named by Murray Gell-Mann (b. 1929) an American theoretical physicist and Nobel Prize winner. Its name was initially a playful tag that Gell-Mann invented, sounding something like "kwork". Later Gell-Mann came across the line "Three quarks for Master Marks" in James Joyce's *Finnegan's Wake,* and the tag became known as a quark. There are six kinds or "flavors" (up, down, strange, charm, bottom, and top) of quarks, and each "flavor" has 3 varieties or "colors" (red, blue, and green). All eighteen types have different electric charges (a basic characteristic of all elementary particles). Three quarks form a proton (having one unit of positive electric charge) or a neutron (zero charge), and two quarks (a quark and an antiquark) form a meson. Like all known particles, a quark has its anti-matter opposite, known as an antiquark (having the same mass but opposite charge).

What substance, other than water, is **less dense as a solid** than as a liquid?

Only bismuth and water share this characteristic. Density (the mass per unit volume or mass/volume) refers to how compact or crowded a substance is. For instance the density of water is 1 gram/cm^3 or 1 kg/l (kilogram/per liter); the density of a rock is 3.3 g/cm^3; pure iron is 7.9 g/cm^3; and the Earth (as a whole) 5.5 g/cm^3 (average). Water as a solid (i.e., ice) floats; which is a good thing, otherwise ice would sink to the bottom of every lake or stream, which would result in great flooding.

9

Why is **liquid water** more dense than ice?

Pure liquid water is most dense at 39.2°F (3.98°C) and decreases in density as it freezes. The water molecules in ice are held in a relatively rigid geometric pattern by their hydrogen bonds, producing an open, porous structure. Liquid water has fewer bonds, therefore; more molecules can occupy the same space, making liquid water more dense than ice.

What does **half-life** mean?

Half-life is the time it takes for the number of radioactive nuclei originally present in a sample to decrease to one half of their original number. Thus, if a sample has a half-life of one year, its radioactivity will be reduced to half its original amount at the end of a year and to one quarter at the end of two years. The half-life of a particular radionuclide is always the same, independent of temperature, chemical combination, or any other condition.

Who made the **first organic compound to be synthesized** from inorganic ingredients?

Urea was synthesized by Friedrich Wohler (1800–1882) in 1828 from ammonia and cyanic acid. This synthesis dealt a deathblow to the vital-force theory. This theory assumed the existence of a mysterious "Vital Force," which intervened in the formation of compounds so that their preparation in the laboratory could hardly be expected.

Who is known as the father of **crystallography**?

The French priest and mineralogist, Rene-Just Haüy (1743-1822), is called the father of crystallography. In 1781 Haüy had a fortunate accident when he dropped a piece of calcite and it broke into small fragments. He noticed that the fragments broke along straight planes that met at constant angles. He hypothesized that each crystal was built up of successive additions of what is now called a unit cell to form a simple geometric shape with constant angles. An identity or difference in crystalline form implied an identity or difference in chemical composition. This was the beginning of the science of crystallography.

What is a **chemical garden** and how is one made?

Mix 4 tablespoons of bluing, 4 tablespoons of salt, and 1 tablespoon household ammonia. Pour this mixture over pieces of coal or brick in a suitable dish or bowl. Put several drops of red or green ink or mercurochrome on various parts of the coal and leave undisturbed for several days.

A crystal garden—a dishful of crystals that grow like plants and look like coral—will begin to appear. How soon the crystals will begin to appear depends on the temperatures and humidity in the room. Before long, crystals will be growing all over the briquettes, on the side of the dish, and down onto the plate. The crystals will be pure white with a snow-like texture.

CHEMICAL ELEMENTS, ETC.

See also: Minerals and Other Materials

Why is **Berzelius** considered one of the founders of science in its modern form?

Swedish chemist Jöns Jakob Berzelius (1779–1848) devised chemical symbols, determined atomic weights, contributed to the atomic theory, and discovered several new elements. Between 1810 and 1816, he described the preparation, purification, and analysis of 2,000 chemical compounds. Then he determined atomic weights for 40 elements. He simplified chemical symbols, introducing a notation (still used today)—letters with numbers—that replaced the pictorial symbols his predecessors used. He discovered cerium (in 1803, with Wilhelm Hisinger), selenium (1818), silicon (1824), and thorium (1829).

Who developed the **periodic table**?

Dmitri Ivanovich Mendeleyev (1834-1907) was a Russian chemist whose name will always be linked with his outstanding achievement, the development of the periodic table. He was the first chemist really to understand that all elements are related members of a single ordered system. He changed what had been a highly fragmented and speculative branch of chemistry into a true, logical science. His nomination for the 1906 Nobel Prize for Chemistry failed by one vote, but his name became recorded in perpetuity 50 years later when element 101 was called mendelevium.

According to Mendeleyev, the properties of the elements, as well as those of their compounds, are periodic functions of their atomic weights (in the 1920s, it was discovered that atomic number was the key rather than weight). Mendeleyev compiled the first true periodic table listing all the 63 (then-known) elements. In order to make the table work, Mendeleyev had to leave gaps, and he predicted that further elements would eventually be discovered to fill them. Three were discovered in Mendeleyev's lifetime: gallium, scandium, and germanium.

There are 94 naturally occurring elements; of the 15 remaining elements (elements 95 to 109), 10 are undisputed. By 1984, over 6.8 million chemical compounds had been produced from these elements; 65,000 of them in common use.

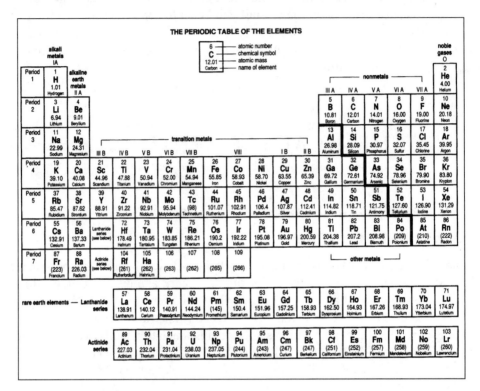

What is meant by **Group I elements**?

These are the elements at the left of the periodic table: lithium (Li, element 3), potassium (K, element 19), rubidium (Rb, element 37), cesium (Cs, element 55), francium (Fr, element 87), and sodium (Na, element 11). They are alkali metals, sometimes called the sodium family of elements. Because of their great chemical reactivity (easily form positive ions), none exist in nature in the elemental state.

Which elements are called **Group II elements**?

These are beryllium (Be, element 4); magnesium (Mg, element 12); calcium (Ca, element 20); strontium (Sr, element 38); barium (Ba, element 56); and radium (Ra, element 88). They are known as the alkaline-Earth metals. Like the alkali metals, they are never found as free elements in nature and are moderately reactive metals. Harder and less volatile than the alkali metals (Group I), these elements all burn in air.

What are the **transition elements?**

The transition elements are the 10 subgroups of elements between group 2 and group 13, starting with period 4. They include gold (Au, element 79), silver (Ag, element 97), platinum (Pt, element 78), iron (Fe, element 26), copper (Cu, element 29), and other metals. The transition elements or metals are so named because they represent a gradual shift from the strongly electropositive elements of Group I and II to the electronegative elements of Groups VI and VII.

What are the names and numbers of the **transuranic chemical elements?**

These elements include 17 established artificial elements. They have atomic numbers higher than that of uranium (U, element 92), which is the last stable element of the periodic system.

Element number	Name	Symbol
93	Neptunium	Np
94	Plutonium	Pu
95	Americium	Am
96	Curium	Cm
97	Berkelium	Bk
98	Californium	Cf
99	Einsteinium	Es
100	Fermium	Fm
101	Mendelevium	Md
102	Nobelium	No
103	Lawrencium	Lr
104	Unnilquadium	Unq
105	Unnilpentium	Unp
106	Unnilhexium	Unh
107	Unnilseptium	Uns
108	Unniloctium	Uno
109	Unnilennium	Une

Which elements are the **"noble metals"?**

The noble metals are gold (Au, element 79), silver (Ag, element 47), mercury (Hg, element 80), and the platinum (Pt, element 78) group (including palladium (Pd, element 46), iridium (Ir, element 77), rhodium (Rh, element 45), ruthenium (Ru, element 44), and osmium (Os, element 76)). The term refers to those metals highly resistant to chemical reaction or oxidation (resistant to corrosion) and is contrasted to "base" metals which are not so resistant. The term has its origins in ancient alchemy whose goals of transformation and perfection were pursued through the different properties of metals and chemicals. The term is not synonymous with "precious metals," although a metal, like platinum, may be both.

What were the **proposed names for elements 104 and 105** in use before the official names were chosen?

Rutherfordium (Rf) and kurchatovium (Ku) were the names proposed by rival groups for element 104 (unnilquadium (Unq)). Hahnium (Ha) and nielsbohrium (Ns) were proposed for element 105 (unnilpentium (Unp)).

Which elements are **liquid at room temperature?**

Mercury ("liquid silver", Hg, element 80) and bromine (Br, element 35) are liquid at room temperature 68° to 70°F (20° to 25°C). Gallium (Ga, element 31) with a melting point of 85.6°F (29.8°C) and cesium (Cs, element 55) with a melting point of 83°F (28.4°C), are liquids at slightly above room temperature.

Which compound contains the **most elements?**

Chemical Abstracts Service searched its database to find the compound with the most elements—ten.

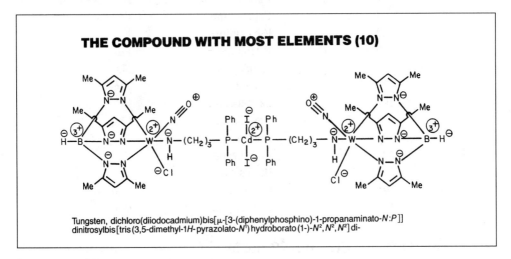

THE COMPOUND WITH MOST ELEMENTS (10)

Tungsten, dichloro(diiodocadmium)bis[μ-[3-(diphenylphosphino)-1-propanaminato-N:P]] dinitrosylbis[tris(3,5-dimethyl-1H-pyrazolato-N^1) hydroborato (1-)-N^2, N^2, N^2] di-

What is **Harkin's rule?**

Atoms having even atomic numbers are more abundant in the universe than are atoms having odd atomic numbers. Chemical properties of an element are determined by its atomic number, which is the number of protons in the atom's nucleus or the number of electrons revolving around this nucleus.

Which chemical element is the **most abundant in the universe?**

Hydrogen (H, element 1) makes up about 75% of the mass of the universe. It is estimated that more than 90% of all atoms in the universe are hydrogen atoms. Most of the rest are helium (He, element 2) atoms.

Which chemical element is the **most abundant on Earth?**

Oxygen (O, element 8) is the most abundant element in the Earth's crust, waters, and atmosphere. It composes 49.5% of the total mass of these compounds.

What is the **second most abundant** chemical element on Earth?

Silicon (Si, element 14) is the second most abundant element. Silicon dioxide and silicates make up about 87% of the materials in the Earth's crust.

Which elements have the **most isotopes?**

The elements with the most isotopes, with 36 each, are xenon (Xe) with 9 stable isotopes (identified from 1920 to 1922) and 27 radioactive isotopes (identified from 1939 to 1981), and cesium (Cs) with one stable isotope (identified in 1921) and 35 radioactive isotopes (identified from 1935 to 1983).

The element with the least number of isotopes is hydrogen (H), with three isotopes, including two stable ones—protium (identified in 1920) and deuterium (identified in 1931)—and one radioactive isotope—tritium (first identified in 1934, but later considered a radioactive isotope in 1939).

Why are the **rare Earth elements** "rare"?

Elements numbered 58 through 71 in the periodic table plus yttrium (Y, element 39) and thorium (Th, element 90) are called "rare earths" because they are difficult to extract from monazite ore, where they occur. The term has nothing to do with scarcity or rarity in nature.

MEASUREMENT, METHODOLOGY, ETC.

Who was the **father of analytical chemistry?**

René Descartes' (1596–1650) reform of science was the establishment of a mechanical philosophy, which sought to explain the properties and actions of bodies in terms of the parts of which they are composed. He proposed the use of mathematical methods to investigate scientific problems. Descartes rejected all untested ancient and medieval authority as opinion that must be subjected to objective scientific analysis to test its validity. Honored in physics and mathematics, he is also considered one of the foremost philosophers of the modern era. Descartes believed in a firm unity of all knowledge, scientific and philosophical, which he symbolized in the metaphor of a tree whose roots are metaphysics, whose trunk is physics, and whose branches are specific topics (such as medicine, mechanics, and morality).

Who is considered the **father of modern chemistry?**

Several contenders share this honor. Robert Boyle (1627–1691), a British natural philosopher, is considered one of the founders of modern chemistry. Best known for his discovery of Boyle's Law (volume of a gas is inversely proportional to its pressure at constant temperature), he was a pioneer in the use of experiments and the scientific method. A founder of the Royal Society, he worked to remove the mystique of alchemy from chemistry to make it a pure science.

The French chemist, Antoine-Laurent Lavoisier (1743–1794) is regarded as another founder of modern chemistry. His wide-ranging contributions include the discrediting of the phlogiston theory of combustion, which had been for so long a stumbling block to a true understanding of chemistry. He established modern terminology for chemical substances and did the first experiments in quantitative organic analysis. He is sometimes credited with having discovered or established the law of conservation of mass in chemical reactions.

Where is Albert Einstein's brain?

Einstein was cremated after his death on April 18, 1955. The doctor who performed the autopsy removed Einstein's brain, part of which now rests in a bottle somewhere in Weston, Missouri. Albert Einstein (1879–1955) was the principal founder of modern theoretical physics; his theory of relativity (speed of light is a constant and not relative to the observer or source of light), and the relationship of mass and energy ($E = MC^2$), fundamentally changed human understanding of the physical world. His stature as a scientist, together with his strong humanitarian stance on major political and social issues, made him one of the outstanding men of the twentieth century.

Who is generally regarded as the father of quantum mechanics?

The German mathematical physicist, Werner Karl Heisenberg (1901-1976), is regarded as the father of quantum mechanics (theory of small-scale physical phenomena). His theory of uncertainty in 1927 overturned traditional classical mechanics and electromagnetic theory regarding energy and motion when applied to sub-atomic particles such as electrons and parts of atomic nuclei. The theory states that it is impossible to specify precisely both the position and the simultaneous momentum (mass x volume) of a particle, but they could only be predicted. This meant that a result of an action can only be expressed in terms of probability that a certain effect will occur, not certainty.

What is the **Kelvin** temperature scale?

Temperature is the level of heat in a gas, liquid, or solid. The freezing and boiling points of water are used as standard reference levels in both the metric (centigrade or Celsius) and the English system (Fahrenheit). In the metric system, the difference between freezing and boiling is divided into 100 equal intervals called degree Celsius or degree centigrade (°C). In the English system, the intervals are divided into 180 units, with 1 unit called degree Fahrenheit (°F). But temperature can be measured from absolute zero (no heat, no motion); this principle defines thermodynamic temperature and establishes a method to measure it upward. This scale of temperature is called the Kelvin temperature scale, after its inventor, William Thomson, Lord Kelvin (1824–1907), who devised it in 1848. The Kelvin (symbol K) has the same magnitude as the degree Celsius (the difference between freezing and boiling water is 100 degrees), but the two temperatures differ by 273.15 degrees (absolute zero, which is -273.15°C on the Celsius scale). Below is a comparison of the three temperatures:

Characteristic	K°	C°	F°
Absolute zero	0	-273.15	-459.67
Freezing point of water	273.15	0	32
Normal human body temperature	310.15	37	98.6
Boiling point of water	373.15	100	212

To convert Celsius to Kelvin: Add 273.15 to the temperature (K = C + 273.15).

Who invented the **thermometer**?

The Greeks of Alexandria knew that air expanded as it was heated, and it is known that Hero of Alexandria (1st century C.E.) and Philo of Byzantium made simple thermometers or "thermoscopes," but they were not real thermometers. In 1592, Galileo **17**

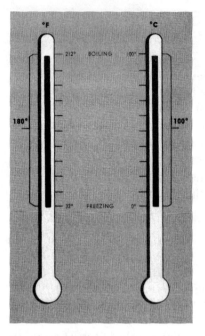
Fahrenheit and Celsius scales

(1564–1642) made a kind of thermometer which also functioned as a barometer, but in 1612, his friend Santorio Santorio (1561–1636) first adapted the air thermometer (a device in which a colored liquid was driven down by the expansion of air) to measure the body's temperature change during illness and recovery. Still, it was not until 1713 that Daniel Fahrenheit (1686–1736) began developing a thermometer having a fixed scale. He worked out his scale from two "fixed" points: the melting point of ice and the heat of the healthy human body. He realized that the melting point of ice was a constant temperature, whereas the freezing point of water varied. Fahrenheit put his thermometer into a mixture of ice, water, and salt (which he marked off as 0°) and using this as a starting point, marked off melting ice at 32° and blood heat at 96°. In 1835, it was discovered that normal blood measured 98.6°F. Sometimes, Fahrenheit used spirit of wine as the liquid in the thermometer tube, but more often he used specially purified mercury. Later, the boiling point of water (212°F) became the upper fixed point.

What was unusual about the original Celsius temperature scale?

In 1742, the Swedish astronomer Anders Celsius (1701–1744) set the freezing point of water at 100°C and the boiling point of water at 0°C. It was Carolus Linnaeus (1707–1778), who reversed the scale, but a later textbook attributed the modified scale to Celsius and the name has remained.

How are Celsius temperatures converted Into Fahrenheit temperatures?

To convert Fahrenheit to Celsius: Subtract 32 from the temperature and multiply the difference by 5; then divide the product by 9 ($C = \frac{5}{9} [F - 32]$). To convert Celsius to Fahrenheit: Multiply the temperature by 1.8, then add 32 ($F = \frac{9}{5} C + 32$ or $F = 1.8C + 32$).

Some useful comparisons of the two scales:

Temperature	Fahrenheit	Celsius or Centigrade
Absolute zero	-459.67	-273.15
Point of equality	-40.0	-40.0
Zero fahrenheit	0.0	-17.8
Freezing point of water	32.0	0.0
Normal human blood temperature	98.4	36.9
100 degrees F	100.0	37.8
Boiling point of water (at standard pressure)	212.0	100.0

What is STP?

The abbreviation *STP* is often used for *s*tandard *t*emperature and *p*ressure. As a matter of convenience, scientists have chosen a specific temperature and pressure as standards for comparing gas volumes. The standard temperature is 0°C (273°K) and the standard pressure is 760 torr (1 atmosphere).

How did the electrical term "ampere" originate?

It was named for Andre Marie Ampere (1775-1836), the physicist who formulated the basic laws of the science of electrodynamics. The *ampere* (symbol A), often abbreviated as "amp," is the unit of electric current, defined at the constant current, that, maintained in two straight parallel infinite conductors placed one meter apart in a vacuum, would produce a force between the conductors of 2×10^{-7} Newton per meter. For example, the amount of current flowing through a 100-watt light bulb is 1 amp; through a toaster, 10 amps; a TV set, 3 amps; a car battery, 50 amps (while cranking). Newton (symbol N) is defined as a unit of force needed to accelerate one kilogram by one meter second^{-2}, or $1N = 1Kg^{MS-2}$.

How did the electrical "volt" originate?

The unit of voltage is the volt, named after Alessandro Volta (1745-1827), the Italian scientist who built the first battery. Voltage measures the force or "oomph" with which electrical charges are pushed through a material. Some common voltages are 1.5 volts for a flashlight battery; 12 volts for a car battery; 115 volts for ordinary household receptacles; and 230 volts for a heavy-duty household receptacle.

What is a mole in chemistry?

A mole (symbol Mol), a fundamental measuring unit for the amount of a substance, refers to either a gram atomic weight or a gram molecular weight of a substance. It is **19** the quantity of a substance that contains 6.02 x 10²³ atoms, molecules, or formula units of that substance. This number is called Avogadro's number or constant, after Amedeo Avogadro (1776-1856) who is considered to be one of the founders of physical science.

How does a **gram atomic weight** differ from a **gram formula weight**?

A gram atomic weight is the amount of an *element* (substance made up of atoms having the same atomic number) equal to its atomic weight (the number of protons) in grams. A gram formula weight is an amount of a *compound* (a combination of elements) equal to its formula weight in grams.

SPACE

UNIVERSE

What was the "Big Bang"?

The Big Bang theory is the explanation most commonly accepted by astronomers for the origin of the universe. It proposes that the universe began as the result of an explosion—the Big Bang—15 billion to 20 billion years ago. Two observations form the basis of this cosmology. First, as Edwin Hubble (1889–1953) demonstrated, the universe is expanding uniformly, with objects at greater distances receding at greater velocities. Secondly, the Earth is bathed in a glow of radiation that has the characteristics expected from a remnant of a hot primeval fireball. This radiation was discovered by Arno A. Penzias (b. 1933) and Robert W. Wilson (b. 1936) of Bell Telephone Laboratories. In time, the matter created by the Big Bang came together in huge clumps to form the galaxies. Smaller clumps within the galaxies formed stars. Parts of at least one clump became a group of planets—our solar system.

How old is the Universe?

The universe is believed to be somewhere between 15 billion and 20 billion years old. The figure is derived from the concept that the universe has been expanding at the same rate since its birth at the "Big Bang." The rate of expansion is a ratio known as Hubble's constant. It is calculated by dividing the speed at which the galaxy is moving away from the Earth by its distance from the Earth. By inverting Hubble's Constant, that is dividing the distance of a galaxy by its recessional speed, the age of the universe can be calculated. The estimates of both the velocity and distance of galaxies from the

Earth are subject to uncertainties and not all scientists accept that the universe has always expanded at the same rate. Therefore, many still hold the age of the universe open to question.

Who is **Stephen Hawking**?

Hawking (b. 1943), a British physicist and mathematician, is considered to be the greatest theoretical physicist of the late twentieth century. In spite of being severely handicapped by *a*myotrophic *l*ateral *s*clerosis (ALS), he has made major contributions to scientific knowledge about black holes and the origin and evolution of the universe though his research into the nature of space-time and its anomalies. For instance, Hawking proposed that a black hole could emit thermal radiation and predicted that a black hole would disappear after all its mass has been converted into radiation (called *Hawking's radiation*). A current objective of Hawking is to synthesize quantum mechanics and relativity theory into a theory of quantum gravity. He is also the author of several books including the popular best-selling work *A Brief History of Time*.

What are **quasars**?

The name quasar is short for *quasi*-stellar radio source. Quasars appear to be stars, but they have large red shifts in their spectra indicating that they are receding from the Earth at great speeds, some up to 90% of the speed of light. Their exact nature is still unknown, but many believe quasars to be the cores of distant galaxies, the most distant objects yet seen. Quasars were first identified in 1963 by astronomers at the Palomar Observatory.

What is a **syzygy**?

A syzygy is a configuration that occurs when three celestial bodies lie in a straight line, such as the sun, Earth, and moon during a solar or lunar eclipse. The particular case when a planet is on the opposite side of the Earth from the sun is called an opposition.

STARS

What is a **binary star**?

A binary star is a pair of stars revolving around a common center of gravity. About 50% of all stars are members of either binary star systems or multiple star systems which contain more than two stars.

The bright star Sirius, about 8.6 light years away, is composed of two stars, one about 2.3 times the mass of the sun, the other a white dwarf star about 980 times the mass of Jupiter. Alpha Centauri, the nearest star to Earth after the sun, is actually three stars. Alpha Centauri A and Alpha Centauri B, two sunlike stars, orbit each other and Alpha Centauri C, a low mass red star, orbits around the others.

What is a **pulsar**?

A pulsar is a rotating neutron star that gives off sharp regular pulses of radio waves at rates ranging from 0.001 to 4 seconds. Stars burn by fusing hydrogen into helium. When they use up their hydrogen, their interiors begin to contract. During this contraction energy is released and the outer layers of the star are pushed out. These layers are large and cool; the star is now a red giant. A star with more than twice the mass of the sun will continue to expand, becoming a supergiant. At that point, it may blow up, in an explosion called a supernova. After a supernova, the remaining material of the star's core may be so compressed that the electrons and protons become neutrons. A star 1.4 to 4 times the mass of the sun can be compressed into a neutron star only about 12 miles (20 kilometers) across. Neutron stars rotate very fast. The neutron star at the center of the Crab Nebula spins 30 times per second.

A pulsar is formed by the collapse of a star with 1.4 to 4 times the mass of the sun. Some of these neutron stars emit radio signals from their magnetic poles in a direction that reaches Earth. These signals were first detected by Jocelyn Bell (b. 1943) of Cambridge University in 1967. Because of their regularity some people speculated that they were extraterrestrial beacons constructed by alien civilizations. This theory was eventually ruled out and the rotating neutron star came to be accepted as the explanation for these pulsating radio sources, or pulsars.

What is a **black hole**?

When a star with a mass greater than about four times that of the sun collapses even the neutrons cannot stop the force of gravity. There is nothing to stop the contraction and the star collapses forever. The material is so dense that nothing—not even light—can escape. The American physicist John Wheeler, in 1967, gave this phenomenon the name "black hole." Since no light escapes from a black hole, it cannot be observed directly. However, if a black hole existed near another star, it would draw matter from the other star into itself, producing x-rays as a result. In the constellation of Cygnus, there is a strong x-ray source, named Cygnus X-1. It is near a star and the two revolve around each other. The unseen x-ray source has the gravitational pull of at least 10 suns and is believed to be a black hole. Another type of black hole, a primordial black hole, may also exist dating from the time of the "Big Bang" when regions of gas and dust were highly compressed.

What does the **color of a star** indicate?

The color of a star gives an indication of its temperature and age. Stars are classified by their spectral type. From oldest to youngest and hottest to coolest, the types of stars are

Type	Color	Temperature	
		Farenheit	**Celsius**
O	Blue	45,000–75,000	25,000–40,000
B	Blue	20,800–45,000	11,000–25,000
A	Blue-White	13,500–20,000	7,500–11,000
F	White	10,800–13,500	6,000–7,500
G	Yellow	9,000–10,800	5,000–6,000
K	Orange	6,300–9,000	3,500–5,000
M	Red	5,400–6,300	3,000–3,500

Each type is further subdivided on a scale of 0-9. The sun is a type G2 star.

Which stars are the **brightest**?

The brightness of a star is called its magnitude. Apparent magnitude is how bright a star appears to the naked eye. The lower the magnitude, the brighter the star. On a clear night, stars of about magnitude +6 can be seen with the naked eye. Large telescopes can detect objects as faint as +27. Very bright objects have negative magnitudes; the sun is -26.8.

Star	Constellation	Apparent Magnitude
Sirius	Canis Major	-1.47
Canopus	Carina	-0.72
Arcturus	Boötes	-0.06
Rigil Kentaurus	Centaurus	+0.01
Vega	Lyra	+0.04
Capella	Auriga	+0.05
Rigel	Orion	+0.14
Procyon	Canis Minor	+0.37
Betelgeuse	Orion	+0.41
Achernar	Eridanus	+0.51

What is the **Milky Way**?

The Milky Way is a hazy band of light that can be seen encircling the night sky. This light comes from the stars that make up the Milky Way galaxy, the galaxy to which the sun and the Earth belong. Galaxies are huge systems of stars separated from one another by largely empty space. Astronomers estimate that the Milky Way galaxy contains at least 100 billion stars and is about 100,000 light years in diameter. The galaxy is shaped like a phonograph record with a central bulge, or nucleus, and spiral arms curving out from the center.

What is the **Big Dipper**?

The Big Dipper is a group of seven stars which are part of the constellation Ursa Major. They appear to form a sort of spoon with a long handle. The group is known as The Plough in Great Britain. The Big Dipper is almost always visible in the Northern Hemisphere. It serves as a convenient reference point when locating other stars; for example, an imaginary line drawn from the two end stars of the dipper leads to Polaris, the North Star.

Where is the **North Star**?

If an imaginary line is drawn from the North Pole into space, there is a star called Polaris, or the North Star, less than one degree away from the line. As the Earth rotates on its axis, Polaris acts as a pivot-point around which all the stars visible in the Northern Hemisphere appear to move, while Polaris itself remains motionless.

What is the "**summer triangle**"?

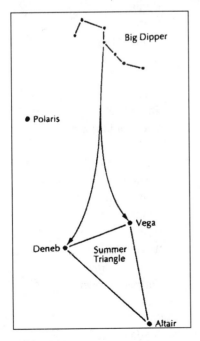

Summer Triangle

The summer triangle is the triangle formed by the stars Deneb, Vega, and Altair as seen in the summer Milky Way.

How many **constellations** are there and how were they named?

Constellations are groups of stars that seem to form some particular shape, that of a person, animal or object. They only appear to form this shape and be close to each other from Earth; in actuality the stars in a constellation are often very distant from each other. There are 88 recognized constellations in the heavens whose boundaries were defined in the 1920s by the International Astronomical Union.

Different cultures in all parts of the world had their own constellations. However, because modern science is predominantly a product of western culture, many of the constellations represent characters from Greek and Roman mythology. When Europeans began to explore the Southern Hemisphere in the 16th and 17th centuries, they derived some of the new star patterns from the technological wonders of their time, such as the microscope.

Names of constellations are usually given in Latin. Individual stars in a constellation are usually designated with Greek letters in the order of brightness; the brightest star is alpha, the second brightest is beta, and so on. The genitive, or possessive, form of the constellation name is used, thus Alpha Orionis is the brightest star of the constellation Orion. The table below lists the constellations and what they represent.

Constellation	Genitive	Abbreviation	Meaning
Andromeda	Andromedae	And	Chained Maiden
Antlia	Antliae	Ant	Air Pump
Apus	Apodis	Aps	Bird of Paradise
Aquarius	Aquarii	Aqr	Water Bearer
Aquila	Aquilae	Aql	Eagle
Ara	Arae	Ara	Altar
Aries	Arietis	Ari	Ram
Auriga	Aurigae	Aur	Charioteer
Boötes	Boötis	Boo	Herdsman
Caelum	Caeli	Cae	Chisel
Camelopardalis	Camelopardalis	Cam	Giraffe
Cancer	Cancri	Cnc	Crab
Canes Venatici	Canum Venaticorum	CVn	Hunting Dogs
Canis Major	Canis Majoris	CMa	Big Dog
Canis Minor	Canis Minoris	CMi	Little Dog
Capricornus	Capricorni	Cap	Goat
Carina	Carinae	Car	Ship's Keel
Cassiopeia	Cassiopeiae	Cas	Queen of Ethiopia
Centaurus	Centauri	Cen	Centaur
Cepheus	Cephei	Cep	King of Ethiopia
Cetus	Ceti	Cet	Whale
Chamaeleon	Chamaeleonis	Cha	Chameleon
Circinus	Circini	Cir	Compass
Columba	Columbae	Col	Dove
Coma Berenices	Comae Berenices	Com	Berenice's Hair
Corona Australis	Coronae Australis	CrA	Southern Crown
Corona Borealis	Coronae Borealis	CrB	Northern Crown
Corvus	Corvi	Crv	Crow
Crater	Crateris	Crt	Cup
Crux	Crucis	Cru	Southern Cross
Cygnus	Cygni	Cyg	Swan

Constellation	Genitive	Abbreviation	Meaning
Delphinus	Delphini	Del	Dolphin
Dorado	Doradus	Dor	Goldfish
Draco	Draconis	Dra	Dragon
Equuleus	Equulei	Equ	Little Horse
Eridanus	Eridani	Eri	River Eridanus
Fornax	Fornacis	For	Furnace
Gemini	Geminorum	Gem	Twins
Grus	Gruis	Gru	Crane
Hercules	Herculis	Her	Hercules
Horologium	Horologii	Hor	Clock
Hydra	Hydrae	Hya	Hydra, Greek monster
Hydrus	Hydri	Hyi	Sea Serpent
Indus	Indi	Ind	Indian
Lacerta	Lacertae	Lac	Lizard
Leo	Leonis	Leo	Lion
Leo Minor	Leonis Minoris	LMi	Little Lion
Lepus	Leporis	Lep	Hare
Libra	Librae	Lib	Scales
Lupus	Lupi	Lup	Wolf
Lynx	Lyncis	Lyn	Lynx
Lyra	Lyrae	Lyr	Lyre or Harp
Mensa	Mensae	Men	Table Mountain
Microscopium	Microscopii	Mic	Microscope
Monoceros	Monocerotis	Mon	Unicorn
Musca	Muscae	Mus	Fly
Norma	Normae	Nor	Carpenter's Square
Octans	Octanis	Oct	Octant
Ophiuchus	Ophiuchi	Oph	Serpent Bearer
Orion	Orionis	Ori	Orion, the Hunter
Pavo	Pavonis	Pav	Peacock
Pegasus	Pegasi	Peg	Winged Horse
Perseus	Persei	Per	Perseus, a Greek hero
Phoenix	Phoenicis	Phe	Phoenix
Pictor	Pictoris	Pic	Painter
Pisces	Piscium	Psc	Fish
Piscis Austrinus	Piscis Austrini	PsA	Southern Fish
Puppis	Puppis	Pup	Ship's Stern
Pyxis	Pyxidis	Pyx	Ship's Compass
Reticulum	Reticuli	Ret	Net
Sagitta	Sagittae	Sge	Arrow
Sagittarius	Sagittarii	Sgr	Archer
Scorpius	Scorpii	Sco	Scorpion

Constellation	Genitive	Abbreviation	Meaning
Sculptor	Sculptoris	Scl	Sculptor
Scutum	Scuti	Sct	Shield
Serpens	Serpentis	Ser	Serpent
Sextans	Sextantis	Sex	Sextant
Taurus	Tauri	Tau	Bull
Telescopium	Telescopii	Tel	Telescope
Triangulum	Trianguli	Tri	Triangle
Triangulum	Triangli Australis	TrA	Southern Australe Triangle
Tucana	Tucanae	Tuc	Toucan
Ursa Major	Ursae Majoris	UMa	Big Bear
Ursa Minor	Ursae Minoris	UMi	Little Bear
Vela	Velorum	Vel	Ship's Sail
Virgo	Virginis	Vir	Virgin
Volans	Volantis	Vol	Flying Fish
Vulpecula	Vulpeculae	Vul	Little Fox

Which star is the **closest to Earth**?

The sun, at a distance of 93 million miles (150 million kilometers), is the closest star to the Earth. After the sun, the closest stars are the members of the triple star system known as Alpha Centauri (Alpha Centauri A, Alpha Centauri B, and Alpha Centauri C, sometimes called Proxima Centauri). They are 4.3 light years away.

What is the **sun** made of?

The sun is an incandescent ball of gases. Its mass is 1.8×10^{27} tons or 1.8 octillion tons (a mass 330,000 times as great as the Earth).

Element	% of mass
Hydrogen	73.46
Helium	24.85
Oxygen	0.77
Carbon	0.29
Iron	0.16
Neon	0.12
Nitrogen	0.09
Silicon	0.07
Magnesium	0.05
Sulfur	0.04
Other	0.10

How **hot** is the sun?

The center of the sun is about 27,000,000°F (15,000,000°C). The surface, or photosphere, of the sun is about 10,000°F (5,500°C). Magnetic anomolies in the photosphere cause cooler regions which appear to be darker than the surrounding surface. These sunspots are about 6,700°F (4,000°C). The sun's layer of lower atmosphere, the chromosphere, is only a few thousand miles thick. At the base, the chromosphere is about 7,800°F (4,300°C), but its temperature rises with altitude to the corona, the sun's outer layer of atmosphere, which has a temperature of about 1,800,000°F (1,000,000°C).

When will the **sun die**?

The sun is approximately 4.5 billion years old. About 5 billion years from now, the sun will burn all of its hydrogen fuel into helium. When this process occurs, the sun will change from the yellow dwarf we know it as to a red giant. Its diameter will extend well beyond the orbit of Venus, and even possibly beyond the orbit of Earth. In either case, the Earth will be burned to a cinder and will be incapable of supporting life.

What is the **ecliptic**?

Ecliptic refers to the apparent yearly path of the sun through the sky with respect to the stars. In the spring, the ecliptic in the Northern Hemisphere is angled high in the evening sky. In fall, the ecliptic lies much closer to the horizon.

Why does the **color of the sun** vary?

Sunlight contains all the colors of the rainbow which blend to form white light, making sunlight appear white. At times, some of the color wavelengths become scattered, especially blue, and the sunlight appears colored. When the sun is high in the sky, some of the blue rays are scattered in the Earth's atmosphere. At such times, the sky looks blue and the sun appears to be yellow. At sunrise or sunset when the light must follow a longer path through the Earth's atmosphere, the sun looks red (red having the longest wavelengths).

How long does it take **light from the sun** to reach the Earth?

Sunlight takes about 8 minutes and 20 seconds to reach the Earth, travelling at 186,282 miles (299,792 kilometers) per second.

How long is a **solar cycle**?

The solar cycle is the periodic change in the number of sun spots. The cycle is taken as the interval between successive minima and is about 11.1 years. During an entire cycle, solar flares, sunspots, and other magnetic phenomena move from intense activity to relative calm and back again. The solar cycle is one area of study to be carried out by up to 10 ATLAS space missions beginning in March 1992 to probe the chemistry and physics of the atmosphere. These studies of the solar cycle will yield a more detailed picture of the Earth's atmosphere and its response to changes in the sun.

What is the **Maunder minimum**?

It was the time between 1645 and 1715 when sunspots were very scarce or missing entirely. From studying old solar records in 1890, E. Walter Maunder observed that sunspot activity during these years was at a minimum. This observation was confirmed in 1976 by John Eddy.

When do **solar eclipses** happen?

A solar eclipse occurs when the moon passes between the Earth and the sun and all three bodies are aligned in the same plane. When the moon completely covers the sun and the umbra, or dark part of the moon's shadow, reaches the Earth, a total eclipse occurs. A total eclipse happens only along a narrow path 100 to 200 miles (160 to 320 kilometers) wide called the track of totality. Just before totality, the only parts of the sun that are visible are a few points of light called Baily's beads shining through valleys on the moon's surface. Sometimes, a last bright flash of sunlight is seen—the diamond ring effect. During totality, which averages 2.5 minutes, but may last up to 7.5 minutes, the sky is dark and stars and other planets are easily seen. The corona, the sun's outer atmosphere, is also visible.

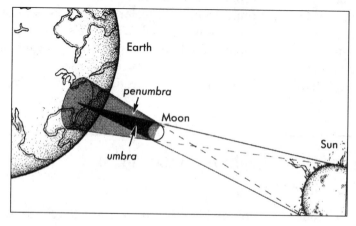

Solar eclipse

If the moon does not appear large enough in the sky to completely cover the sun, it appears silhouetted against the sun with a ring of sunlight showing around it. This is an annular eclipse. Because the sun is not completely covered, its corona cannot be seen and although the sky may darken, it will not be dark enough to see the stars.

During a partial eclipse of the sun, the penumbra of the moon's shadow strikes the Earth. A partial eclipse can also be seen on either side of the track of totality of an annular or total eclipse. The moon will cover part of the sun and the sky will not darken noticeably during a partial eclipse.

PLANETS AND MOONS

See also: The Earth

How old is the **solar system**?

It is is currently believed to be 4.5 billion years old. The Earth and the rest of the solar system formed from an immense cloud of gas and dust. Gravity and rotational forces caused the cloud to flatten into a disc and much of the cloud's mass to drift into the center. This material became the sun. The left-over parts of the cloud formed small bodies called planetesimals. These planetesimals collided with each other, gradually forming larger and larger bodies, some of which became the planets. This process is thought to have taken about 25 million years.

How far are the planets from the sun?

The planets revolve around the sun in elliptical orbits, with the sun at one focus of the ellipse. Thus, a planet is at times closer to the sun than at other times. The distances given below are the average distance from the sun, starting with Mercury, the planet closest to the sun, and moving outward.

Planet	Average distance Miles	Kilometers
Mercury	35,983,000	57,909,100
Venus	67,237,700	108,208,600
Earth	92,955,900	149,598,000
Mars	141,634,800	227,939,200
Jupiter	483,612,200	778,298,400
Saturn	888,184,000	1,427,010,000
Uranus	1,782,000,000	2,869,600,000
Neptune	2,794,000,000	4,496,700,000
Pluto	3,666,000,000	5,913,490,000

How long do the **planets** take to go **around the sun**?

Planet	Period of revolution Earth days	Earth years
Mercury	88	0.24
Venus	224.7	0.62
Earth	365.26	1.00
Mars	687	1.88
Jupiter	4,332.6	11.86
Saturn	10,759.2	29.46
Uranus	30,685.4	84.01
Neptune	60,189	164.8
Pluto	90,777.6	248.53

What are the **diameters** of the planets?

Planet	Diameter Miles	Kilometers
Mercury	3,031	4,878
Venus	7,520	12,104
Earth	7,926	12,756
Mars	4,221	6,794
Jupiter	88,846	142,984
Saturn	74,898	120,536
Uranus	31,763	51,118
Neptune	31,329	50,530
Pluto	1,423	2,290

Note: All diameters are as measured at the planet's equator.

What are the **colors** of the planets?

Planet	Color
Mercury	Orange
Venus	Yellow
Earth	Blue, brown, green
Mars	Red
Jupiter	Yellow, red, brown, white
Saturn	Yellow
Uranus	Green
Neptune	Blue
Pluto	Yellow

Which planets have **rings**?

Jupiter, Saturn, Uranus, and Neptune all have rings. Jupiter's rings were discovered by *Voyager 1* in March, 1979. The rings extend 80,240 miles (129,130 kilometers) from the center of the planet. They are about 4,300 miles (7,000 kilometers) in width and less than 20 miles (30 kilometers) thick. A faint inner ring is believed to extend to the edge of Jupiter's atmosphere.

Saturn has the largest, most spectacular set of rings in the solar system. That the planet is surrounded by a ring system was first recognized by the Dutch astronomer Christiaan Huygens (1629–1695) in 1659. Saturn's rings are 169,800 miles (273,200 kilometers) in diameter, but less than 10 miles (16 kilometers) thick. There are six different rings, the largest of which appear to be divided into thousands of ringlets. The rings appear to be composed of pieces of water ice ranging in size from tiny grains to blocks several tens of yards in diameter.

In 1977 when Uranus occulted (passed in front of) a star, scientists observed that the light from the star flickered or winked several times before the planet itself covered the star. The same flickering occurred in reverse order after the occultation. The reason for this was determined to be a ring around Uranus. Nine rings were initially identified and *Voyager 2* observed two more in 1986. The rings are thin, narrow, and very dark.

Voyager 2 also discovered a series of at least four rings around Neptune in 1989. Some of the rings appear to have arcs, areas where there is a higher density of material than at other parts of the ring.

What is the **gravitational force** on each of the planets, the moon, and the sun relative to the Earth?

If the force on the Earth is taken as 1, the comparative forces are:

Sun	27.9
Mercury	0.37
Venus	0.88
Earth	1.00
Moon	0.16
Mars	0.38
Jupiter	2.64
Saturn	1.15
Uranus	0.93
Neptune	1.22
Pluto	0.06

Weight comparisons can be made by using this table. If a person weighed 100 pounds (45.36 kilograms) on Earth, then the weight of the person on the moon would be 16 pounds (7.26 kilograms) or 100 x 0.16.

Is a day the same on all the planets?

No. A day, the period of time it takes for a planet to make one complete turn on its axis, varies from planet to planet. Venus, Uranus, and Pluto display retrograde motion, that is to say, they rotate in the opposite direction from the other planets. The table below lists the length of the day for each planet.

Planet	Earth days	Length of day	
		Hours	Minutes
Mercury	58	15	30
Venus	243		32
Earth		23	56
Mars		24	37
Jupiter		9	50
Saturn		10	39
Uranus		17	14
Neptune		16	03
Pluto	6	09	18

Which planets are called "inferior" planets and which are "superior" planets?

An inferior planet is one whose orbit is nearer to the sun than Earth's orbit is. Mercury and Venus are the inferior planets. Superior planets are those whose orbits around the sun lie beyond that of Earth. Mars, Jupiter, Saturn, Uranus, Neptune, and Pluto are the superior planets. The terms have nothing to do with the quality of an individual planet.

What are the Jovian and terrestrial planets?

Jupiter, Saturn, Uranus, and Neptune are the Jovian (the adjectival form for the word "Jupiter"), or Jupiter-like, planets. They are giant planets, composed primarily of light elements such as hydrogen and helium.

Mercury, Venus, Earth, and Mars are the terrestrial (derived from *terra,* the Latin word for "earth"), or Earth-like, planets. They are small in size, have solid surfaces, and are composed of rocks and iron. Pluto appears to be a terrestrial planet as well, but it may have a different origin from the other planets.

What is unique about the rotation of the planet Venus?

Unlike Earth and most of the other planets, Venus rotates in a retrograde, or opposite, direction with relation to its orbital motion about the sun. It rotates so slowly that there are only two sunrises and sunsets each Venusian year. Uranus and Pluto's rotation are also retrograde.

Is it true that the **Earth is closer to the sun** in winter than in summer in the Northern Hemisphere?

Yes. However, the Earth's axis, the line around which the planet rotates, is tipped 23.5 degrees with respect to the plane of revolution around the sun. When the Earth is closest to the sun (its perihelion, about January 3), the Northern Hemisphere is tilted away from the sun. This causes winter in the Northern Hemisphere while the Southern Hemisphere is having summer. When the Earth is farthest from the sun (its aphelion, around July 4), the situation is reversed, with the Northern Hemisphere tilted towards the sun. At this time, it is summer in the Northern Hemisphere and winter in the Southern Hemisphere.

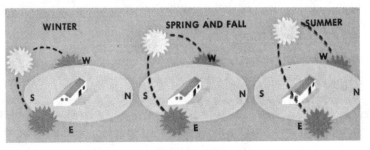

Seasons in the Northern Hemisphere

Is it true that the **rotation speed of the Earth** varies?

The rotation speed is at its maximum in late July and early August and at its minimum in April; the difference in the length of the day is about 0.0012 seconds. Since about 1900 the Earth's rotation has been slowing at a rate of approximately 1.7 seconds per year. In the geologic past the Earth's rotational period was much faster; days were shorter and there were more days in the year. About 350 million years ago, the year had 400-410 days; 280 million years ago, a year was 390 days long.

What is the **circumference of the Earth**?

The Earth is an oblate ellipsoid—a sphere slightly flattened at the poles and bulging at the equator. The distance around the Earth at the equator is 24,902 miles (40,075 kilometers). The distance around the Earth through the poles is 24,860 miles (40,008 kilometers).

Is there **life on Mars**?

Experiments carried out by the Viking Lander in July 1976 indicated there is no evidence of life on Mars. There were three experiments conducted on the composition of the Martian soil and atmosphere which formed the basis for the decision.

Is it true that Pluto is not always the **outermost planet** in the solar system?

Pluto's very eccentric orbit carried it inside Neptune's orbit on January 23, 1979. It will remain there until March 15, 1999. During this time, Neptune is the outermost planet in the solar system. However, because they are so far apart, the planets are in no danger of colliding with one another.

Pluto, discovered in 1930 by American astronomer Clyde Tombaugh (b. 1906), is the smallest planet in the solar system. It is composed of rock and ice, with methane ice on the surface and a thin methane atmosphere. Pluto's single moon, Charon, discovered by James Christy in 1978, has a diameter of 741 miles (1,192 kilometers). This makes Charon, at half the size of Pluto, a very large moon relative to the planet. Some astronomers consider Pluto and Charon to be a double planet system.

What is **Planet X**?

Astronomers have observed perturbations, or disturbances, in the orbits of Uranus and Neptune since the discoveries of both planets. They speculated that Uranus and Neptune were being influenced by the gravity of another celestial body. Pluto, discovered in 1930, does not appear to be large enough to cause these disturbances. The existence of another planet, known as Planet X, orbiting beyond Pluto, has been proposed. As yet there have been no sightings of this tenth planet but the search continues. There is a possibility that the unmanned space probes *Pioneer 10 & 11* and *Voyager 1 & 2*, now heading out of the solar system, will be able to locate this elusive object.

How many **moons** does each planet have?

Planet	Number of moons	Names of moons
Mercury	0	
Venus	0	
Earth	1	The Moon (sometimes called Luna)
Mars	2	Phobos, Deimos
Jupiter	16*	Metis, Adrastea, Amalthea, Thebe, Io, Europa, Ganymede, Callisto, Leda, Himalia, Lysithia, Elara, Ananke, Carme, Pasiphae, Sinope

Planet	Number of moons	Names of moons
Saturn	18*	Atlas, 1981S13 (unnamed as yet), Prometheus, Pandora, Epimetheus, Janus, Mimas, Enceladus, Tethys, Telesto, Calypso, Dione, Helene, Rhea, Titan, Hyperion, Iapetus, Phoebe
Uranus	15	Cordelia, Ophelia, Bianca, Cressida, Desdemona, Juliet, Portia, Rosalind, Belinda, Puck, Miranda, Ariel, Umbriel, Titania, Oberon
Neptune	8	Naiad, Thalassa, Despina, Galatea, Larissa, Proteus, Triton, Nereid
Pluto	1	Charon

*Several other satellites have been reported but not confirmed.

How far is the moon from the Earth?

Since the moon's orbit is elliptical, its distance varies from about 221,463 miles (356,334 kilometers) at perigee (closest approach) to Earth, to 251,968 miles (405,503 kilometers) at apogee (farthest point), with the average distance being 238,857 miles (384,392 kilometers).

What are the diameter and circumference of the moon?

The moon's diameter is 2,159 miles (3,475 kilometers)and its circumference is 6,790 miles (10,864 kilometers). The moon is 27% the size of the Earth.

Why does the moon always keep the same face toward the Earth?

Only one side of the moon is seen because it always rotates in exactly the same length of time that it takes to revolve about the Earth. This combination of motions (called "captured rotation") means that it always keeps the same side toward the Earth.

What are the phases of the moon?

The phases of the moon are changes in the moon's appearance during the month which are caused by the moon's turning different portions of its illuminated hemisphere towards the Earth. When the moon is between the Earth and the sun, its daylight side is turned away from the Earth, so it is not seen. This is called the new moon. As the moon continues its revolution around the Earth, more and more of its surface becomes visible. This is called the waxing crescent phase. About a week after the new moon, half the moon is visible—the first quarter phase. During the next week, more than half of the moon is seen; this is called the waxing gibbous phase. Finally, about two weeks after the new moon, the moon and sun are on opposite sides of the Earth. The side of the moon facing the sun is also facing the Earth, and all the moon's illuminated side is seen as a full moon. In the next two weeks the moon goes through the same phases, but in reverse from a waning gibbous to third or last quarter to waning crescent phase. Gradually, less and less of the moon is visible until a new moon occurs again.

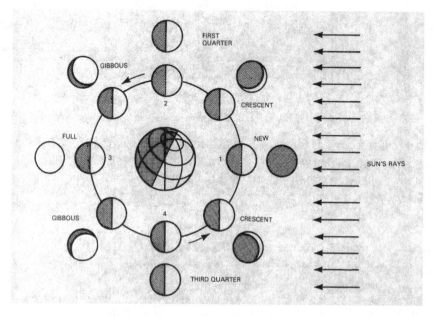

Is there ever a **full moon** twice in one month?

Yes, every two or three years a month has 2 full moons. For instance, this phenomenon occur's on September 1 and September 30, 1993; July 1 and July 30, 1996; January 2 and January 31, 1999; and March 2 and March 31, 1999. In 1999 February has no full moon.

What is the difference between a **hunter's moon** and a **harvest moon**?

The harvest moon is the full moon nearest the autumnal equinox (on or about September 23). It is followed by a period of several successive days when the moon rises soon after sunset. In the Southern Hemisphere the harvest moon is the full moon closest to the vernal equinox (on or about March 21). This gives farmers extra hours of light for harvesting crops. The next full moon after the harvest moon is called the hunter's moon.

Is it possible for a calendar month to pass **without a full moon?**

Yes. February is the only month that may not have a full moon. This occurred in 1866, 1893, 1915, and 1980, and will happen again in 1999 and 2066.

Why do **lunar eclipses** happen?

A lunar eclipse occurs only during a full moon when the moon is on one side of the Earth and the sun is on the opposite side and all three bodies are aligned in the same plane. In this alignment the Earth blocks the sun's rays to cast a shadow on the moon. In a total lunar eclipse the moon seems to disappear from the sky, when the whole moon passes through the umbra, or total shadow created by the Earth. A total lunar eclipse may last up to 1 hour and 40 minutes. If only part of the moon enters the umbra, a partial eclipse occurs. A penumbral eclipse takes place if all or part of the moon passes through the penumbra (partial shadow or "shade") without touching the umbra. It is difficult to detect this type of eclipse from Earth. From the moon one could see that the Earth blocked part of the sun only.

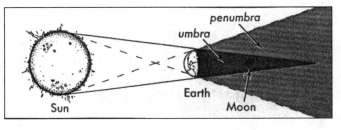

Lunar eclipse

Astronomers have discovered that **the moon has a tail. What does this mean?**

A glowing 15,000 mile (24,000 kilometer) tail of sodium atoms streams from the moon. The faint, orange glow of sodium cannot be seen by the naked eye but is detectable by instruments. Astronomers are not certain of the source of these sodium atoms.

What are the **craters** on the moon that are named for the famous Curie family?

Curie—named for Pierre Curie (1859–1906), French chemist and Nobel prize winner.

Sklodowska—the maiden name of Marie Curie (1867–1934), French physical chemist and Nobel Prize winner.

39

Joliot—named for physicist Frederic Joliot–Curie (1900–1958), Pierre and Marie's son-in-law and Nobel Prize winner.

Is the moon really blue during a blue moon?

A blue moon is the second full moon in a single month. The term does not refer to the color of the moon. It occurs, on average, every 2.72 years. Since 29.53 days pass between full moons (a synodial month), there is never a blue moon in February.

A bluish-looking moon results from effects of the Earth's atmosphere. For example, the phenomenon was widely observed on September 26, 1950, due to Canadian forest fires which had scattered high-altitude dust.

COMETS, METEORITES, ETC.

Where are the asteroids?

The asteroids, also called the minor planets, are smaller than any of the nine major planets in the solar system and are not satellites of any major planet. The term asteroid means "starlike" because asteroids appear to be points of light when seen through a telescope.

Most asteroids are located between Mars and Jupiter, between 2.1 and 3.3 AUs from the sun. Ceres, the first to be discovered, and the largest, was found by Giuseppe Piazzi (1746–1826) on January 1, 1801, and has a diameter of 582 miles (936 kilometers). A second asteroid, Pallas, was discovered in 1802. Since then, over 18 thousand asteroids have been identified and astronomers have established orbits for about 5 thousand of them. Some of these have diameters of only 0.62 miles (1 kilometer). Originally, astronomers thought the asteroids were remnants of a planet that had been destroyed; now they believe asteroids to be material that never became a planet, possibly because it was affected by Jupiter's strong gravity.

Not all asteroids are in the main asteroid belt. Three groups reside in the inner solar system. The Aten asteroids have orbits that lie primarily inside Earth's orbit. However, at their farthest point from the sun, these asteroids may cross Earth's orbit. The Apollo asteroids cross Earth's orbit; some come even closer than the moon. The Amor asteroids cross the orbit of Mars and some come close to Earth's orbit. The

Trojan asteroids move in virtually the same orbit as Jupiter but at points 60 degrees ahead or 60 degrees behind the planet. In 1977 Charles Kowal discovered an object now known as Chiron orbiting between Saturn and Uranus. Originally cataloged as an asteroid, Chiron was later observed to have a coma (a gaseous halo) and it may be reclassified as a comet.

An **asteroid** came close to hitting the Earth sometime in 1989. How much **damage** might it have done?

Asteroid 1989 FC passed within 434,000 miles (700,000 kilometers) of the Earth on March 22, 1989. The impact, had it hit the Earth, would have delivered the energy equivalent of more than 1 million tons of exploding TNT and created a crater up to 4.3 miles (7 kilometers) across.

What was the **Tunguska Event**?

On June 30, 1908, a violent explosion occurred in the atmosphere over the Podkamennaya Tunguska river, in a remote part of central Siberia. The blast's consequences were similar to an H-bomb going off, leveling thousands of square miles of forest. The shock of the explosion was heard over 600 miles (960 kilometers) away. A number of theories have been proposed to account for this event.

Some people thought that a large meteorite or a piece of anti-matter had fallen to Earth. But a meteorite, composed of rock and metal, would have created a crater and none was found at the impact site. There are no high radiation levels in the area which would have resulted from the collision of anti-matter and matter. Two other theories include a mini-black hole striking the Earth and the crash of an extraterrestrial spaceship. However, a mini-black hole would have passed through the Earth and there is no record of a corresponding explosion on the other side of the world. As for the spaceship, no wreckage of such a craft was ever found.

The most likely cause of the explosion was the entry into the atmosphere of a piece of a comet which would have produced a large fireball and blast wave. Since a comet is composed primarily of ice, the fragment would have melted during its passage through the Earth's atmosphere, leaving no impact crater and no debris. Since the Tunguska Event coincided with the Earth's passage through the orbit of Comet Encke, the explosion could have been caused by a piece of that comet.

From where do **comets** originate?

According to a theory developed by Dutch astronomer Jan Oort, there is a large cloud of gas, dust, and comets orbiting beyond Pluto out to perhaps 100,000 astronomical **41** units (AU). Occasional passage of a star close to this cloud disturbs some of the comets from their orbits. Some fall inwards towards the sun.

Comets, sometimes called "dirty snowballs," are made up mostly of ice, with some dust mixed in. When a comet moves closer to the sun, the dust and ice of the core, or nucleus, heats up, producing a tail of material which trails along behind it. The tail is pushed out by the solar wind and almost always points away from the sun.

Most comets have highly elliptical orbits that carry them around the sun and then fling them back out to the outer reaches of the solar system, never to return. Occasionally however, a close passage by a comet near one of the planets can alter a comet's orbit making it stay in the middle or inner solar system. Such a comet is called a short-period comet because it passes close to the sun at regular intervals. The most famous short-period comet is Comet Halley which reaches perihelion (the point in its orbit that is closest to the sun) about every 76 years. Comet Encke, with an orbital period of 3.3 years, is another short-period comet.

When will **Halley's comet** return?

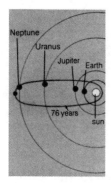

Halley's comet orbit

Halley's comet returns about every 76 years. It was most recently seen in 1985/1986 and is predicted to appear again in 2061, then in 2134. Every appearance of what is now known as Comet Halley has been noted by astronomers since the year 239 B.C.E.

The comet is named for Edmund Halley (1656–1742), England's second Astronomer Royal. In 1682 he observed a bright comet and noted that it was moving in an orbit similar to comets seen in 1531 and 1607. He concluded that the three comets were actually one and the same and that the comet had an orbit of 76 years. In 1705 Halley published *A Synopsis of the Astronomy of Comets* in which he predicted that the comet seen in 1531, 1607, and 1682 would return in 1758. On Christmas night, 1758, a German farmer and amateur astronomer named Johann Palitzsch spotted the comet in just the area of the sky that Halley had foretold.

Prior to Halley, comets appeared at irregular intervals and were often thought to be harbingers of disaster and signs of divine wrath. Halley proved that they are natural objects subject to the laws of gravity. This was a great victory for science in the war against superstition.

When do **meteor showers** occur?

There are a number of groups of meteoroids circling the sun just as the Earth is. When Earth's orbit intercepts the path of one of these swarms of meteoroids, some of them enter Earth's atmosphere. When friction with the air causes a meteoroid to burn up, the streak, or shooting star, that is produced is called a meteor. Large numbers of meteors can produce a spectacular shower of light in the night sky. Meteor showers are named for the constellation that occupies the area of the sky from which they originate. Listed below are the ten meteor showers and the dates during the year during which they can be seen.

Name of Shower	Dates
Quadrantids	January 1–6
Lyrids	April 19–24
Eta Aquarids	May 1–8
Perseids	July 25–August 18
Orionids	October 16–26
Taurids	October 20–November 20
Leonids	November 13–17
Phoenicids	December 4–5
Geminids	December 7–15
Ursids	December 17–24

How does a **meteorite** differ from a **meteoroid**?

A meteorite is a natural object of extraterrestrial origin that survives passage through the Earth's atmosphere and hits the Earth's surface. A meteorite is often confused with a meteoroid or a meteor. A meteoroid is a small object in outer space, generally less than 30 feet (10 meters) in diameter. A meteor (sometimes called a shooting star) is the flash of light seen when an object passes through Earth's atmosphere and burns as a result of heating caused by friction. A meteoroid becomes a meteor when it enters the Earth's atmosphere; if any portion of a meteoroid lands on Earth, it is a meteorite.

There are three kinds of meteorites. Irons contain 85–95% iron; the rest of their mass is mostly nickel. Stony irons are relatively rare meteorites composed of about 50% iron and 50% silicates. Stones are made up mostly of silicates and other stony materials.

What are the **largest meteorites** that have been found in the world?

Name	Location	Weight	
		Tons	Tonnes
Hoba West	Namibia	66.1	60
Ahnighito (The Tent)	Greenland	33.5	30.4
Bacuberito	Mexico	29.8	27
Mbosi	Tanzania	28.7	26
Agpalik	Greenland	22.2	20.1

Name	Location	Weight	
		Tons	Tonnes
Armanty	Outer Mongolia	22	20
Willamette	Oregon, USA	15.4	14
Chupaderos	Mexico	15.4	14
Campo del Cielo	Argentina	14.3	13
Mundrabilla	Western Australia	13.2	12
Morito	Mexico	12.1	11

The famous Willamette (Oregon) iron, listed above, displayed at the American Museum of Natural History in New York, is the largest specimen found in the United States. It is 10 feet (3.048 meters) long and 5 feet (1.524 meters) high.

OBSERVATION AND MEASUREMENT

Who is considered the father of systematic astronomy?

The Greek scientist, Hipparchus (fl. 146–127 B.C.E.), is considered to be the father of systematic astronomy. He measured as accurately as possible the directions of objects in the sky. He compiled the first catalog of stars, containing about 850 entries, and designated each star's celestial coordinates, indicating its position in the sky. Hipparchus also divided the stars according to their apparent brightness or magnitudes.

What is a light year?

A light year is a measure of distance, not time. It is the distance that light, which travels in a vacuum at the rate of 186,000 miles (300,000 kilometers) per second, can travel in a year (365¼ days). This is equal to 5,870 billion miles or 9,460 billion kilometers.

Besides the light year, what other units are used to measure distances in astronomy?

The astronomical unit (AU) is often used to measure distances within the solar system. One AU is equal to the average distance between the Earth and the sun or 92,955,630 miles (149,597,870 kilometers). The parsec is equal to 3.26 light years, or about 19,180 million miles (30,820 million kilometers).

How are new celestial objects named?

Many stars and planets have names that date back to antiquity. The *International Astronomical Union* (IAU), the professional astronomers organization, has attempted, in this century, to standardize names given to newly discovered celestial objects and their surface features.

Stars are generally called by their traditional names, most of which are of Greek, Roman, or Arabic origin. They are also called after the constellation in which they appear, designated in order of brightness by Greek letters. Thus Sirius is also called alpha Canis Majoris, which means it is the brightest star in the constellation Canis Major. Other stars are called by catalog numbers which include the star's coordinates. To many astronomers' horror, several commercial star registries exist, and for a fee, you can submit a star name to them. These names are not officially recognized by the IAU.

The IAU has made some recommendations for naming the surface features of the planets and their satellites. For example, features on Mercury are named for composers, poets, and writers; the features of Venus for women; Saturn's moon Mimas for people and places in Arthurian legend.

Comets are named for their discoverers. Newly discovered asteroids are first given a temporary designation consisting of the year of discovery plus two letters. The first letter indicates the half month of discovery and the second the order of discovery in that half month. Thus asteroid 1991BA was the first asteroid (A) discovered in the second half of January (B) in 1991. After an asteroid's orbit is determined it is given a permanent number and its discoverer is given the honor of naming it. Asteroids have been named after such diverse things as mythological figures (Ceres, Vesta), an airline (Swissair), and the Beatles (Lennon, McCartney, Harrison, Starr).

What is an **astrolabe**?

Invented by the Greeks or Alexandrians in about 100 B.C.E. or before, an astrolabe is a two-dimensional working model of the heavens, with sights for observations. It consists of two concentric flat disks, one fixed, representing the observer on Earth, the other moving which can be rotated to represent the appearance of the celestial sphere at a given moment. Given latitude, date, and time, the observer can read off the altitude and azimuth of the sun, the brightest stars and the planets. By measuring the altitude of a particular body, one can find the time. The astrolabe can also be used to find times of sunrise, sunset, twilight, or the height of a tower or depth of a well. It was replaced by the sextant and other more accurate instruments.

Who invented the **telescope**?

Hans Lippershey (ca. 1570–ca. 1619), a German-Dutch lens grinder and spectacle maker, is generally credited with inventing the telescope in 1608 because he was the first scientist to apply for a patent. Two other inventers, Zacharias Janssen and Jacob Metius, also developed telescopes. Modern historians consider Lippershey and Janssen as the two likely candidates for the title of inventor of the telescope, with Lippershey possessing the strongest claim.

Who is the **Hubble** for whom the space telescope is named?

Edwin Powell Hubble (1889–1953) was an American astronomer known for his studies of galaxies. His study of nebulae, or clouds, the faint, unresolved luminous patches in the sky, showed that some of them were large groups of many stars. Hubble classified galaxies by their shapes as being spiral, elliptical, or irregular.

Hubble's Law establishes a relationship between the velocity of recession of a galaxy and its distance. The speed at which a galaxy is moving away from the solar system (measured by its redshift, the shift of its light to longer wavelengths, presumed to be caused by the Doppler effect) is directly proportional to the galaxy's distance from the solar system.

The Hubble Space Telescope was deployed by the space shuttle *Discovery* on April 25, 1990. The telescope, which would be free of distortions caused by the Earth's atmosphere, was designed to see deeper into space than any telescope on land. However, on June 27, 1990, the National Aeronautics and Space Administration announced that the telescope had a defect in one of its mirrors which prevented it from properly focusing. Although other instruments, such as one designed to make observations in ultraviolet light, are still operating, nearly 40% of the telescope's experiments will have to be postponed. There is a possibility that special lenses for the telescope's instruments can compensate for the mirror's flaw; however, it will be several years until such repairs can be carried out by space shuttle astronauts.

EXPLORATION

How probable is it that **intelligent life** exists on other planets?

The possibility of intelligent life depends on several factors. An estimation can be calculated by using an equation developed originally by American astronomer Frank Drake (b. 1930). Drake's equation reads $N = N^* f_p n_e f_1 f_i f_c f_L$. This means that the number of advanced civilizations (N) is equal to

N^*, the number of stars in the Milky Way galaxy, times

f_p, the fraction of those stars that have planets, times

n_e, the number of planets capable of supporting life, times

f_1, the fraction of planets suitable for life on which life actually arises, times

f_i, the fraction of planets where intelligent life evolves, times

f_c, the fraction of planets with intelligent life which develops a technically advanced civilization, times

f_L, the fraction of time which a technical civilization lasts.

The equation is obviously subjective and the answer depends on whether optimistic or pessimistic numbers are assigned to the various factors. However, the galaxy is so large that the possibility of life elsewhere must be considered.

Is anyone looking for **extraterrestrial life**?

SETI (the *S*earch for *E*xtra*t*errestrial *I*ntelligence) actively began in 1960 when American astronomer Frank Drake (b. 1930) spent three months at the National Radio Astronomy Observatory in Green Bank, West Virginia, searching for radio signals coming from the nearby stars Tau Ceti and Epsilon Eridani. Although no signals were detected and although scientists interested in SETI have often been ridiculed, support for the idea of seeking out intelligent life in the universe has grown.

Project Sentinel, which used a radio dish at Harvard University's Oak Ridge Observatory in Massachusetts, was capable of monitoring 128,000 channels at a time. This project was upgraded in 1985 to META (*M*egachannel *E*xtra*t*errestrial *A*ssay), thanks in part to a donation by filmmaker Steven Spielberg. Project META is capable of receiving 8.4 million channels. NASA began a ten-year search in October 1992 using radio telescopes in Arecibo, Puerto Rico, and Barstow, California.

Scientists are searching for radio signals which stand out from the random noises caused by natural objects. Such signals might repeat at regular intervals or contain mathematical sequences. There are millions of radio channels and a lot of sky to be examined. The search for extraterrestrial intelligence will probably continue for some time.

What is meant by the phrase **"greening of the galaxy"**?

It is an expression describing the spreading of human life, technology, and culture through interstellar space and eventually across the entire Milky Way galaxy, the Earth's home galaxy.

When was the **Outer Space Treaty** signed?

The United Nations Outer Space Treaty was signed on January 23, 1967. The Treaty provides a framework for the exploration and sharing of outer space. It governs the outer space activities of nations that wish to exploit and make use of space, the moon, and other celestial bodies. It is based on a humanist and pacifist philosophy and on the principle of the non-appropriation of space and the freedom that all nations have to explore and use space. Many countries have signed this agreement, including those from the western alliance, the former Eastern bloc, and non-aligned countries.

Space law, or those rules governing the space activities of various countries, international organizations, and private industries, has been evolving since 1957 when the General Assembly of the United Nations created the Committee on the Peaceful Uses of Outer Space (COPUOS). One of its subcommittees was instrumental in drawing up the 1967 Outer Space Treaty.

What is a "close encounter of the third kind"?

UFO expert J. Allen Hynek (1910–1986) developed the following scale to describe encounters with extraterrestrial beings or vessels:

Close Encounter of the First Kind—sighting of a UFO at close range with no other physical evidence.

Close Encounter of the Second Kind—sighting of a UFO at close range, but with some kind of proof, such as a photograph, or an artifact from a UFO.

Close Encounter of the Third Kind—sighting of an actual extraterrestrial being.

Close Encounter of the Fourth Kind—abduction by a UFO.

Who was the first man in space?

Yuri Gagarin (1934–1968), a Soviet cosmonaut, became the first man in space when he made a full orbit of the Earth in *Vostok I,* April 12, 1961. Gagarin's flight lasted only 1 hour and 48 minutes, but as the first man in space, he became an international hero. Partly because of this Soviet success, U.S. President John F. Kennedy (1917-1963), announced on May 25, 1961, that the United States would land a man on the moon before the end of the decade. The United States took its first step toward that goal when it launched the first American into orbit on February 20, 1962. Astronaut John H. Glenn, Jr. (b. 1921), completed three orbits in *Friendship 7* and travelled about 81,000 miles (130,329 kilometers). Prior to this, on May 5, 1961, Alan B. Shepard, Jr. (b. 1923), became the first American to man a spaceflight, aboard *Freedom 7.* This suborbital flight reached an altitude of 116.5 miles (187.45 kilometers).

Did physical changes occur in the astronauts' bodies when they were in space?

The astronauts grew 1½ to 2¼ inches taller. This height increase was due to spinal lengthening and straightening. Their waist measurements decreased by several inches, due to an upward shift of the internal organs in the body. The calves of their legs became smaller because the muscles of the legs forced blood and other fluids toward the upper part of the body, thus decreasing the girth measurement around the thighs

and calves.

What did NASA mean when it said **Voyager 1 and 2** would take a "grand tour" of the planets?

Once every 176 years the giant outer planets—Jupiter, Saturn, Uranus and Neptune—align themselves in such a pattern that a spacecraft launched from Earth to Jupiter at just the right time might be able to visit the other three planets on the same mission. A technique called "gravity assist" used each planet's gravity as a power boost to point *Voyager* toward the next planet. 1977 was the opportunity year for the "grand tour."

What is the message attached to the **Voyager** spacecraft?

Voyager 1 (launched September 5, 1977) and *Voyager 2* (launched August 20, 1977) by NASA were unmanned space probes designed to explore the outer planets and then travel out of the solar system. A gold-coated copper phonograph record containing a message to any possible extraterrestrial civilization that they may encounter is attached to each spacecraft. The record contains both video and audio images of Earth and the civilization which sent this message to the stars.

The record begins with 118 pictures. These show the Earth's position in the galaxy; a key to the mathematical notation used in other pictures; the sun; other planets in the solar system; human anatomy and reproduction; various types of terrain (seashore, desert, mountains); examples of vegetation and animal life; people of both sexes and of all ages and ethnic types engaged in a number of activities; structures (from grass huts to the Taj Mahal to the Sydney Opera House) showing diverse architectural styles; and means of transportation, including roads, bridges, cars, planes, and space vehicles.

The pictures are followed by greetings from Jimmy Carter, then President of the United States, and Kurt Waldheim, then Secretary General of the United Nations. Brief messages in 54 languages, ranging from ancient Sumerian to English, are included as is a "song" of the humpback whales.

The next section is a series of sounds common to the Earth. These include thunder, rain, wind, fire, barking dogs, footsteps, laughter, human speech, the cry of an infant, and the sounds of a human heartbeat and human brainwaves.

The record concludes with approximately 90 minutes of music, "Earth's Greatest Hits." The musical selections were drawn from a broad spectrum of cultures and include such diverse pieces as a Pygmy girl's initiation song; bagpipe music from Azerbaijan; the Fifth Symphony, First Movement by Ludwig von Beethoven; and "Johnny B. Goode" by Chuck Berry.

It will be tens, even hundreds of thousands of years before either *Voyager* comes close to another star and perhaps the message will never be heard. But it is a sign of humanity's hope to encounter life elsewhere in the universe.

Which astronauts have **walked on the moon?**

Twelve astronauts have walked on the moon. Each Apollo flight had a crew of three. One crew member remained in orbit in the command service module (CSM) while the other two actually landed on the moon.

Apollo 11, July 16–24, 1969
> Neil A. Armstrong
> Edwin E. Aldrin, Jr.
> Michael Collins (CSM pilot, did not walk on the moon)

Apollo 12, November 14–24, 1969
> Charles P. Conrad
> Alan L. Bean
> Richard F. Gordon, Jr. (CSM pilot, did not walk on the moon)

Apollo 14, January 31–February 9, 1971
> Alan B. Shepard, Jr.
> Edgar D. Mitchell
> Stuart A. Roosa (CSM pilot, did not walk on the moon)

Apollo 15, July 26–August 7, 1971
> David R. Scott
> James B. Irwin
> Alfred M. Worden (CSM pilot, did not walk on the moon)

Apollo 16, April 16–27, 1972
> John W. Young
> Charles M. Duke, Jr.
> Thomas K. Mattingly, II (CSM pilot, did not walk on the moon)

Apollo 17, December 7–19, 1972
> Eugene A. Cernan
> Harrison H. Schmitt
> Ronald E. Evans (CSM pilot, did not walk on the moon)

Who made the first **golf shot on the moon?**

Alan B. Shepard, Jr. (b. 1923), commander of *Apollo 14,* launched on January 31, 1971, made the first golf shot. He attached a six iron to the handle of the contingency sample return container, dropped a golf ball on the moon, and took a couple of one-handed swings. He missed with the first, but connected with the second. The ball, he reported, sailed for miles and miles.

Which United States **manned space flight** was the longest?

It was Skylab SL-4, which lasted 84 days, 1 hour, 15 minutes, and 31 seconds from November 16, 1973 to February 8, 1974. During this time, the three-man crew, Gerald

Carr, Edward Gibson, and William Pogue, observed Comet Kohoutek and set records for time spent on experiments in every discipline from medical investigations to materials science. The longest single Soviet stay in space was accomplished by cosmonauts Vladimir Titov (b. 1947) and Musa Manarov (b. 1951) during *Soyuz TM4.* Their space flight aboard the space station Mir lasted 366 days (December 21, 1987–December 21, 1988).

When and what was the first **animal sent into orbit**?

A dog named Laika, aboard the Soviet *Sputnik 2,* launched November 3, 1957, was the first animal sent into orbit. This event followed the successful Soviet launch on October 4, 1957, of *Sputnik 1,* the first man-made satellite ever placed in orbit. Laika was a very small female dog and became the first living creature to go into orbit. She was placed in a pressurized compartment within a capsule that weighed 1103 pounds (500 kilograms). After a few days in orbit, she died, and *Sputnik 2* reentered the Earth's atmosphere on April 14, 1958. Some sources list the dog as a Russian samoyed laika named "Kudyavka" or "Limonchik."

What were the **first monkeys and chimpanzees in space**?

On a United States Jupiter flight on December 12, 1958, a squirrel monkey named Old Reliable was sent into space, but not into orbit. The monkey drowned during recovery.

On another Jupiter flight, on May 28, 1959, two female monkeys were sent 300 miles (482.7 kilometers) high. Able was a six-pound (2.7 kilograms) rhesus monkey and Baker was an 11-ounce (0.3 kilograms) squirrel monkey. Both were recovered alive.

A chimpanzee named Ham was used on a Mercury flight on January 31, 1961. Ham was launched to a height of 157 miles (253 kilometers) into space, but did not go into orbit. His capsule reached a maximum speed of 5,857 miles (9,426 kilometers) per hour and landed 422 miles (679 kilometers) downrange in the Atlantic Ocean where he was recovered unharmed.

On November 29, 1961, the United States placed a chimpanzee, named Enos, into orbit and recovered him alive after two complete orbits around the Earth. Like the Soviets, who usually used dogs, the United States had to obtain information on the effects of space flight on living beings before they could actually launch a human into space.

Who were the first man and woman to **walk in space**?

On March 18, 1965, the Soviet cosmonaut Alexei Leonov (b. 1934) became the first person to walk in space when he spent 10 minutes outside his *Voskhod 2* spacecraft. The first woman to walk in space was Soviet cosmonaut Svetlana Savitskaya (b. 1947) who, during her second flight aboard the *Soyuz T-12* (July 17, 1984), performed 3½ hours of extra-vehicular activity.

The first American to walk in space was Edward White II (1930–1967) from the spacecraft *Gemini 4* on June 3, 1965. Kathryn D. Sullivan (b. 1951) became the first American woman to walk in space when she spent 3½ hours outside the Challenger orbiter during the space shuttle mission 41G on October 11, 1984.

American astronaut Bruce McCandless II (b. 1937) performed the first untethered space walk from the space shuttle *Challenger* on February 7, 1984, using an MMU (*m*anual *m*aneuvering *u*nit) back-pack.

What were the first words spoken by an astronaut after touchdown of the lunar module on the **Apollo 11** flight and by an astronaut standing on the moon?

On July 20, 1969, at 4:17:43 p.m. Eastern Daylight Time (20:17:43 Greenwich Mean Time), Neil A. Armstrong (b. 1930) and Edwin E. Aldrin, Jr. (b. 1930), landed the lunar module *Eagle* in the moon's Sea of Tranquility, and Armstrong radioed: "Houston, Tranquility Base here. The *Eagle* has landed." Several hours later, when Armstrong descended the lunar module ladder and made the small jump between the *Eagle* and the lunar surface, he announced: "That's one small step for man, one giant leap for mankind." The article "a" was missing in the live voice transmission, and was later inserted in the record to amend the message to "one small step for *a* man."

Who are the **American women** who have gone **into** space?

As of June 1992, the following American women have flown in space:

Sally K. Ride
June 18, 1983	STS-7	Challenger
October 5, 1984	41G	Challenger

Judith Resnick
August 31, 1984	41D	Discovery
January 28, 1986	51L	Challenger

Kathryn D. Sullivan
October 5, 1984	41G	Challenger
April 24, 1990	STS-31	Discovery
March 24, 1992	STS-45	Atlantis

Anna L. Fischer
November 8, 1984   51A   Discovery

Margaret R. Seddon
April 12, 1985   51D   Discovery
June 5, 1981   STS-40   Columbia

Shannon W. Lucid
June 17, 1985   51G   Discovery
October 18, 1989   STS-34   Atlantis
August 2, 1991   STS-43   Atlantis

Bonnie J. Dunbar
October 30, 1985   61A   Challenger
January 9, 1990   STS-32   Columbia
June 25, 1992   STS-50   Columbia

Mary L. Cleave
November 26, 1985 61B   Atlantis
May 4, 1989   51L   Atlantis

Christa McAuliffe
January 28, 1986   51L   Challenger

Ellen S. Baker
October 18, 1989   STS-34   Atlantis
June 25, 1992

Kathryn C. Thornton
November 22, 1989 STS-33   Discovery
May 7, 1992   STS-49   Endeavour

Marsha S. Irvins
January 9, 1990   STS-32   Columbia

Linda M. Godwin
April 5, 1991   STS-40   Columbia

Millie Hughes-Fulford
June 5, 1991   STS-40   Columbia

Who was the first **black American in space?**

Guion S. Bluford, Jr. (b. 1942), became the first African American to fly in space during the Space Shuttle *Challenger* mission STS-8 (August 30–September 5, 1983). Astronaut Bluford, who holds a Ph.D. in aerospace engineering, made a second shuttle flight aboard *Challenger* mission STS-61-A/Spacelab D1 (October 30-November 6, 1985). The first black man to fly in space was the Cuban cosmonaut, Arnaldo Tamayo-Mendez, who flew aboard *Soyuz 38* and spent eight days aboard the Soviet space station *Salyut 6* during September 1980. Dr. Mae C. Jemison became the first African American woman in space on September 12, 1992, aboard the Space Shuttle *Endeavour,* mission Spacelab-J. **53**

How many successful space flights were **launched in 1991,** and by what countries?

A total of 88 flights which achieved Earth orbit or beyond were made in 1991:

Country or Organization	Number of Launches
U.S.S.R.	59
United States	18
European Space Agency	8
Japan	2
Peoples' Republic of China	1

In 1990, there were 116 such launches, with 75 by the U.S.S.R.; 27 (including 7 commercial launches) by the United States; 5 by the European Space Agency; 5 by the People's Republic of China; and 1 by Israel.

Who has the most **time accumulated in space?**

Soviet cosmonaut Sergei Krikalev (b. 1958) has logged 465 days traveling in space. He accumulated this total on only two missions; during the Soyuz TM-7 mission (November 26, 1988–April 27, 1989), he spent 152 days in space, and during Soyuz TM-12 (May 18, 1991–March 25, 1992), he spent 313 days in space. During both flights, he and other cosmonauts remained aboard the orbiting *MIR* space station.

What was the first **meal on the moon?**

American astronauts Neil A. Armstrong (b. 1930) and Edwin E. Aldrin, Jr. (b. 1930) ate four bacon squares, three sugar cookies, peaches, pineapple-grapefruit drink, and coffee before their historic moonwalk on July 20, 1969.

Who were the first married couple to go into space together?

Astronauts Jan Davis and Mark Lee were the first married couple in space. They flew aboard the space shuttle *Endeavor* on an eight-day mission which began on September 12, 1992. Ordinarily NASA bars married couples from flying together. An exception was made for Davis and Lee because they had no children and had begun training for the mission long before they got married.

Who was the first **woman in space?**

Valentina V. Tereshkova-Nikolaeva (b. 1937), a Soviet cosmonaut, was the first woman in space. She was aboard the *Vostok 6,* launched June 16, 1963. She spent three days circling the Earth, completing 48 orbits. Although she had little cosmonaut training, she had been an accomplished parachutist and was especially fit for the rigors of space travel.

The United States space program did not put a woman in space until 20 years later when, on June 18, 1983, Sally K. Ride (b. 1951) flew aboard the space shuttle *Challenger* mission STS-7. In 1987, she moved to the administrative side of NASA and was instrumental in issuing the "Ride Report" which recommended future missions and direction for NASA. She retired from NASA in August 1987 to become a research fellow at Stanford University after serving on the Presidential Commission that investigated the *Challenger* disaster. At present, she is the director of the California Space Institute at the University of California San Diego.

When was the first United States **satellite** launched?

Explorer 1, launched January 31, 1958, by the U.S. Army was the first United States satellite launched into orbit. This 31-pound (14.06-kilogram) satellite carried instrumentation that led to the discovery of the Earth's radiation belts, which would be named after University of Iowa scientist James A. Van Allen. It followed four months after the world's first satellite, the Soviet Union's *Sputnik 1.* On October 3, 1957, the Soviet Union placed the large, 184-pound (83.5-kilogram) satellite into low Earth orbit. It carried instrumentation to study the density and temperature of the upper atmosphere, and its launch was the event that opened the age of space.

What is the mission of the **Galileo** spacecraft?

Galileo, launched October 18, 1989, will circle the sun three times and require almost six years to reach Jupiter, looping past Venus once and the Earth twice. The anticipated arrival date at Jupiter is currently late 1995. The *Galileo* spacecraft was designed to make a detailed study of Jupiter and its rings and moons over a period of years. It will release a probe into the jovian atmosphere to analyze the different layers. This probe's range will be limited by its ability to withstand atmospheric pressure (up to several Earth atmospheres). The United States and the former U.S.S.R. have launched many spacecraft to the planets, and only Pluto remains to be explored. The surfaces of the moon, Mars, and Venus have all been examined by unmanned landers, and nearly 1103 pounds (500 kilograms) of rock samples from the moon have been brought back to Earth. For the next two decades of solar exploration, the objectives will be to observe, at close quarters, objects which have never been visited by spacecraft, and to increase the amount of data on the planets and their moons which have already been examined. No explorations of Pluto are planned for the foreseeable future.

Who was the "father" of the **Soviet space program?**

Sergei P. Korolev (1907–1966) made enormous contributions to the development of Soviet manned space flight, and his name is linked with their most significant space achievements. Trained as an aeronautical engineer, he directed the Moscow group studying the principles of rocket propulsion, and in 1946 took over the Soviet program to develop long-range ballistic rockets. Under Korolev, the Soviets used these rockets for space projects and launched the world's first satellite in October 4, 1957. Besides a vigorous unmanned interplanetary research program, Korolev's goal was to place men in space, and following tests with animals his manned space flight program was initiated when Yuri Gagarin (1934–1968) was successfully launched into Earth orbit.

How many **fatalities** have occurred during space-related missions?

The fourteen astronauts and cosmonauts listed below died in space-related accidents.

Date	Astronaut/Cosmonaut	Mission
January 27, 1967	Roger Chaffee (U.S.)	Apollo 1
January 27, 1967	Edward White II (U.S.)	Apollo 1
January 27, 1967	Virgil "Gus" Grissom (U.S.)	Apollo 1
April 24, 1967	Vladimir Komarov (U.S.S.R.)	Soyuz 1
June 29, 1971	Viktor Patsayev (U.S.S.R.)	Soyuz 11
June 29, 1971	Vladislav Volkov (U.S.S.R.)	Soyuz 11
June 29, 1971	Georgi Dobrovolsky (U.S.S.R.)	Soyuz 11
January 28, 1986	Gregory Jarvis (U.S.)	STS 51L
January 28, 1986	Christa McAuliffe (U.S.)	STS 51L
January 28, 1986	Ronald McNair (U.S.)	STS 51L
January 28, 1986	Ellison Onizuka (U.S.)	STS 51L
January 28, 1986	Judith Resnik (U.S.)	STS 51L
January 28, 1986	Francis Scobee (U.S.)	STS 51L
January 28, 1986	Michael Smith (U.S.)	STS 51L

Chaffee, Grissom, and White died in a cabin fire during a test firing of the *Apollo 1* rocket.

Komarov was killed in *Soyuz 1* when the capsule's parachute failed.

Dobrovolsky, Patsayev, and Volkov were killed during the *Soyuz II*'s re-entry when a valve accidently opened and released their capsule's atmosphere.

Jarvis, McAuliffe, McNair, Onizuka, Resnik, Scobee, and Smith died when the space shuttle *Challenger* STS 51L exploded 73 seconds after lift-off.

In addition, 19 other astronauts and cosmonauts have died of non-space related causes. Fourteen of these died in air crashes (including Yuri Gagarin), four died of natural causes, and one died in an auto crash.

What was the worst **disaster in the U.S. space program** and what caused it?

Challenger mission STS 51L was launched on January 28, 1986, but exploded only 73 seconds after lift-off. The entire crew of seven was killed and the *Challenger* was completely destroyed. The investigation of the *Challenger* tragedy was performed by the Rogers Commission, established and named for its chairman, former Secretary of State William Rogers.

The consensus of the Rogers Commission (which studied the accident for several months) and participating investigative agencies is that the accident was caused by a failure in the joint between the two lower segments of the right solid rocket motor. The specific failure was the destruction of the seals that are intended to prevent hot gases from leaking through the joint during the propellant burn of the rocket motor. The evidence assembled by the Commission indicated that no other element of the space shuttle system contributed to this failure.

Although the Commission did not affix blame to any individuals, the public record made clear that the launch should not have been made that day. The weather was unusually cold at Cape Canaveral and temperatures had dipped below freezing during the night. Test data had suggested that the seals (called O-rings) around the solid rocket booster joints lost much of their effectiveness in very cold weather.

What is "Earthshine"?

A spacecraft in orbit around the Earth is illuminated by sunlight and "earthshine." Earthshine consists of sunlight reflected by the Earth and thermal radiation emitted by the Earth's surface and atmosphere.

58

EARTH

AIR

See also: Climate and Weather

What is the **composition** of the **Earth's atmosphere**?

The Earth's atmosphere, apart from water vapor and pollutants, is composed of 78% nitrogen, 20% oxygen, and less than 1% each of argon and carbon dioxide. It also has traces of hydrogen, neon, helium, krypton, xenon, methane, and ozone. Earth's original atmosphere was probably composed of ammonia and methane; 20 million years ago the air started to contain a broader variety of elements.

How many **layers** does the **Earth's atmosphere** contain?

The atmosphere, the "skin" of gas that surrounds the Earth, consists of five layers that are differentiated by temperature:

The troposphere—is the lowest level; it averages about 7 miles (11 kilometers) in thickness, varying from 5 miles (8 kilometers) at the poles to 10 miles (16 kilometers) at the equator. Most clouds and weather form in this layer. Temperature decreases with altitude in the troposphere.

The stratosphere—ranges between 7 miles (11 kilometers) to 30 miles (48 kilometers) above the Earth's surface. The ozone layer, important because it absorbs most of the sun's harmful ultraviolet radiation, is located in this band. Temperatures rise slightly with altitude to a maximum of about 32°F (0°C).

The mesosphere—(above the stratosphere) extends from 30 miles (48 kilometers) to 55 miles (85 kilometers) above the Earth. Temperatures decrease with altitude to -130°F (-90°C).

The thermosphere—(also known as the hetereosphere) is between 55 miles (85 kilometers) to 435 miles (700 kilometers). Temperatures in this layer range to 2696°F (1475°C).

The exosphere—beyond the thermosphere, applies to anything above 435 miles (700 kilometers). In this layer, temperature no longer has any meaning.

The ionosphere—is a region of the atmosphere that overlaps the others, reaching from 30 miles (48 kilometers) to 250 miles (402 kilometers). In this region, the air becomes ionized (electrified) from the sun's ultraviolet rays, etc. This area affects the transmission and reflection of radio waves. It is divided into three regions: the D region (at 35 to 55 miles, or 56 to 88 kilometers), the E Region (Heaviside-Kennelly Layer, 55 to 95 miles, or 56 to 153 kilometers), and the F Region (Appleton Layer, at 95 to 250 miles, or 153 to 402 kilometers).

What are the **Van Allen belts**?

The Van Allen belts (or zones) are two regions of highly charged particles above the Earth's equator trapped by the magnetic field, which surrounds the Earth. Also called the magnetosphere, the first belt extends from a few hundred to about 2,000 miles (3,200 kilometers) above the Earth's surface and the second is between 9,000 and 12,000 miles (14,500 to 19,000 kilometers). The particles, mainly protons and electrons, come from the solar wind and cosmic rays. The belts are named in honor of James Alfred Van Allen (b.1914), the American physicist who discovered them in 1958 and 1959, with the aid of radiation counters carried aboard the artificial satellites, *Explorer I* (1958) and *Pioneer 3* (1959).

Why is the **sky blue**?

The sunlight interacting with the Earth's atmosphere makes the sky blue. In outer space the astronauts see blackness because outer space has no atmosphere. Sunlight consists of light waves of varying wavelengths, each of which is seen as a different color. The minute particles of matter and molecules of air in the atmosphere intercept and scatter the white light of the sun. A larger portion of the blue color in white light is scattered, more so than any other color because the blue wavelengths are the shortest. When the size of atmospheric particles are smaller than the wavelengths of the colors, selective scattering occurs—the particles only scatter one color and the atmosphere will appear to be that color. Blue wavelengths especially are affected, bouncing off the air particles to become visible. This is why the sun looks yellow (yellow equals white minus blue). At sunset, the sky changes color because as the sun drops to the horizon, sunlight has more atmosphere to pass through and loses more of its blue wavelengths (the shortest of all the colors). The orange and red, having the longer wavelengths and making up more of sunlight at this distance, are most likely to be scattered by the air particles.

What is the **weight of one cubic foot of dry air** at one atmosphere of barometric pressure?

Temperature (Fahrenheit)	Weight per cubic foot (pounds)
50°	0.07788
60°	0.07640
70°	0.07495

What is the **density of air?**

The density of dry air is 1.29 grams per liter at 32°F (0°C) at average sea level. Average sea level, measured by the height of a column of mercury in a barometer, is 29.92 inches (760 millimeters).

PHYSICAL CHARACTERISTICS, ETC.

See also: Space—Planets and Moons

How much does the **Earth weigh?**

It is estimated to weigh 6 sextillion, 588 quintillion short tons (6.6 sextillion short tons) or 5.97×10^{24} kilograms with the Earth's mean density being 5.515 times that of water (the standard). This is calculated from using the parameters of an ellipsoid adopted by the International Astronomical Union in 1964 and recognized by the International Union of Geodesty and Geophysics in 1967 in which the area of the ellipsoid is 196,938,800 square miles.

What is the **interior of the Earth** like?

The Earth is divided into a number of layers. The topmost layer is the crust, which contains about 0.6% of the Earth's volume. The depth of the crust varies from 3.5 to 5 miles (5 to 9 kilometers) beneath the oceans to 50 miles (80 kilometers) beneath some mountain ranges. The crust is formed primarily of rocks such as granite and basalt.

Between the crust and the mantle is a boundary known as the Mohorovicvicv discontinuity (or Moho for short), named for the Croatian seismologist, Andrija Mohorovičič (1857–1936), who discovered it in 1909. Below the Moho is the mantle, extending down about 1,800 miles (2,900 kilometers). The mantle is composed mostly of oxygen, iron, silicon, and magnesium, and accounts for about 82% of the Earth's volume. Although mostly solid, the upper part of the mantle, called the asthenosphere, is partially liquid.

61

The core-mantle boundary, also called the Gutenberg discontinuity for the German-American seismologist, Beno Gutenberg (1889–1960), separates the mantle from the core. Made up primarily of nickel and iron, the core contains about 17% of the Earth's volume. The outer core is liquid and extends from the base of the mantle to a depth of about 3,200 miles (5,155 kilometers). The solid inner core reaches from the bottom of the outer core to the center of the Earth, about 3,956 miles (6,371 kilometers) deep. The temperature of the inner core is estimated to be about 7,000°F (3,850°C).

Which elements are contained in the **Earth's crust**?

The most abundant elements in the Earth's crust are listed in the table below. In addition, nickel, copper, lead, zinc, tin, and silver account for less than 0.02% with all other elements comprising 0.48%.

Element	Percentage
Oxygen	47.0
Silicon	28.0
Aluminum	8.0
Iron	4.5
Calcium	3.5
Magnesium	2.5
Sodium	2.5
Potassium	2.5
Titanium	0.4
Hydrogen	0.2
Carbon	0.2
Phosphorous	0.1
Sulfur	0.1

How does the **temperature of the Earth** change as one goes deeper underground?

The Earth's temperature increases with depth. Measurements made in deep mines and drill–holes indicate that the rate of temperature increase varies from place to place in the world, ranging from 59° to 167°F (15 to 75°C) per kilometer in depth. Actual temperature measurements cannot be made beyond the deepest drill–holes which are a little more than 6.2 miles (10 kilometers) deep. Estimates show that the temperatures at the Earth's center can reach values of 5,000°F (2,760°C) or higher.

What are the **highest and lowest** points on Earth?

The highest point on land is the top of Mt. Everest (in the Himalayas on the Nepal-Tibet border) at 29,028 feet (8,848 meters) above sea level, plus or minus 10 feet (3 meters) because of snow. This height was established by the Surveyor General of India in 1954 and accepted by the National Geographic Society. Prior to that the height was taken to be 29,002 feet (8,840 meters). Satellite measurements taken in 1987 indicate that Mt. Everest is 29,864 feet (9,102 meters) high but this measurement has not been adopted by the National Geographic Society.

The lowest point on land is the Dead Sea between Israel and Jordan which is 1,312 feet (399 meters) below sea level. The lowest point on the Earth's surface is thought to be in the Marianas Trench in the western Pacific Ocean extending from southeast of Guam to northwest of the Marianas Islands. It has been measured as 36,198 feet (11,034 meters) below sea level.

What is the highest and lowest **elevation in the United States**?

Mt. McKinley, Alaska, at 20,320 feet (6,194 meters), is the highest point in the United States and North America. Mt. Whitney, California, at 14,494 feet (4,421 meters), is the highest point in the continental United States. Death Valley, California, at 282 feet (86 meters) below sea level, is the lowest in the United States and in the western hemisphere.

How much of the **Earth's surface** is land and how much is water?

Approximately 30% of the Earth's surface is land. This is about 57,259,000 square miles (148,300,000 square kilometers). The area of the Earth's water surface is approximately 139,692,000 square miles (361,800,000 square kilometers), or 70% of the total surface area.

WATER

What is an **aquifer**?

Some rocks of the upper part of the Earth's crust contain many small holes, or pores. When these holes are large or are joined together so that water can flow through them easily, the rock is considered to be permeable. A large body of permeable rock in which water is stored and flows through is called an aquifer (from the Latin for "water" and "to bear"). Sandstones and gravels are excellent examples of permeable rock.

As water reservoirs, aquifers provide about 60% of American drinking water. The huge Ogallala Aquifer, underlying about two million acres of the Great Plains, is a major source of water for the central United States. It has been estimated that after oceans (containing 1,370 million cubic kilometers of water), aquifers, with an estimated 50 million cubic kilometers, are the second largest store of water. Water is purified as it is filtered through the rock, but it can be polluted by spills, dumps, acid rain, and other causes. In addition, recharging of water by rainfall often cannot keep up with the volume removed by heavy pumping. The Ogallala Aquifer's supply of water could be depleted by 25% by the year 2020.

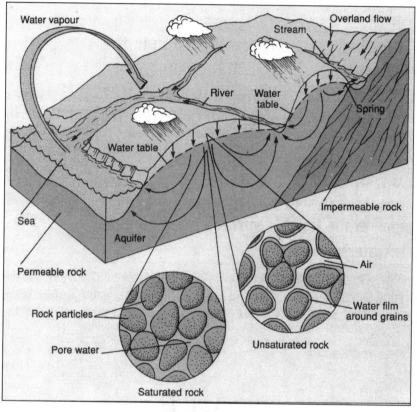

Aquifer

What is the **chemical composition** of the **ocean**?

The ocean contains every known naturally occurring element plus various gases, chemical compounds, and minerals.

Constituent	Concentration (parts per million)
Chloride	18,980
Sodium	10,560
Sulfate	2,560
Magnesium	1,272
Calcium	400
Potassium	380
Bicarbonate	142
Bromide	65
Strontium	13
Boron	4.6
Fluoride	1.4

Why is the **sea blue**?

There is no single cause for the colors of the sea. What is seen depends in part on when and from where the sea is observed. Eminent authority can be found to support almost any explanation. Some explanations include absorption and scattering of light by pure water; suspended matter in sea water; the atmosphere; and color and brightness variations of the sky. For example, one theory is that when sunlight hits seawater, part of the white light, composed of different wavelengths of various colors, is absorbed, and some of the wavelengths are scattered after colliding with the water molecules. In clear water, red and infrared light are greatly absorbed but blue is least absorbed, so that the blue wavelengths are reflected out of the water. The blue effect requires a minimum depth of 10 feet (3 meters) of water.

What causes **waves** in the ocean?

The most common cause of surface waves is air movement (the wind). Waves within the ocean can be caused by tides, interactions among waves, submarine earthquakes or volcanic activity, and atmospheric disturbances. Wave size depends on wind speed, wind duration, and the distance of water over which the wind blows. The longer the distance the wind travels over water, or the harder it blows, the higher the waves. As the wind blows over water it tries to drag the surface of the water with it. The surface cannot move as fast as air, so it rises. When it rises, gravity pulls the water back, carrying the falling water's momentum below the surface. Water pressure from below pushes this swell back up again. The tug of war between gravity and water pressure constitutes wave motion. Capillary waves are caused by breezes of less than two knots. At 13 knots the waves grow taller and faster than they grow longer, and their steepness cause them to break, forming whitecaps. For a whitecap to form, the wave height must be 1/7 the distance between wave crests. Tidal waves are really seismic waves or tsunamis, and are caused by earthquakes or volcanoes. These are very long waves—100

65

to 200 miles (161 to 322 kilometers) with high speeds (500 miles, or 805 kilometers, per hour)—that, when approaching shallow water, can grow into a 100-foot–high (30.5–meter) wave as its wavelength is reduced abruptly.

How far can **sunlight penetrate** into the **ocean?**

Because seawater is relatively transparent, approximately 5% of sunlight can penetrate clean ocean water to a depth of 262 feet (80 meters). When the water is turbid (cloudy) due to currents, mixing of silt, increased growth of algae, or other factors, the depth of penetration is reduced to less than 164 feet (50 meters).

How **deep** is the ocean?

The average depth of the ocean floor is 13,124 feet (4,000 meters). The average depth of the four major oceans is given below:

Ocean	Average depth	
	Feet	Meters
Pacific	13,740	4,188
Atlantic	12,254	3,735
Indian	12,740	3,872
Arctic	3,407	1,038
Average overall	13,124	4,000

There are great variations in depth because the ocean floor is often very rugged. The greatest depth variations occur in deep, narrow depressions known as trenches along the margins of the continental plates. The deepest measurements made—36,000 feet (11,033 meters), deeper than the height of the world's tallest mountains—was taken in Mariana Trench east of the Mariana Islands. In January 1960, the French oceanographer Jacques Piccard, together with the United States Navy Lieutenant David Walsh, took the bathyscaphe *Trieste* to the bottom of the Mariana Trench.

Ocean	Deepest point	Depth	
		Feet	Meters
Pacific	Mariana Trench	36,200	11,033
Atlantic	Puerto Rico Trench	28,374	8,648
Indian	Java Trench	25,344	7,725
Arctic	Eurasia Basin	17,881	5,450

What is a **tidal bore?**

It is a wave which moves inland or upriver as the incoming tidal current surges against the flow of river water. A tidal bore occurs in a long narrow estuary (the area of a river mouth in which tides have an effect) or in a river that flows into an estuary.

Where are the world's **highest tides**?

The Bay of Fundy (New Brunswick, Canada) has the world's highest tides. They average about 45 feet (14 meters) high in the northern part of the bay, far surpassing the world average of 2.5 feet (0.8 meter).

What is the difference between an **ocean** and a **sea**?

There is no neatly defined distinction between ocean and sea. One definition terms the ocean as a great body of interconnecting salt water that covers 71% of the Earth's surface. There are four major oceans—the Arctic, Atlantic, Indian, and Pacific—but some sources do not include the Arctic Ocean, calling it a marginal sea. The terms "ocean" and "sea" are often used interchangeably but a sea is generally considered to be smaller than an ocean. The name is often given to salt-water areas on the margins of an ocean, such as the Mediterranean Sea.

How **salty** is **seawater**?

Sea water is, on average, 3.3 to 3.7% salt. The amount of salt varies from place to place. In areas where large quantities of fresh water are supplied by melting ice, rivers, or rainfall, such as the Arctic or Antarctic, the level of salinity is lower. Areas such as the Persian Gulf and the Red Sea have salt contents over 4.2%. If all the salt in the ocean were dried, it would form a mass of solid salt the size of Africa. Most of the ocean salt comes from processes of dissolving and leaking from the solid Earth over hundreds of millions of years. Some is the result of salty volcanic rock that flows up from a giant rift that runs through all the ocean's basins.

Is the **Dead Sea** really dead?

The Dead Sea, on the boundary between Israel and Jordan, is called "dead" because nothing can live in it. It has a salt content of 25%. Because it is the lowest body of water on the Earth's surface, any water that flows into the Dead Sea has no outflow. As the water evaporates, dissolved minerals are left in the sea.

How much salt is in **brackish water**?

Brackish water has a saline (salt) content between that of fresh water and sea water. It is neither fresh nor salty, but somewhere in between. Brackish waters are usually **67** regarded as those containing 0.5 to 30 parts per thousand salt, while the average saltiness of seawater is 35 parts per thousand.

Where is the world's **deepest lake**?

Lake Baikal, located in southeast Siberia, is 5,314 feet (1,620 meters) deep at its maximum depth, making it the deepest lake in the world.

Where are the **five largest lakes** in the world located?

Location	Area		Length		Depth	
	Square miles	Square kma	Miles	Kma	Feet	Meters
Caspian Sea[b], Asia-Europe	143,244	370,922	760	1,225	3,363	1,025
Superior, North America	31,700	82,103	350	560	1,330	406
Victoria, Africa	26,828	69,464	250	360	270	85
Aralb, Asia	24,904	64,501	280	450	220	67
Huron, North America	23,010	59,600	206	330	750	229

[a]-Kilometers
[b]-Salt water lake

What is the **bearing capacity of ice** on a lake?

The following chart indicates the maximum safe load. It applies only to clear lake ice that has not been heavily travelled. For early winter slush ice, ice thickness should be doubled for safety.

Ice thickness		Examples	Maximum safe load	
Inches	Centimeters		Tons	Kilograms
2	5	One person on foot		
3	7.6	Group in single file		
7.5	19	Car or snowmobiles	2	907.2
8	20.3	Light truck	2½	1,361
10	25.4	Medium truck	3½	1,814.4
12	30.5	Heavy truck	9	7,257.6
15	38		10	9,072
20	50.8		25	22,680

Which of the **Great Lakes** is the largest?

Lake	Surface area		Maximum depth	
	Square miles	Square kilometers	Feet	Meters
Superior	31,700	82,103	1,333	406
Huron	23,010	59,600	750	229
Michigan	22,300	57,757	923	281
Erie	9,910	25,667	210	64
Ontario	7,540	19,529	802	244

What is an **oxbow lake**?

An oxbow or ox-bow lake is a crescent-shaped lake lying alongside a winding river, formed when a meander (a bend or loop in the river) becomes separated from the main stream. Erosion and deposition tend to accentuate the meander, with the faster-flowing water on the outside edges of the curves eroding the banks, while the slower-moving water on the inside edges deposits silt on the opposite banks. Over time, this process widens the loop of the meander, while narrowing the neck of land dividing the straight path of the river, until the neck vanishes and the river runs past the isolated meander. Without a current to keep it clear, the narrow horseshoe-shaped lake silts up. During the comparatively brief period after the meander is cut off and before it fills with silt, it is an oxbow lake.

A different explanation for the mechanics involved in cutting off a meander has been suggested by laboratory experiments. A stream needs a minimum slope or gradient (horizontal distance divided by height change) to be able to flow and to transport sediment; a meander decreases the gradient of a stream by increasing the distance that it must cover before dropping to a lower height. If a stream's gradient falls below the minimum required, it will tend to cut across the meanders in order to continue flowing.

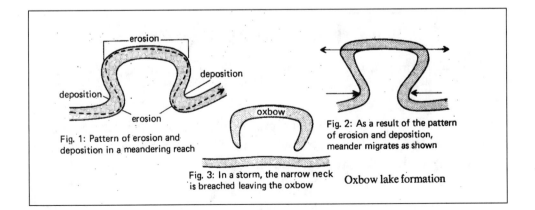

Fig. 1: Pattern of erosion and deposition in a meandering reach

Fig. 2: As a result of the pattern of erosion and deposition, meander migrates as shown

Fig. 3: In a storm, the narrow neck is breached leaving the oxbow

Oxbow lake formation

What is a **yazoo**?

A yazoo is a tributary of a river that runs parallel to the river, being prevented from joining the river because the river has built up high banks. The name is derived from the Yazoo River, a tributary of the Mississippi River, which demonstrates this effect.

Which **rivers flow northward**?

The major rivers that flow north are the Ob, Yenesei, and Lena in Asia; the Nile in Africa; the Rhine in Europe; the Amazon in South America; the Mackenzie and Yukon Rivers in North America.

In the United States, the St. John's River in Florida flows north and the Willamette River in Oregon also flows northward for almost all its length. The Monongahela River, originating in West Virginia, winds northeast into Pennsylvania and then flows northward to Pittsburgh.

What are the **longest rivers** in the world?

River	Length	
	Miles	**Kilometers**
Nile (Egypt)	4,145	6,670
Amazon (Brazil)	4,000	6,404
Chang jiang–Yangtze (China)	3,964	6,378
Mississippi-Missouri (river system) (U.S.)	3,740	6,021
Yenisei-Angara (river system) (U.S.S.R.)	3,442	5,540

When will Niagara Falls disappear?

The water dropping over Niagara Falls digs great plunge pools at the base, undermining the shale cliff and causing the hard limestone cap to cave in. Niagara has eaten itself seven miles (11 kilometers) upstream since it was formed 10 thousand years ago. At this rate, it will disappear into Lake Erie in 22,800 years. The Niagara River connects Lake Erie with Lake Ontario, and marks the U.S.–Canada boundary (New York–Ontario).

What is the world's **highest waterfall**?

Angel Falls, named after the explorer and bush pilot, Jimmy Angel, on the Carrao tributary in Venezuela is the highest waterfall in the world. It has a total height of 3,212 feet (979 meters) with its longest unbroken drop being 2,648 feet (807 meters).

It is difficult to determine the height of a waterfall because many are composed of several sections rather than one straight drop. The highest waterfall in the United States is Yosemite Falls on a tributary of the Merced River in Yosemite National Park, California, with a total drop of 2425 feet (739 meters). There are three sections to the Yosemite Falls: Upper Yosemite is 1,430 feet (435 meters); Cascades (middle portion), 675 feet (205 meters); and Lower Yosemite, 320 feet (97 meters).

LAND

See also: Space—Planets and Moons

Are there **tides in the solid part of the Earth** as well as in its waters?

The solid Earth is distorted about 4.5 to 14 inches (11.4 to 35.6 centimeters) by the gravitational pull of the sun and moon. It is the same gravitational pull that creates the tides of the waters. When the moon's gravity pulls water on the side of the Earth near to it, it pulls the solid body of the Earth on the opposite side away from the water to create bulges on both sides, and causing high tides. These occur every 12½ hours. Low tides occur in those places from which the water is drained to flow into the two high–tide bulges. The sun causes tides on the Earth that are about 33 to 46% as high as those due to the moon. During a new moon or a full moon when the sun and moon are in a straight line, the tides of the moon and the sun reinforce each other, to make high tides higher; these are called spring tides. At the quarter moons, the sun and moon are out of step (at right angles), the tides are less extreme than usual; these are called neap tides. Smaller bodies of water, such as lakes, have no tides because the whole body of water is raised all at once, along with the land beneath it.

Do the **continents move**?

In 1912, a German geologist, Alfred Lothar Wegener (1880–1930), theorized that the continents had drifted or floated apart to their present locations and that once all the continents had been a single land mass near Antarctica, which is called Pangaea (from the Greek word meaning *all-earth*). Pangaea then broke apart some 200 million years ago into two major continents called Laurasia and Gondwanaland. These two continents continued drifting and separating until the continents evolved their present shapes and positions. Wegener's theory was discounted but it has since been found that the continents do move sideways (not drift) at an estimated 0.75 inch (1.9 centimeters) annually because of the action of plate tectonics. American geologist, William Maurice Ewing (1906–1974) and Harry Hammond Hess (1906–1969) proposed that the Earth's crust is not a solid mass, but composed of eight major and seven minor plates that can move apart, slide by each other, collide, or override each other. Where these plates meet are major areas of mountain-building, earthquakes, and volcanoes.

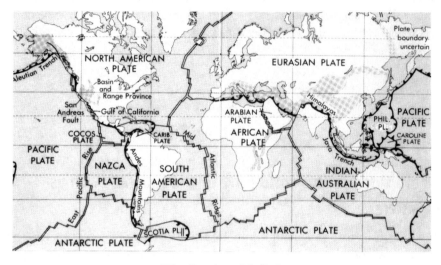

Lithosphere plates of the Earth

How much of the Earth's surface is **covered with ice**?

About 10.4% of the world's land surface is glaciated, or permanently covered with ice. Approximately 6,020,000 square miles (15,600,000 square kilometers) are covered by ice in the form of ice sheets, ice caps, or glaciers. An ice sheet is a body of ice that blankets an area of land, completely covering its mountains and valleys. Ice sheets have an area of over 19,000 square miles (50,000 square kilometers); ice caps are smaller. Glaciers are larger masses of ice that flow, under the force of gravity, at a rate of between ten and one thousand feet (three and 300 meters) per year. Glaciers on steep slopes flow faster. For example, the Quarayoq Glacier in Greenland averages 65 to 80 feet (20 to 24 meters) per day. The area of glaciation in parts of the world are:

Place	Area	
	Square miles	Square kilometers
Antarctia	5,250,000	12,588,000
North Polar Regions (Greenland, Northern Canada, Arctic Ocean islands)	799,000	2,070,000

Place	Area	
	Square miles	Square kilometers
Asia	44,000	115,800
Alaska & Rocky Mountains	29,700	76,900
South America	10,200	26,500
Iceland	4,699	12,170
Alpine Europe	3,580	9,280
New Zealand	391	1,015
Africa	5	12

How much of the Earth's surface is **permanently frozen**?

About one fifth of the Earth's land is permafrost, or ground that is permanently frozen. This classification is based entirely on temperature and disregards the composition of the land. It can include bedrock, sod, ice, sand, gravel, or any other type of material in which the temperature has been below freezing for over 2 years. Nearly all permafrost is thousands of years old.

Where are the **northernmost** and **southernmost points of land**?

The most northern point of land is Cape Morris K. Jesup on the northeastern extremity of Greenland. It is at 83 degrees, 39 minutes north latitude and is 440 miles (708 kilometers) from the North Pole. However, the *Guinness Book of Records 1992* reports that, an islet of 100 feet across, called Oodaq, is more northerly at 83 degrees, 40 minutes north latitude and 438.9 miles (706 kilometers) from the North Pole. The southernmost point of land is the South Pole (since the South Pole, unlike the North Pole, is on land).

In the United States, the northernmost point of land is Point Barrow, Alaska (71 degrees, 23 minutes north latitude), and the southernmost point of land is Ka Lae or South Cape (18 degrees, 55 minutes north latitude) on the island of Hawaii. In the 48 contiguous states, the northernmost point is Northwest Angle, Minnesota (49 degrees, 23 minutes north latitude); the southernmost point is Key West, Florida (24 degrees, 33 minutes north latitude).

How thick is the ice that covers **Antarctica**?

The ice that covers Antarctica is 15,700 feet (4,785 meters) in depth at its thickest point. This is about ten times taller than the Sears Tower in Chicago, the world's tallest building. However, the average thickness is 7,100 feet (2,164 meters).

Who was the **first person on Antarctica**?

Historians are unsure who first set foot on Antarctica, the fifth largest continent covering 10% of the Earth's surface with its area of 5.4 million square miles (14 million square kilometers). In 1771–1775 British Captain James Cook (1728–1779) circumnavigated the continent. American explorer Nathaniel Palmer (1799–1877) discovered Palmer Peninsula in 1820, without realizing that this was a continent. In this same year Fabian Gottlieb von Bellingshausen (1779–1852) sighted the Antarctic continent. American sealer John Davis went ashore at Hughes Bay on February 7, 1821. In 1823, sealer James Weddell (1787–1834) travelled further south (74 degrees south) than anyone and entered what is now called the Weddell Sea. In 1840, American Charles Wilkes (1798–1877), who followed the coast for 1,500 miles, announced the existence of Antarctica as a continent. In 1841, Sir James Clark Ross (1800–1862) discovered Victoria Land, Ross Island, Mount Erebus, and the Ross Ice Shelf. In 1895, the whaler Henryk Bull landed on the Antarctic continent. Norwegian explorer Roald Amundsen (1872–1928) was the first to reach the South Pole on December 14, 1911.

When was the **Ice Age**?

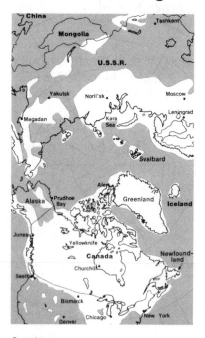

Great ice age

Ice ages, or glacial periods, have occurred at irregular intervals for over 2.3 billion years. During an ice age, sheets of ice cover large portions of the continents. The exact reasons for the changes in the Earth's climate are not known, although some think they are caused by changes in the Earth's orbit around the sun.

The Great Ice Age occurred during the Pleistocene Epoch which began about 2 million years ago and lasted until 11,000 years ago. At its height, about 27% of the world's present land area was covered by ice. In North America, the ice covered Canada and moved southward to New Jersey; in the Midwest, it reached as far south as St. Louis. Small glaciers and ice caps also covered the western mountains. Greenland was covered in ice as it is today. In Europe, ice moved down from Scandinavia into Germany and Poland; the British Isles and the Alps also had ice caps. Glaciers also covered the northern plains of Russia, the plateaus of Central Asia, Siberia, and the Kamchetka Peninsula.

The glaciers' effect on the United States can still be seen. The drainage of the Ohio River and the position of the Great Lakes were influenced by the glaciers. The rich soil of the Midwest is mostly glacial in origin. Rainfall in areas south of the glaciers formed large lakes in Utah, Nevada, and California. The Great Salt Lake in Utah is a remnant of one of these lakes. The large ice sheets locked up a lot of water; sea level fell about 450 feet (137 meters) below what it is today. As a result, some states, such as Florida, were much larger during the ice age.

The glaciers of the last ice age retreated about 11,000 years ago. Some believe that the ice age is not over yet; the glaciers follow a cycle of advance and retreat many times. There are still areas of the Earth covered by ice and this may be a time in between glacial advances.

What is a **moraine**?

A moraine is a mound, ridge, or any other distinct accumulation of unsorted, unstratified material or drift, deposited chiefly by direct action of glacier ice.

Where are the world's **largest deserts**?

Desert	Location	Area	
		Square miles	Square kilometers
Sahara	North Africa	3,500,000	9,065,000
Gobi	Mongolia-China	500,000	1,295,000
Kalahari	Southern Africa	225,000	582,800
Great Sandy	Australia	150,000	338,500
Great Victoria	Australia	150,000	338,500

The Sahara desert is three times the size of the Mediterranean Sea. In the United States, the largest desert is the Mojave Desert in southern California with an area of 15,000 square miles (38,900 square kilometers).

A desert is an area that receives little precipitation and has little plant cover. Many deserts form a band north and south of the equator at about 20 degrees latitude because moisture-bearing winds do not release their rain over these areas. As the moisture-bearing winds from the higher latitudes approach the equator, their temperatures increase and they rise higher and higher in the atmosphere. When the winds arrive over the equatorial areas and come in contact with the colder parts of the Earth's atmosphere, they cool down, and release all their water to create the tropical rain forests near the equator.

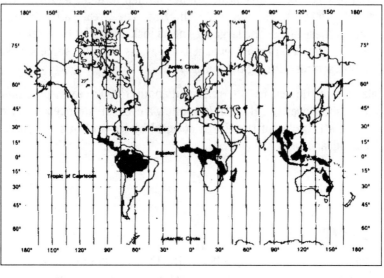

Rain forest locations

How tall is a **sand dune**?

Dunes can range from 3 feet (1 meter) to greater than 650 feet (200 meters) in height. Star dunes (having one central peaked mound with radiating ridges) can grow as high as 1,000 feet (300 meters).

What is a **hoodoo**?

A hoodoo is a fanciful name for a grotesque rock pinnacle or pedestal, usually of sandstone, that is the result of weathering in a semi-arid region. An outstanding example of hoodoos occurs in the Wasatch Formation at Bryce Canyon, Utah.

Are all **craters** part of a volcano?

No, not all craters are of volcanic origin. A crater is a nearly circular area of deformed sedimentary rocks, with a central vent–like depression. Some craters are caused by the collapse of the surface when underground salt or limestone dissolves. The withdrawal of groundwater and the melting of glacial ice can also cause the surface to collapse, forming a crater.

Craters are also caused by large meteorites, comets, and asteroids which hit the Earth. A notable impact crater is Meteor Crater near Winslow, Arizona. It is 4,000 feet

(1,219 meters) in diameter, 600 feet (183 meters) deep and is estimated to have been formed 30,000 to 50,000 years ago.

Meteor Crater, Arizona

How is "speleothem" defined?

Speleothem is a term given to those cave features that form after a cave itself has formed. They are secondary mineral deposits that are created by the solidification of fluids or from chemical solutions. These mineral deposits usually contain calcium carbonate ($CaCO_3$) or limestone, but gypsum or silica may also be found. Stalactites, stalagmites, soda straws, cave coral, boxwork and cave pearls are all types of speleothems.

What is a tufa?

It is a general name for calcium carbonate ($CaCO_3$) deposits or spongy porous limestone found at springs in limestone areas, or in caves as massive stalactite or stalagmite deposits. Tufa, derived from the Italian word for "soft rock," is formed by the precipitation of calcite from the water of streams and springs.

How does a stalactite differ from a stalagmite?

A stalactite is a conical or cylindrical calcite ($CaCO_3$) formation hanging from a cave roof. It forms from the centuries-long buildup of mineral deposits resulting from the seepage of water from the limestone rock above the cave. This water containing calcium bicarbonate evaporates, losing some carbon dioxide, to deposit small quantities of calcium carbonate (carbonate of lime), which eventually forms a stalactite.

A stalagmite is a stone formation which develops upward from the cave floor and resembles an icicle upside down. Formed from water containing calcite which drips from the limestone walls and roof of the cave, it sometimes joins a stalactite to form a column.

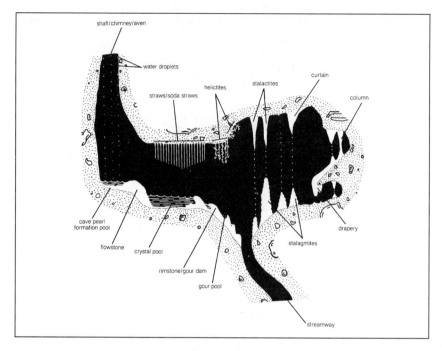

Cave formations

What is the difference between **spelunking** and **speleology**?

Spelunking, or sport caving, is exploring caves as a hobby or for recreation. Speleology is the scientific study of caves and related phenomena, such as the world's deepest cave, Réseau Jean Bernard, Haute Savoie, France with a depth of 5,256 feet (1,602 meters), or the world's longest cave system, Mammoth Cave in Kentucky, with a length of 348 miles (560 kilometers).

Where is the **southernmost city** in the world?

Punta Arenas (formerly Magallanes), Chile, is farther south than any other settlement of sufficient size and having enough commercial importance to deserve being called a city. It lies on the extreme southern end of Patagonia on the Strait of Magellan at 53 degrees 10 minutes south latitude. The world's southernmost village is Puerto Williams (population about 350), Tierra del Fuego, Chile, being 680 miles (1,094 kilometers) north of Antarctica.

What and where is the **continental divide** of North America?

The continental divide is a continuous line that runs north and south the length of North America. About 50 miles west of Denver, Colorado, U.S. Highway Route 40 crosses Berthoud Pass at an altitude of 11,314 feet (3,450 meters) above sea level. At the crest of this pass is the continental divide. West of this divide, all water flows eventually to the Pacific Ocean. East of this line, the waters will flow toward the Atlantic Ocean.

Which **natural attractions** in the United States are the most popular?

1. The Grand Canyon, Arizona
2. Yellowstone National Park, Wyoming
3. Niagara Falls, New York
4. Mount McKinley, Alaska
5. California's "Big Trees": the sequoias and redwoods
6. Hawaii's volcanoes
7. Florida's Everglades

How long is the **Grand Canyon**?

The Grand Canyon, cut out by the Colorado River over a period of 15 million years in the northwest corner of Arizona, is the largest land gorge in the world. It is 4 to 13 miles (6.4 to 21 kilometers) wide at its brim, 4,000 to 5,500 feet (1,219 to 1,676 meters) deep, and 217 miles (349 kilometers) long, extending from the mouth of the Little Colorado River to Grand Wash Cliffs (and 277 miles, 600 feet or 445.88 kilometers if Marble Canyon is included).

However, it is not the deepest canyon in the United States; that distinction belongs to Kings Canyon, which runs through the Sierra and Sequoia National Forests near East Fresno, California, with its deepest point being 8,200 feet (2,500 meters). Hell's Canyon of the Snake River between Idaho and Oregon is the deepest United States canyon in low-relief territory. Also called the Grand Canyon of the Snake, it plunges 7,900 feet (2,408 meters) down from Devil Mountain to the Snake River.

What are the **LaBrea tar pits**?

The tar pits are located in an area of Los Angeles, California formerly known as Rancho LaBrea. Heavy, sticky tar oozed out of the earth, the scum from great petroleum reservoirs far underground. The pools were cruel traps for uncounted numbers of animals. Today the tar pits are a part of Hancock Park. In the park, many fossil remains are displayed along with life-sized reconstructions of these prehistoric species.

What is the composition of the Rock of Gibraltar?

It is composed of gray limestone, with a dark shale overlay on parts of the western slopes. Located on a peninsula at the southern extremity of Spain, the Rock of Gibraltar is a mountain at the east end of the Strait of Gibraltar, the narrow passage between the Atlantic Ocean and the Mediterranean Sea. "The Rock" is 1,398 feet (426 meters) tall at its highest point.

From what type of stone was **Mount Rushmore** National Monument carved?

Granite. The monument, in the Black Hills of southwestern South Dakota, depicts the 60-foot-high (18-meter-high) faces of four United States presidents: George Washington, Thomas Jefferson, Abraham Lincoln, and Theodore Roosevelt. Sculptor Gutzon Borglum (1867–1941) designed the monument, but died before the completion of the project; his son, Lincoln, finished it. From 1927 to 1941, 360 people, mostly construction workers, drillers, and miners, "carved" the figures using dynamite.

VOLCANOES AND EARTHQUAKES

How many **kinds of volcanoes** are there?

Volcanoes are usually cone-shaped hills or mountains built around a vent connecting to reservoirs of molten rock, or magma, below the surface of the Earth. At times the molten rock is forced upwards by gas pressure until it breaks through weak spots in the Earth's crust. The magma erupts forth as lava flows or shoots into the air as clouds of lava fragments, ash, and dust. The accumulation of debris from eruptions cause the volcano to grow in size. There are four kinds of volcanoes:

Cinder cones are built of lava fragments. They have slopes of 30 degrees to 40 degrees and seldom exceed 1,640 feet (500 meters) in height. Sunset Crater in Arizona and Paricutin in Mexico are examples of cinder cones.

Composite cones are made of alternating layers of lava and ash. They are characterized by slopes of up to 30 degrees at the summit, tapering off to 5 degrees at the base. Mount Fuji in Japan and Mount St. Helens in Washington are composite cone volcanoes.

Shield volcanoes are built primarily of lava flows. Their slopes are seldom more than 10 degrees at the summit and 2 degrees at the base. The Hawaiian Islands are clusters of shield volcanoes. Mauna Loa is the world's largest active volcano, rising 13,653 feet (4,161 meters) above sea level.

Lava domes are made of viscous, pasty lava squeezed like toothpaste from a tube. Examples of lava domes are Lassen Peak and Mono Dome in California.

Where is the "Circle of Fire"?

The belt of volcanoes bordering the Pacific Ocean is often called the "Circle of Fire" or the "Ring of Fire". The Earth's crust is composed of 15 pieces, called plates, which "float" on the partially molten layer below them. Most volcanoes, earthquakes, and mountain building occur along the plate boundaries. The Circle of Fire marks the boundary between the plate underlying the Pacific Ocean and the surrounding plates. It runs up the west coast of the Americas from Chile to Alaska (through the Andes Mountains, Central America, Mexico, California, the Cascade Mountains, and the Aleutian Islands) then down the east coast of Asia from Siberia to New Zealand (through Kamchatka, the Kurile Islands, Japan, the Philippines, Celebes, New Guinea, the Solomon Islands, New Caledonia, and New Zealand). Of the 850 active volcanoes in the world, over 75% of them are part of the Circle of Fire.

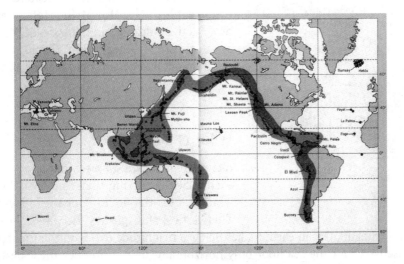

Circle or Ring of Fire

Which **volcanoes** have been the **most destructive?**

The five most destructive eruptions from volcanoes since 1700 are as follows:

Volcano	Date of eruption	Number killed	Lethal agent
Mt. Tambora, Indonesia	April 5, 1815	92,000	2,000 directly by the volcano, 80,000 from starvation afterwards
Karkatoa, Indonesia	Aug. 26, 1883	36,417	90% killed by a tsunami
Mt. Pelee, Martinque	Aug. 30, 1902	29,025	Pyroclastic flows
Nevada del Ruiz, Colombia	Nov. 13, 1985	23,000	Mud flow
Unzen, Japan	1792	14,300	70% killed by cone collapse; 30% by a tsunami

When did **Mount St. Helens** erupt?

Cascade Range volcanoes

Mount St. Helens, located in southwestern Washington state in the Cascades mountain range, erupted on May 18, 1980. Sixty-one people died as a result of the eruption. This was the first known eruption in the 48 contiguous United States to claim a human life. Geologists call Mount St. Helens a composite volcano (a steep-sided, often symmetrical cone constructed of alternating layers of lava, flows, ash, and other volcanic debris). Composite volcanoes tend to erupt explosively. Mount St. Helens and the other active volcanoes in the Cascade Mountains are a part of the "Ring of Fire" (Pacific zone having frequent and destructive volcanic activity).

What is a **tsunami?**

A tsunami is a giant wave set in motion by a large earthquake occurring under or near the ocean which causes the ocean floor to shift vertically. This vertical shift pushes the water ahead of it, starting a tsunami. Ocean earthquakes below a magnitude of 6.5 on the Richter scale, and those that shift the sea floor only horizontally, do not produce these destructive waves.

Who invented the **ancient Chinese earthquake detector?**

Chinese quake detector

The detector, invented by Zhang Heng (78–139 C.E.) around 132 C.E., was a copper-domed urn with dragons' heads circling the outside, each containing a bronze ball. Inside the dome was suspended a pendulum that would swing when the ground shook, and knock a ball from the mouth of a dragon into the waiting open mouth of a bronze toad below. The ball made a loud noise and signaled the occurrence of an earthquake. Knowing which ball had been released, one could determine the direction of the earthquake's epicenter (the point on the Earth's surface directly above the quake's point of origin).

How does a **seismograph** work?

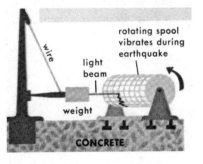

Seismograph device

A seismograph records earthquake waves. When an earthquake occurs, three types of waves are generated. The vibrations are detected by a delicately balanced beam; the movements of the beam are recorded on a moving tape. The three types of waves arrive at different times after the quake. By comparing the arrival times of the waves at different seismographs throughout the world, it is possible to determine exactly where the earthquake took place.

What is the **Richter scale?**

On a machine called a seismograph, the Richter scale measures the magnitude of an earthquake, i.e., the size of the ground waves generated at the earthquake's source. The scale was devised by American geologist Charles W. Richter (1900–1985) in 1935. Every increase of one number means a tenfold increase in magnitude.

Richter Scale	
Magnitude	**Possible effects**
1	Detectable only by instruments
2	Barely detectable, even near the epicenter
3	Felt indoors
4	Felt by most people; slight damage
5	Felt by all; damage minor to to moderate
6	Moderately destructive
7	Major damage
8	Total and major damage

Another widely used system of measurement was developed by Guiseppe Mercalli (1850–1914) in 1902 and modified by Harry Wood and Frank Neumann in the 1930s. This scale, the Mercalli scale, measures the effects of the shock.

When did the **most severe earthquake** in American history occur?

The New Madrid earthquakes (a series of quakes starting on December 16, 1811, and lasting until March 1812) are considered to be the most severe earthquake event in United States history. It shook more than two-thirds of the United States and was felt in Canada. It changed the level of land by as much as 20 feet, altered the course of the Mississippi River, and created new lakes, such as Lake St. Francis west of the Mississippi and Reelfoot Lake in Tennessee. Because the area was so sparsely populated, no known loss of life occurred.

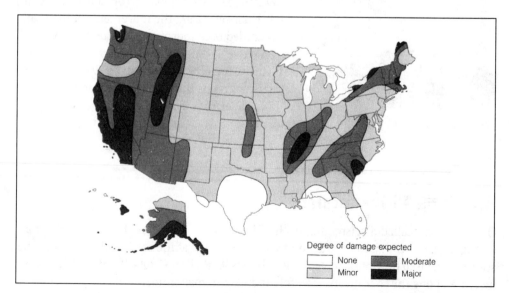

Degree of damage expected

None   Moderate
Minor   Major

Of what magnitude was the **earthquake** that hit **San Francisco** on April 18, 1906 on the Richter scale?

Causing 700 deaths and damage estimated at 500 million dollars, the earthquake measured 8.3 on the Richter scale. The October 17, 1989, earthquake in San Francisco measured 7.1 on the Richter scale, killed 67 people, and caused billions of dollars worth of damage.

OBSERVATION AND MEASUREMENT

What are the major eras, periods, and epochs in geologic time?

Modern dating techniques have given a range of dates as to when the various geologic time periods have started, as they are listed below:

Era	Period	Epoch	Beginning date (Est. millions of years)
Cenozoic	Quaternary	Holocene	10,000 years ago
		Pleistocene	1.9
	Tertiary	Pliocene	6
		Miocene	25
		Oligocene	38
		Eocene	55
		Paleocene	65
Mesozoic	Cretaceous		135
	Jurassic		200
	Triassic		250
Paleozoic	Permian		285
	Carboniferous (divided into Mississippian and Pennsylvanian periods by some in the U.S.)		350
	Devonian		410
	Silurian		425
	Ordovician		500
	Cambrian		570
Precambrian	Proterozoic		2500
	Archeozoic		3800
	Azoic		4600

What is **magnetic declination**?

It is the angle between magnetic north and true north at a given point on the Earth's surface. It varies at different points on the Earth's surface and at different times of the year.

Which direction does a compass needle point at the north magnetic pole?

The compass needle would be attracted by the ground and point straight down.

What is a **Foucault pendulum**?

An instrument devised by Jean Foucault (1819–1868) in 1851 to prove that the Earth rotates on an axis, the pendulum consisted of a heavy ball suspended by a very long fine wire. Sand beneath the pendulum recorded the plane of rotation of the pendulum over time.

A reconstruction of Foucault's experiment is located in Portland, Oregon, at its Convention Center. It swings from a cable 90 feet (27.4 meters) long, making it the longest pendulum in the world.

What is the **prime meridian**?

The north-south lines on a map run from the North Pole to the South Pole and are called "meridians," a word that meant "noon," for when it is noon on one place on the line, it is noon at any other point as well. The lines are used to measure longitudes, or how far east or west a particular place might be, and they are 69 miles (111 kilometers) apart at the equator. The east-west lines are called parallels, and unlike meridians, are all parallel to each other. They measure latitude, or how far north or south a particular place might be. There are 180 lines circling the Earth, one for each degree of latitude. The degrees of both latitude and longitude are divided into 60 minutes, further divided into 60 seconds each.

The prime meridian is the meridian of 0 degrees longitude, used as the origin for measurement of longitude. The meridian of Greenwich, England, is used almost universally for this purpose.

Who is regarded as the father of American **geology**?

Born in Scotland, the American William Maclure (1763–1840) was a member of a commission set up to settle claims between the United States and France during 1803–1807. In 1809 he made a geographical chart of the United States in which the land areas were divided by rock types. In 1817 he revised and enlarged this map. Maclure wrote the first English-language articles and books on United States geology.

When were **relief maps** first used?

The Chinese were the first to use relief maps where the contours of the terrain were represented in models. Relief maps in China go back at least to the third century B.C.E. Some early maps were modeled in rice or carved in wood. It is likely that the idea of making relief maps was transmitted from the Chinese to the Arabs and then to Europe. The earliest known relief map in Europe was a map showing part of Austria, made in 1510 by Paul Dox.

What is the **Piri Re'is map**?

In 1929, a map was found in Constantinople that caused great excitement. Painted on parchment and dated in the Moslem year 919 (1513 according to the Christian calendar), it was signed by an admiral of the Turkish navy known as Piri Re'is. This map appears to be one of the earliest maps of America and it shows South America and Africa in there correct relative longitudes. The mapmaker also indicated that he had used a map drawn by Columbus for the western part. It was an exciting statement because for several centuries geographers had been trying to find a "lost map of Columbus" supposedly drawn by him in the West Indies.

Who was the first person to map the **Gulf Stream**?

In his travels to and from France as a diplomat, Benjamin Franklin (1706–1790) noticed a difference in speed in the two directions of travel between France and America. He was the first to study ships' reports seriously to determine the cause of the speed variation. As a result, he found that there was a current of warm water coming from the Gulf of Mexico that crossed the North Atlantic Ocean in the direction of Europe. In 1770, Franklin mapped it.

What are **Landsat maps**?

They are images of the Earth taken at an altitude of 567 miles (912.3 kilometers) by an orbiting Landsat satellite, or ERTS (*Earth Resources Technology Satellite*). The Landsats were originally launched in the 1970s. Rather than cameras, the Landsats use multispectral scanners, which detect visible green and blue wavelengths, and four infrared and near-infrared wavelengths. These scanners can detect differences between soil, rock, water, and vegetation; types of vegetation; states of vegetation (e.g., healthy/unhealthy or underwatered/well-watered); and mineral content. The differences are especially accurate when multiple wavelengths are compared using multispectral scanners. Even visible light images have proved useful—some of the earliest Landsat images showed that some small Pacific islands were up to 10 miles (16 kilometers) away from their charted positions.

The results are displayed in "false-color" maps, where the scanner data is represented in shades of easily distinguishable colors—usually, infrared is shown as red, red as green, and green as blue. The maps are used by farmers, oil companies, geologists, foresters, foreign governments, and others interested in land management. Each image covers an area approximately 115 miles (185 kilometers) square. Maps are offered for sale by the United States Geological Survey.

Other systems that produce similar images include the French SPOT satellites, the Russian Salyut and Mir manned space stations, and NASA's Airborne Imaging Spectrometer, which senses 128 infrared bands. NASA's Jet Propulsion Laboratories are developing instruments which will sense 224 bands in infrared, which will be able to detect specific minerals absorbed by plants.

CLIMATE AND WEATHER

TEMPERATURE

Is the **world** actually **getting warmer**?

The year 1990 was the warmest year since 1880, with an average global temperature of 59.8°F (15.45°C). Six of the seven warmest years of the twentieth century have occurred during the 1980s. The average annual temperature for the contiguous United States (during the time period 1895–1990) is 52.5°F (11.4°C). The United States began the century with cool temperatures; the 1920s through 1950s were warm; the 1960s and 1970s were cool; and the 1980s were warm.

How much has the **Earth warmed** in the past century?

A United Nations committee of 300 leading scientists examined numerous data sets and concluded that the Earth has warmed (0.3°C to 0.6°C) during the past century. This data could mean that the global warming is a result of the ozone layer depletion, or it could be part of a larger weather fluctuation pattern.

What are the **highest** and **lowest recorded** **temperatures** on Earth?

The highest temperature in the world was recorded as 136°F (58°C) at Al Aziziyah (el-Azizia), Libya, on September 13, 1922; the highest temperature recorded in the United States was 134°F (56.7°C) in Death Valley, California, on July 10, 1913. The temperatures of 140°F (60°C) at Delta, Mexico, in August, 1953 and 136.4°F (58°C) at San Luis, Mexico, on August 11, 1933 are not internationally accepted. The lowest temperature was -128.6°F (-89.6°C) at Vostok Station in Antarctica on July 21, 1983. The record cold temperature for an inhabited area was -90.4°F (-68°C) at Oymyakon, Siberia (population 4,000), on February 6, 1933. This temperature tied with the readings at Verkhoyansk, Siberia, on January 3, 1885, February 5 and 7, 1892. The lowest temperature reading in the United States was -79.8°F (-62.1°C) on January 23, 1971, in Prospect Creek, Alaska; for the contiguous 48 states, the coldest temperature was -69.7°F (-56.5°C) at Rogers Pass, Montana, on January 20, 1954.

Why are the hot humid days of summer called **dog days**?

This period of extremely hot, humid, sultry weather that traditionally occurs in the Northern Hemisphere in July and August received its name from the dog star Sirius of the constellation *Canis Major*. At this time of year, Sirius, the brightest visible star, rises in the east at the same time as the sun. Ancient Egyptians believed that the heat of this brilliant star added to the sun's heat to create this hot weather. Sirius was blamed for the withering droughts, sickness, and discomfort that occurred during this time. Traditional dog days start on July 3 and end on August 11.

Which place has the **maximum amount of sunshine** in the United States?

Yuma, Arizona, has an annual average of 90% of sunny days or over 4,000 sunny hours per year. St. Petersburg, Florida, had 768 consecutive sunny days from February 9, 1967, to March 17, 1969. On the other extreme, the South Pole has no sunshine for 182 days annually, and the North Pole has none for 176 days.

Why was 1816 known as the **"year without a summer"**?

The eruption of Mount Tambora, a volcano in Indonesia, in 1815 threw billions of cubic yards of dust over 15 miles into the atmosphere. Because the dust penetrated the stratosphere, wind currents spread it throughout the world. As a consequence of this volcanic activity, in 1816 normal weather patterns were greatly altered. Some parts of Europe and the British Isles experienced average temperatures 2.9°F to 5.8°F (1.6°C to 3.2°C) below normal. In New England heavy snow fell between June 6 and June 11 and frost occurred every month of 1816. Crop failures were experienced in Western Europe and Canada as well as in New England. By 1817, the excess dust had settled and the climate returned to more normal conditions.

How can the **temperature** be determined from the frequency of **cricket chirps**?

Count the number of cricket chirps in 14 seconds and add 40.

AIR PHENOMENA

What is a **bishop's ring**?

It is a ring around the sun, usually with a reddish outer edge. It is probably due to dust particles in the air, as is seen after all great volcanic eruptions.

How often does an **aurora** appear?

Because it depends on solar winds (electrical particles generated by the sun) and sunspot activity, the frequency of an aurora cannot be determined. Auroras usually appear two days after a solar flare (a violent eruption of particles on the sun's surface) and reach their peak two years into the 11–year sunspot cycle. The auroras, occurring in the polar regions, are broad displays of usually colored light at night. The northern polar aurora is called Aurora Borealis or Northern Lights and the southern polar aurora is called the Aurora Australis.

When and by whom were **clouds** first classified?

The French naturalist Jean Lamarck (1744–1829) proposed the first system for classifying clouds in 1802. However his work did not receive wide acclaim. A year later the Englishman Luke Howard (1772–1864) developed a cloud classification system that has been generally accepted and is still used today. Clouds are distinguished by their general appearance ("heap clouds" and "layer clouds") and by their height above the ground. Latin names and prefixes are used to describe these characteristics. The shape names are *cirrus* (curly or fibrous), *stratus* (layered), *cumulus* (lumpy or piled). The prefixes denoting height are *cirro* (high clouds with bases above 20,000 feet or 6.1 kilometers) and *alto* (mid-level clouds from 6,000 to 20,000 feet or 1830 to 6,100 kilometers). There is no prefix for low clouds. *Nimbo* or *nimbus* is also added as a name or prefix to indicate that the cloud produces precipitation.

What are the four major **cloud groups** and their types?

1. High Clouds—composed almost entirely of ice crystals. The bases of these clouds start at 16,500 feet (5,000 meters) and reach 45,000 feet (13,700 meters).

Types of clouds

Cirrus (from Latin, "lock of hair")—are thin feather-like crystal clouds in patches or narrow bands. The large ice crystals that often trail downward in well-defined wisps are called "mares tails."

Cirrostratus—is a thin, white cloud layer that resembles a veil or sheet. This layer can be striated or fibrous. Because of the ice content, these clouds are associated with the halos that surround the sun or moon.

Cirrocumulus—are thin clouds that appear as small white flakes or cotton patches and may contain super-cooled water.

2. Middle Clouds—composed primarily of water. The height of the cloud bases range from 6,500 to 23,000 feet (2,000 to 7,000 meters).

Altostratus—appears as a bluish or grayish veil or layer of clouds that can gradually merge into altocumulus clouds. The sun may be dimly visible through it, but flat, thick sheets of this cloud type can obscure the sun.

Altocumulus—is a white or gray layer or patches of solid clouds with rounded shapes.

3. Low Clouds—composed almost entirely of water that may at times be super-cooled; at subfreezing temperatures, snow and ice crystals may be present as well. The bases of these clouds start near the Earth's surface and climb to 6,500 feet (2,000 meters) in the middle latitudes.

Stratus—are gray uniform sheet-like clouds with a relatively low base or they can be patchy, shapeless, low gray clouds. Thin enough for the sun to shine through, these clouds bring drizzle and snow.

Stratocumulus—are globular rounded masses that form at the top of the layer.

Nimbostratus—are seen as a gray or dark relatively shapeless massive cloud layer containing rain, snow, and ice pellets.

4. Clouds with Vertical Development—contain super-cooled water above the freezing level and grow to great heights. The cloud bases range from 1,000 feet (300 meters) to 10,000 feet (3,000 meters).

Cumulus—are detached, fair weather clouds with relatively flat bases and dome-shaped tops. These usually do not have extensive vertical development and do not produce precipitation.

Cumulonimbus—are unstable large vertical clouds with dense boiling tops that bring showers, hail, thunder, and lightning.

What **color** is **lightning**?

Lightning is often seen as white or white-yellow, although it may appear to have other colors depending on the background.

How **hot** is **lightning**?

The temperature of the air around a bolt of lightning is about 54,000°F (30,000°C), which is six times hotter than the surface of the sun, yet many times people survive a bolt of lightning. American park ranger Roy Sullivan was hit by lightning seven times between 1942 and 1977. In cloud–to–ground lightning, its energy seeks the shortest route to Earth, which could be through a person's shoulder, down the side of the body through the leg to the ground. As long as the lightning does not pass across the heart or spinal column, the victim usually does not die.

Does lightning ever **strike in the same place**?

It is not true that lightning does not strike twice in the same place. In fact, tall buildings, such as the Empire State Building, in New York city, can be struck several times during the same storm. During one storm, lightning struck the Empire State Building 12 times.

How **long** is a **lightning stroke**?

The visible length of the streak of lightning depends on the terrain and can vary greatly. In mountainous areas where clouds are low, the flash can be as short as 300 yards (170 meters); whereas in flat terrain, where clouds are high, the bolt can measure as long as 4 miles (6.5 kilometers). The usual length is about 1 mile (1.6 kilometers), but streaks of lightning up to 20 miles (32 kilometers) have been recorded. The stroke channel is very narrow—perhaps as little as ½ inch (1.3 centimeters). It is surrounded by a "corona envelope" or a glowing discharge that can be as wide as 10 to 20 feet (3 to 6 meters) in diameter. The speed of lightning can vary from 100 to 1,000 miles (160 to 1,600 kilometers) per second for the downward leader track; the return stroke is 87,000 miles (140,000 kilometers) per second (almost half the speed of light).

How many **volts** are in **lightning**?

A stroke of lightning discharges from 10 to 100 million volts of electricity. An average lightning stroke has 30,000 amperes.

How is the **distance of a lightning flash** calculated?

Count the number of seconds between seeing a flash of lightning and hearing the sound of the thunder. Divide the number by 5 to determine the number of miles away that the lightning flashed.

What is **ball lightning**?

Ball lightning is a rare form of lightning in which a persistent and moving luminous white or colored sphere is seen. It can last from a few seconds to several minutes, and travels at about a walking pace. Spheres have been reported to vanish harmlessly, or to pass into or out of rooms leaving, in some cases, sign of their passage such as a hole in a window pane. Sphere dimensions vary but are most commonly from 4 to 8 inches (10 to 20 centimeters).

Other types of lightning include the common streak lightning (a single or multiple zigzagging line from cloud to ground); forked lightning (lightning formed two branches simultaneously); sheet lightning (a shapeless flash covering a broad area); ribbon lightning (streak lightning blown sideways by the wind to make it appear like parallel successive strokes); bead or chain lightning (a stroke interrupted or broken into evenly-spaced segments or beads); and heat lightning (lightning seen along the horizon during hot weather and believed to be a reflection of lightning occurring beyond the horizon).

When does the green flash phenomenon occur?

On rare occasions, the sun may look bright green for a moment, as the last tip of the sun is setting. This green flash occurs because the red rays of light are hidden below the horizon and the blue are scattered in the atmosphere. The green rays are seldom seen because of dust and pollution in the lower atmosphere. It may best be seen when the air is cloudless and when a distant, well–defined horizon exists, as on an ocean.

What are **fulgurites**?

Fulgurites (from the Latin word *fulgur,* meaning lightning) are petrified lightning, created when lightning strikes an area of dry sand. The intense heat of the lightning melts the sand surrounding the stroke into a rough glassy tube forming a fused record of its path. These tubes may be ½ to 2 inches (1.5 to 5 centimeters) in diameter, and up to 10 feet (3 meters) in length. They are extremely brittle and break easily. The inside walls of the tube are glassy and lustrous while the outside is rough, with sand particles adhering to it. Fulgurites are usually tan or black in color, but translucent white ones have been found.

What is **Saint Elmo's fire**?

Saint Elmo's fire has been described as a corona from electric discharge produced on high grounded metal objects, chimney tops, and ship masts. Since it often occurs during thunderstorms, the electrical source may be lightning. Another description refers to this phenomenon as weak static electricity formed when an electrified cloud touches a high exposed point. Molecules of gas in the air around this point become ionized and glow. The name originated with sailors who were among the first to witness the display of spearlike or tufted flames on the tops of their ships' masts. Saint Elmo (which is a corruption of Saint Ermo) is the patron saint of sailors, so they named the fire after him.

What is the order of **colors in a rainbow**?

Red, orange, yellow, blue, indigo, and violet are the colors of the rainbow, but these are not necessarily the sequence of colors that an observer might see. Rainbows are formed when raindrops reflect sunlight. As sunlight enters the drops, the different wavelengths of the colors that compose sunlight are refracted at different lengths to produce a spectrum of color. Each observer sees a different set of raindrops at a slightly different angle. Drops at different angles from the observer send different wave lengths (i.e., different color) to the observer's eyes. Since the color sequence of the rainbow is the result of refraction, the color order depends on how the viewer sees this refraction from the viewer's angle of perception.

What is the origin of the **Brown Mountain Lights** of North Carolina?

For a period of some thirty years the "lights" seen at Brown Mountain could not be explained. In 1922, the United States Geological Survey studied the mystery. The area has extraordinary atmospheric conditions and automobile, locomotive, and fixed lights from miles away are reflected in the atmosphere.

WIND

What is the **Coriolis effect**?

The 19th century French engineer Gaspard C. Coriolis (1792–1843) discovered that the rotation of the Earth deflects streams of air. Because the Earth spins to the east, all moving objects in the Northern Hemisphere tend to turn somewhat to the right of a straight path while those in the Southern Hemisphere turn slightly left. The Coriolis effect explains the lack of northerly and southerly winds in the tropics and polar regions; the northeast and southeast trade winds and the polar easterlies all owe their westward deflection to the Coriolis effect.

When was the **jet stream** discovered?

A jet stream is a flat and narrow tube of air that moves more rapidly than the surrounding air. Discovered by World War II bomber pilots flying over Japan and the Mediterranean Sea, jet streams have become important with the advent of airplanes capable of cruising at over 30,000 feet (9,000 meters). The currents of air flow from west to east and are usually a few miles deep, up to 100 miles (160 kilometers) wide and well over 1,000 miles (1,600 kilometers) in length. The air current must flow at over 57.5 miles (90 kilometers) per hour.

There are two polar jet streams, one in each hemisphere. They meander between 30 and 70 degrees latitude, occur at altitudes of 25,000 to 35,000 feet (7,600 to 10,600 meters), and achieve maximum speeds of over 230 miles (360 kilometers) per hour. The subtropical jet streams, again one per hemisphere, wander between 20 and 50 degrees latitude. They are found at altitudes of 30,000 to 45,000 feet (9,000 to 13,700 meters), and have speeds of over 345 miles (550 kilometers) per hour.

Why are the **horse latitudes** called by that name?

They are two high pressure belts characterized by low winds, about 30 degrees north and south of the equator. Dreaded by early sailors, these areas have undependable winds with periods of calm. In the Northern Hemisphere, particularly near Bermuda, sailing ships carrying horses from Spain to the New World were often becalmed. When water supplies ran low, these animals were the first to be rationed water. Dying from thirst or tossed overboard, the animals were sacrificed to conserve water for the men. Explorers and sailors reported that the seas were "strewn with bodies of horses." Supposedly this is why the areas are called the "horse latitudes." The term might be rooted in the complaints by sailors who were paid in advance and received no overtime when the ships slowly transversed this area. Supposedly during this time they were "working off a dead horse."

What are **halcyon days**?

This term is often used to refer to a time of peace or prosperity. In sea tradition, it is the two-week period of calm weather before and after the shortest day of the year, about December 21. The phrase is taken from halcyon, the name the ancient Greeks gave to the kingfisher. According to legend, the halcyon built its nest on the surface of the ocean and was able to quiet the winds while its eggs were hatching.

Is Chicago the **windiest city**?

In 1990 Chicago ranked 21st in the list of 68 windy cities with an average wind speed of 10.3 miles (16.6 kilometers) per hour. Cheyenne, Wyoming, with an average wind speed of 12.9 miles (20.8 kilometers) per hour, ranks number one, closely followed by Great Falls, Montana, with an average wind speed of 12.8 miles (20.6 kilometers) per hour. The highest surface wind ever recorded was on Mount Washington, New Hampshire, at an elevation of 6288 feet (1.9 kilometers). On April 12, 1934 its wind was 231 miles (371.7 kilometers) per hour and its average wind speed was 35 miles (56.3 kilometers) per hour.

What is meant by the **wind chill factor**?

The wind chill factor or wind chill index is a number, which expresses the cooling effect of moving air at different temperatures. It indicates in a general way how many calories of heat are carried away from the surface of the body.

COOLING POWER OF WIND EXPRESSED AS "EQUIVALENT CHILL TEMPERATURE"																							
WIND SPEED		TEMPERATURE (°F)																					
CALM	CALM	40	35	30	25	20	15	10	5	0	-5	-10	-15	-20	-25	-30	-35	-40	-45	-50	-55	-60
KNOTS	MPH	EQUIVALENT CHILL TEMPERATURE																					
3-6	5	35	30	25	20	15	10	5	0	-5	-10	-15	-20	-25	-30	-35	-40	-45	-50	-55	-65	-70
7-10	10	30	20	15	10	5	0	-10	-15	-20	-25	-35	-40	-45	-50	-60	-65	-70	-75	-80	-90	-95
11-15	15	25	15	10	0	-5	-10	20	-25	-30	-40	-45	-50	-60	-65	-70	-80	-85	-90	-100	-105	-110
16-19	20	20	10	5	0	-10	-15	-25	-30	-35	-45	-50	-60	-65	-75	-80	-85	-95	-100	-110	-115	-120
20-23	25	15	10	0	-5	-15	-20	-30	-35	-45	-50	-60	-65	-75	-80	-90	-95	-105	-110	-120	-125	-135
24-28	30	10	5	0	-10	-20	-25	-30	-40	-50	-55	-65	-70	-80	-85	-95	-100	-110	-115	-125	-130	-140
29-32	35	10	5	-5	-10	-20	-30	-35	-40	-50	-60	-65	-75	-80	-90	-100	-105	-115	-120	-130	-135	-145
33-36	40	10	0	-5	-15	-20	-30	-35	-45	-55	60	-70	-75	-85	-95	-100	-110	-115	-125	-130	-140	-150
WINDS ABOVE 40 HAVE LITTLE ADDITIONAL EFFECTS.		LITTLE DANGER					INCREASING DANGER (Flesh may freeze within 1 minute)						GREAT DANGER (Flesh may freeze within 30 seconds)										
DANGER OF FREEZING EXPOSED FLESH FOR PROPERLY CLOTHED PERSONS																							

Wind chill index

Scientists have devised an equivalent temperature scale which makes it easy to determine the wind chill factor. The accompanying chart is used for that purpose. Find the wind speed in the column on the left and find the temperature reading in the horizontal row at the top. The intersection of these two points gives the corresponding wind chill factor. For example, a temperature of 10°F along with a wind speed of 29 to 32 miles per hour corresponds to a wind chill factor of -35°F.

Is there a **formula** for computing the **wind chill**?

The formula is

$$Te = 33 - \frac{(10.45 + V - V(33 - T)}{22.04}$$

Te = Wind chill equivalent temperature in degrees Celsius
V = Wind speed in meters per second
T = Air temperature in degrees Celsius

Who is associated with **developing the concept of wind chill**?

The Antarctic explorer Paul A. Siple coined the term in his dissertation "Adaptation of the Explorer to the Climate of Antarctica," submitted in 1939. Siple was the youngest member of Admiral Byrd's Antarctica expedition in 1928–1930 and later made other trips to the Antarctic as part of Byrd's staff and for the United States Department of the Interior assigned to the United States Antarctic Expedition. He also served in many other endeavors related to the study of cold climates.

How does a **cyclone** differ from a **hurricane** or **tornado**?

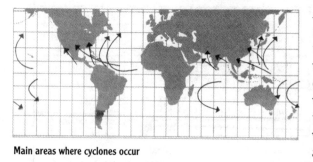
Main areas where cyclones occur

All three wind phenomena are rotating winds that spiral in toward a low–pressure center as well as upward. Their differences lie in their size, wind velocity, rate of travel, and duration. Generally, the faster the winds spin, the shorter (in time) and smaller (in size) the event becomes.

A cyclone has rotating winds from 10 to 60 miles (16 to 97 kilometers) per hour; can be up to 1,000 miles (1,600 kilometers) in diameter, travels about 25 miles (40 kilometers) per hour, and lasts from one to several weeks. A hurricane (or typhoon, as it is called in the Pacific Ocean area) has winds that vary from 75 to 200 miles (120 to 320 kilometers) per hour, moves between 10 to 20 miles (16 to 32 kilometers) per hour, can have a diameter up to 600 miles (960 kilometers), and can exist from several days to more than a week. A tornado can reach a rotating speed of 300 miles (400 kilometers) per hour, travels between 25 to 40 miles (40 to 64 kilometers) per hour and generally lasts minutes, although some have existed for 5–6 hours. Its diameter can range from 300 yards (274 meters) to a mile (1.6 kilometers) and its average path length is 16 miles (26 kilometers), with a maximum of 300 miles (483 kilometers).

Typhoons, hurricanes, and cyclones tend to breed in low–altitude belts over the oceans generally from five degrees to fifteen degrees latitude north or south. A tornado generally forms several thousand feet above the Earth's surface usually during warm, humid weather; many times it is in conjunction with a thunderstorm. Although a tornado can occur in many places, they mostly take place on the continental plains of North America (i.e., from the Plains States eastward to western New York and the southeastern Atlantic states). 82% of tornadoes materialize during the warmest hours of the day (noon to midnight), while 23% of all tornado activity occurs between 4 p.m. and 6 p.m.

What is the **Beaufort Scale?**

The Beaufort scale was devised in 1805 by a British Admiral, Sir Francis Beaufort (1774–1857), to help mariners in handling ships. It uses a series of numbers from 0 to 17 to indicate wind speeds and applies to both land and sea.

Beaufort number	Name	Wind speed	
		Miles per hour	Kilometers per hour
0	Calm	less than 1	less than 1.5
1	Light air	1–3	1.5–4.8
2	Light breeze	4–7	6.4–11.3
3	Gentle breeze	8–12	12.9–19.3
4	Moderate breeze	13–18	21–29
5	Fresh breeze	19–24	30.6–38.6
6	Strong breeze	25–31	40.2–50
7	Moderate gale	32–38	51.5–61.1
8	Fresh gale	39–46	62.8–74
9	Strong gale	47–54	75.6–86.9
10	Whole gale	55–63	88.5–101.4
11	Storm	64–73	103–117.5
12–17	Hurricane	74 and above	119.1 and above

What is the **Fujita and Pearson Tornado Scale?**

The Fujita and Pearson Tornado Scale, developed by T. Theodore Fujita and Allen Pearson, ranks tornadoes by their wind speed. Sometimes known as the Fujita scale, the ranking ranges from F0 (very weak) to F6 (inconceivable).

F0—Light damage: damage to trees, billboards, and chimneys.

F1—Moderate damage: mobile homes could be pushed off their foundations and cars could be pushed off roads.

F2—Considerable damage: roofs torn off, mobile homes demolished, and large trees uprooted.

F3—Severe damage: even well-constructed homes can be torn apart, trees uprooted, and cars lifted off the ground.

F4—Devastating damage: houses can be leveled, cars thrown, and objects can become flying missiles.

F5—Incredible damage: structures lifted off foundations and carried away; cars become missiles. Less than 2% of tornadoes are in this category.

F6—Maximum tornado winds not expected to exceed 318 miles per hour.

Fujita and Pearson Tornado Scale

Scale	Speed miles per hour	Path length in miles	Path width
0	≤72	≤1.0	≤17 yards
1	73–112	1.0–3.1	18–55 yards
2	113–157	3.2–9.9	56–175 yards
3	158–206	10.0–31.0	176–556 yards
4	207–260	32.0–99.0	0.34–0.9 miles
5	261–318	100–315	1.0–3.1 miles
6	319–380	316–999	3.2–9.9 miles

Which year had the **most tornadoes?**

From the period 1916 (when records started to be kept) to 1989, more tornadoes occurred in 1973 than in any other year. That year 1,102 tornadoes struck in 46 states, killing 87 persons. From 1980 to 1989, an average of 820 tornadoes occurred each year. The largest outbreak of tornadoes occurred on April 3 and 4, 1974. In this "Super Outbreak" 127 tornadoes were recorded in the Plains and midwestern states. Six of these tornadoes had winds greater than 26 miles (420 kilometers) per hour and some of them were the strongest ever recorded.

Which **month** is the **most dangerous for tornadoes** in the United States?

According to one study, May is the most dangerous month for tornadoes in the United States, with 329, while February is the safest with only three. In another study the months December and January were usually the safest, and the months having the greatest number of tornadoes were April, May, and June. In February, tornado frequency begins to increase, depending on the geographical area. February tornadoes tend to occur in the central Gulf states; in March the center of activity moves eastward to the southeastern Atlantic states where tornado activity peaks in April. In May the center of activity focuses on the southern Plains states; in June this moves to the northern Plains and Great Lakes area (into western New York).

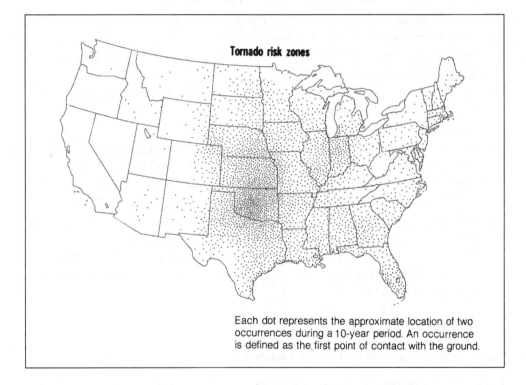

Tornado risk zones

Each dot represents the approximate location of two occurrences during a 10-year period. An occurrence is defined as the first point of contact with the ground.

How are **hurricanes classified**?

The Saffir/Simpson Hurricane Damage-Potential scale assigns numbers 1 through 5 to measure the disaster potential of a hurricane's winds and its accompanying storm surge. The purpose of the scale, developed in 1971 by Herbert Saffir and Robert Simpson, is to assist disaster agencies in gauging the potential significance of these storms in terms of assistance.

101

Saffir/Simpson Hurricane Scale Ranges

Scale number (category)	Barometric pressure (in inches)	Winds (miles per hour)	Surge (in feet)	Damage
1	≥28.94	74–95	4–5	Minimal
2	28.50–28.91	96–110	6–8	Moderate
3	27.91–28.47	111–130	9–12	Extensive
4	27.17–27.88	131–155	13–18	Extreme
5	<27.17	>155	>18	Catastrophic

Damage categories:

Minimal—No real damage to building structures. Some tree, shrubbery, and mobile home damage. Coastal road flooding and minor pier damage.

Moderate—Some roof, window, and door damage. Considerable damage to vegetation, mobile homes and piers. Coastal and low-lying escape routes flood two to four hours before center of storm arrives. Small craft can break moorings in unprotected areas.

Extensive—Some structural damage to small or residential buildings. Mobile homes destroyed. Flooding near coast destroys structures and flooding of homes five feet above sea level as far inland as six miles.

Extreme—Extensive roof, window, and door damage. Major damage to lower floors of structures near the shore, and some roof-failure on small residences. Complete beach erosion. Flooding of terrain 10 feet above sea level as far as six   miles inland requiring massive residential evacuation.

Catastrophic—Complete roof failure to many buildings; some complete building failure with small utility buildings blown away. Major damage to lower floors of all structures 19 feet above sea level located within 500 yards of the shoreline. Massive evacuation of residential areas on low ground 5 to 10 miles from shoreline may be required.

How do **hurricanes** get their **names**?

Since 1950, hurricane names are officially selected from library sources and are decided during the international meetings of the World Meteorological Organization (WHO). The names are chosen to reflect the cultures and languages found in the Atlantic, Caribbean, and Hawaiian regions. The National Hurricane Center near Miami, Florida, selects the name from one of the six listings for Region 4 (Atlantic and Caribbean area) when a tropical storm with rotary action and wind speeds above 39 miles (62 kilometers) per hour develops. Letters Q, U, X, Y, and Z are not included because of the scarcity of names beginning with those letters. Once a storm has done great damage, its name is retired from the six–year list cycle.

1992	1993	1994	1995	1996	1997
Andrew	Arlene	Alberto	Allison	Arthur	Ana
Bonnie	Bret	Beryl	Barry	Bertha	undecided
Charley	Cindy	Chris	Chantal	Cesar	Claudette
Danielle	Dennis	Debby	Dean	Diana	Danny
Earl	Emily	Ernesto	Erin	Edouard	Erika
Frances	Floyd	Florence	Felix	Fran	Fabian
Georges	Gert	Gordon	Gabrielle	Gustav	Grace
Hermine	Harvey	Helene	Humberto	Horrtense	Henri
Ivan	Irene	Isaac	Iris	Isidore	Isabel
Jeanne	Jose	Joyce	Jerry	Josephine	Juan
Karl	Katrina	Keith	Karen	Klaus	Kate
Lisa	Lenny	Leslie	Luis	Lili	Larry
Mitch	Maria	Michael	Marilyn	Marco	Mindy
Nicole	Nate	Nadine	Noel	Nana	Nicholas
Otto	Ophelia	Oscar	Opal	Omar	Odette
Paula	Philippe	Patty	Pablo	Paloma	Peter
Richard	Rita	Rafael	Roxanne	Rene	Rose
Shary	Stan	Sandy	Sebastien	Sally	Sam
Tomas	Tammy	Tony	Tanya	Teddy	Teresa
Virginie	Vince	Valerie	Van	Vicky	Victor
Walter	Wilma	William	Wendy	Wilfred	Wanda

Which United States **hurricanes** have caused the **most deaths**?

The ten deadliest United States hurricanes are listed below.

Hurricane	Year	Deaths
1. Texas (Galveston)	1900	6,000
2. Florida (Lake Okeechobee)	1928	1,836
3. Florida (Keys/S. Texas)	1919	600–900+
4. New England	1938	600
5. Florida (Keys)	1935	408
6. AUDREY (Louisiana/Texas)	1957	390
7. Northeast U.S.	1944	390
8. Louisiana (Grand Isle)	1909	350
9. Louisiana (New Orleans)	1915	275
10. Texas (Galveston)	1915	275

What was the **greatest natural disaster** in United States history?

The greatest natural disaster occurred when a hurricane struck Galveston, Texas on September 8, 1900, and killed over 6,000 people. However, the costliest national disaster to date is when Hurricane Andrew hit Florida on August 24, 1992 and Louisiana on August 25, 1992. Classified as a category four hurricane, Andrew's winds of 135 miles (225 kilometers) per hour gusted up to 165 miles (275 kilometers) per hour. Early warning kept the death-toll low, but property damage is estimated at $20 billion.

SNOW, RAIN, HAIL, ETC.

How does **dew** form?

At night, as the temperature of the air falls, the amount of water vapor the air can hold also decreases. Excess water vapor then condenses as very small drops on whatever it touches.

What is the shape of a **raindrop**?

Although a raindrop has been illustrated as being pear-shaped or tear-shaped, high-speed photographs reveal that a large raindrop has a spherical shape with a hole not quite through it (giving it a doughnut-like shape). Water surface tension pulls the drop into this shape. As a drop larger than 0.08 inches (2 millimeters) in diameter falls, it will become distorted. Air pressure flattens its bottom and its sides bulge. If it becomes larger than 0.25 inches (6.4 millimeters) across, it will keep spreading crosswise as it falls and will bulge more at its sides, while at the same time, its middle will thin into a bow-tie shape. Eventually in its path downward, it will divide into two smaller spherical drops.

How **fast** can **rain** fall?

The speed of rainfall varies with drop size and wind speed. A typical raindrop in still air falls about 600 feet (183 meters) per minute or about 7 miles (11 kilometers) per hour.

What is the **greatest amount of rainfall** ever measured?

Time duration	Amount		Place
	Inches	**Centimeters**	
1 minute	1.5	3.8	Barst, Guadeloupe, West Indies, November 26, 1970
24 hours	73.62	187	Cilaos, La Réunion, Indian Ocean, March 15–16, 1952
24 hours in the United States	19.0	48.3	Alvin, Texas, July 25–26, 1979
calendar month	366.14	930	Cherrapunji, Meghalaya, India, July 1861
12 months	1,041.78	2646	Cherrapunji, Meghalaya, India, between August 1, 1860 and July 31, 1861
12 months in the United States	739	1877	Kukui, Maui, Hawaii, December 1981–December 1982

Where is the **rainiest place on Earth**?

The wettest place in the world is Tutunendo, Colombia, with an average annual rainfall of 463.4 inches (1177 centimeters) per year. The place that has the most rainy days per year is Mount Wai-'ale'ale on Kauai, Hawaii. It has up to 350 rainy days annually.

In contrast the longest rainless period in the world was from October, 1903, to January, 1918, at Arica, Chile—a period of 14 years. In the United States the longest dry spell was 767 days at Bagdad, California, from October 3, 1912, to November 8, 1914.

When do **thunderstorms** occur?

In the United States thunderstorms usually occur in the summertime, especially from May through August. Thunderstorms tend to occur in late spring and summer when large amounts of tropical maritime air move across the United States. Storms usually develop when the surface air is heated the most from the sun (2 to 4 p.m.). Thunderstorms are relatively rare in the New England area, North Dakota, Montana, and other northern states (latitude 60 degrees) where the air is often too cold. Also these storms are rare along the Pacific Ocean because the summers there are too dry for these storms to occur. Florida, the Gulf states, and the southeastern states tend to have the most storms, averaging 70 to 90 annually. The mountainous southwest averages 50 to 70 storms annually. In the world, thunderstorms are most plentiful in the areas between latitude 35 degrees north and 35 degrees south; in these areas there can be as many as 3,200 storms within a 12-hour nighttime period. As many as 1,800 storms can occur at once throughout the world.

105

Lightning does perform a vital function; it returns to the Earth much of the negative charge the Earth loses by leakage into the atmosphere. The annual death toll in the United States from lightning is greater than the annual death toll from tornadoes or hurricanes; 150 Americans die annually from lightning and 250 are injured.

How far away can **thunder** be heard?

Thunder is the crash and rumble associated with lightning. It is caused by the explosive expansion and contraction of air heated by the stroke of lightning. This results in sound waves that can be heard easily six to seven miles (9.7 to 11.3 kilometers) away. Occasionally such rumbles can be heard as far away as 20 miles (32.1 kilometers). The sound of great claps of thunder are produced when intense heat and the ionizing effect of repeated lightning occurs in a previously heated air path. This creates a shock wave that moves at the speed of sound.

What would cause frogs and toads to fall from the sky like a rain shower?

Documented cases of showers of frogs have been recorded since 1794, usually during heavy summer rainstorms. Whirlwinds, waterspouts, and tornadoes are given as the conventional explanation. Extensive falls of fish, birds, and other animals have also been reported.

How large can **hailstones** become?

The average hailstone is about ¼ inch (0.63 centimeters) in diameter. However, hailstones weighing up to 7.5 pounds (3.4 kilograms) are reported to have fallen in Hyderabad state in India in 1939, although scientists think these huge hailstones may be several stones that partly melted and stuck together. On April 14, 1986, hailstones weighing 2.5 pounds (1 kilogram) were reported to have fallen in the Gopalgang district of Bangladesh.

The largest hailstone ever recorded in the United States fell in Coffeyville, Kansas, on September 3, 1970. It measured 5.57 inches (14.1 centimeters) in diameter and weighed 1.67 pounds (0.75 kilograms).

Hail is precipitation consisting of balls of ice. Hailstones usually are made of concentric, or onion-like, layers of ice alternating with partially melted and refrozen snow, structured around a tiny central core. It is formed in cumulonimbus or thunderclouds when freezing water and ice cling to small particles in the air, such as dust. The winds in the cloud blow the particles through zones of different temperatures causing them to accumulate additional layers of ice and melting snow and to increase in size.

What is the difference between **freezing rain** and **sleet**?

Freezing rain is rain that falls as a liquid but turns to ice on contact with a freezing object to form a smooth ice coating called *glaze*. Usually freezing rain only lasts a short time, because it either turns to rain or to snow. Sleet is frozen or partially frozen rain in the form of ice pellets. Sleet forms as falling rain from a warm layer of air passes through a freezing air layer near the Earth's surface to form hard, clear, tiny ice pellets that can hit the ground so fast that they bounce off with a sharp click.

How does **snow** form?

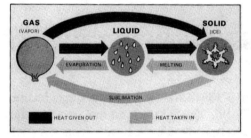

Sublimation of vapor into ice

Snow is not frozen rain. Snow forms by sublimation of water vapor (the turning of water vapor directly into ice, without going through the liquid stage). High above the ground, chilled water vapor turns to ice when its temperature reaches the dew point. The result of this sublimation is a crystal of ice, usually hexagonal. Snow begins in the form of these tiny hexagonal ice crystals in the high clouds; the young crystals are the seeds from which snowflakes will grow. As water vapor is pumped up into the air by updrafts, more water is deposited on the ice crystals, causing them to grow. Soon some of the larger crystals fall to the ground as snowflakes.

How much **water** is **in an inch of snow**?

An average figure is 10 inches (25 centimeters) of snow is equal to 1 inch (2.5 centimeters) of water. Heavy, wet snow has a high water content; 4 to 5 inches (10 to 12.5 centimeters) may contain 1 inch (2.5 centimeters) of water. A dry, powdery snow might require 15 inches (37.5 centimeters) of snow to equal 1 inch (2.5 centimeters) of water.

Are all **snowflakes shaped** alike?

Although no two snowflakes are identical, they all have six sides or six prongs. The many forms that a snow crystal can have are explained by the number of water molecules involved (10^{18}) and the different ways that the molecules can arrange themselves in a three-dimensional hexagonal pattern. Temperature may be the key to the crystal's **107** shape—flat, plate-like forms, long column-like forms, or prismlike forms. Snow crystals are classified into 10 categories in the International Snow Classification. Half are symmetrical and include plates, stellars, columns, needles, and capped columns. Columns and capped columns occur at temperatures colder than 12°F (-10°C). Needles occur at the warmest temperatures. Plates and stellars, the most beautiful of the snow crystals, are rare and occur under special weather conditions. Very plentiful are the asymmetrical snow crystals, the spatial dendrites (three-dimensional irregular stellar- or plate-like crystals occurring in high-moisture clouds) and the irregular crystals (ice prisms called diamond dust).

What causes exceptionally large snowflakes?

Giant snowflakes, 4 to 6 inches (10 to 15 centimeters) in diameter, have been verified. Large flakes are probably aggregations of smaller flakes created by collisions and/or electrostatic attraction.

What is the record for the greatest snowfall in the United States?

The record for the most snow in a single storm is 189 inches (480 centimeters) at Mount Shasta Ski Bowl in California from February 13–19, 1959. For the most snow in a 24 hour day, the record goes to Silver Lake, Colorado, on April 14–15, 1921, with 76 inches (193 centimeters) of snow. The year record is held by Paradise, Mount Rainier, in Washington with 1,224.5 inches (3,110 centimeters) from February 19, 1971, to February 18, 1972. The highest average annual snowfall was 241 inches (612 centimeters) for Blue Canyon, California. In March, 1911, Tamarack, California, had the deepest snow accumulation— over 37.5 feet (11.4 meters).

Is it ever too cold to snow?

No matter how cold the air gets, it still contains some moisture, and this can fall out of the air in the form of very small snow crystals. Very cold air is associated with no snow because these invasions of air from northerly latitudes are associated with clearing conditions behind cold fronts. Heavy snowfalls are associated with relatively mild air in advance of a warm front. The fact that snow piles up, year after year, in Arctic regions illustrates that it is never too cold to snow.

When does frost form?

A frost is a crystalline deposit of small thin ice crystals formed on objects which are at freezing or below freezing temperatures. This phenomenon occurs when atmospheric water vapor condenses directly into ice without first becoming a liquid; this process is called sublimation. Usually frost appears on clear, calm nights especially during early autumn when the air above the Earth is quite moist. A light frost damages only the most tender plants and vines; whereas a heavy frost (a heavy deposit of crystallized water) might kill plants that were not sturdy. A killing frost may do that to sturdy plants. Black frost (or hard frost) occurs in late autumn when both objects and air are at temperatures below freezing. Leaf edges and leaf tips turn black from this frost. Hoar or hoarfrost is the term for frost in Europe. Finally, permafrost is ground permanently frozen which never thaws out completely.

WEATHER PREDICTION

What is **nowcasting**?

Nowcasting is a form of very short-range weather forecasting. The term is sometimes used loosely to refer to any area-specific forecast for the period up to 12 hours ahead that is based on very detailed observational data. However, nowcasting should probably be defined more restrictively as the detailed description of the current weather along with forecasts obtained by extrapolation up to about two hours ahead.

Can **groundhogs** accurately predict the weather?

Over a 60-year period, groundhogs have accurately predicted the weather (i.e., when spring will start) only 28% of the time on Groundhog Day. Groundhog Day was first celebrated in Germany where farmers would watch for a badger to emerge from winter hibernation. If the day was sunny, the sleepy badger would be frightened by this shadow and duck back for another 6 weeks' nap; if it was cloudy he would stay out, knowing that spring had arrived. German farmers who emigrated to Pennsylvania brought the celebration to America. Finding no badgers in Pennsylvania, they chose the groundhog as a substitute.

Is a **halo around the sun or moon** a sign of rain or snow approaching?

The presence of a ring around the sun or, more commonly, the moon in the night sky, betrays very high ice crystals composing cirrostratus clouds. The brighter the ring, the greater the odds of precipitation and the sooner it may be expected. Rain or snow will not *always* fall, but two times out of three, precipitation will start to fall within twelve to eighteen hours. These cirroform clouds are a forerunner of an approaching warm front and an associated low pressure system.

What is **barometric pressure** and what does it mean?

Barometric, or atmospheric, pressure is the force exerted on a surface by the weight of the air above that surface, as measured by an instrument called a barometer. Pressure is greater at lower levels because the air's molecules are squeezed under the weight of the air above. So while the average air pressure at sea level is 14.7 pounds per square inch, at 1,000 feet above sea level, the pressure drops to 14.1 pounds per square inch, and at 18,000 feet the pressure is 7.3 pounds, about half of the figure at sea level. Changes in air pressure bring weather changes. High pressure areas bring clear skies and fair weather; low pressure areas bring wet or stormy weather. Areas of very low pressure have serious storms, such as hurricanes.

Can weather be predicted from the stripes on a **wooly-bear caterpillar**?

It is an old superstition that the severity of the coming winter can be predicted by the width of the brown bands or stripes around the wooly-bear caterpillar in the autumn. If the brown bands are wide, says the superstition, the winter will be mild, but if the brown bands are narrow, a rough winter is foretold. Studies at the American Museum of Natural History in New York City failed to show any connection between the weather and the caterpillar's stripes. This belief is only a superstition; it has no basis in scientific fact.

When did **weather forecasting** begin?

On May 14, 1692, the weekly newspaper, *A Collection for the Improvement of Husbandry and Trade,* gave a seven–day table with pressure and wind readings for the comparable dates of the previous year. Readers were expected to make up their own forecasts from the data. Other journals soon followed with their own weather features. In 1771, a new journal, called the *Monthly Weather Paper,* was completely devoted to weather prediction. In 1861, the British Meteorological Office began issuing daily weather forecasts. The first broadcast of weather forecasts was done by the University of Wisconsin's station 9XM at Madison, Wisconsin, on January 3, 1921.

MINERALS AND OTHER MATERIALS

ROCKS AND MINERALS

See also: Energy

How do **rocks** differ?

Rocks can be conveniently placed into three groups—igneous, sedimentary, and metamorphic.

Igneous rocks, such as granite, pegmatite, rhyolite, obsidian, gabbro, and basalt are formed by the solidification of molten magma that emerges through the Earth's crust via volcanic activity. The nature and properties of the crystals vary greatly, depending in part on the composition of the original magma and partly on the conditions under which the magma solidified. There are thousands of different igneous rock types. For example, granite is formed by slow cooling of molten material (within the Earth). It has large crystals of quartz, feldspars, and mica.

Sedimentary rocks such as brecchia, sandstone, shale, limestone, chert, and coals are produced by the accumulation of "sediments." These are fine rock particles or fragments, skeletons of microscopic organisms, or minerals leached from rocks that have accumulated from weathering. These "sediments" are then redeposited under water and were later compressed in layers over time. The most common sedimentary rock is sandstone, which is predominantly quartz crystals.

Metamorphic rocks, such as marble, slate, schist, gneiss, quartzite, and hornsfel, are formed by the alteration of igneous and sedimentary rocks through heat and/or pressure. These physical and/or chemical changes to rocks can be exemplified by the formation of marble from thermal changes that have occurred to limestone.

What is **petrology** and what does a **petrologist** do?

Petrology is the science of rocks. A petrologist is a person who is interested in the mineralogy of rocks and who also tries to unravel the record of the geological past contained within rocks. From these rocks, a petrologist can learn about past climates and geography, past and present composition, and the conditions which prevail within the interior of the Earth.

How are **fossils formed**?

Fossil scallops of Cretaceous age from north-central Texas.

Fossils are the remains of animals or plants that were preserved in rock before the beginning of recorded history. It is unusual for complete organisms to be preserved; fossils usually represent the hard parts such as bones or shells of animals and leaves, seeds, or woody parts of plants.

Some fossils are simply the bones, teeth, or shells themselves, which can be preserved for a relatively short period of time. Another type of fossil is the imprint of a buried plant or animal that decomposes, leaving a film of carbon that retains the form of the organism.

Some buried material is replaced by silica and other materials that permeate the organism and replace the original material in a process called petrification. Some woods are replaced by agate or opal so completely that even the cellular structure is duplicated. The best examples of this can be found in the Petrified Forest National Park in Arizona.

Molds and casts are other very common fossils. A mold is made from an imprint, such as a dinosaur footprint, in soft mud or silt. This impression may harden, then be covered with other materials. The original footprint will have formed a mold and the sediments filling it will be a cast of the footprint.

How **old** are **fossils**?

The oldest known fossils are of single-celled organisms, blue-green algae, found in 3.2 billion-year-old cherts, shales, and sandstone from the Transvaal of South Africa. Multicellular fossils dating from about 700 million years ago are also known. The largest number of fossils come from the Cambrian era of 590 million years ago, when living organisms began to develop skeletons and hard parts. Since these parts tended to last longer than ordinary tissue, they were more likely to be preserved in the clay to become fossilized.

What is a **geode**?

A geode is a hollow, stone-like formation lined inside with inward projecting, small crystals. Geodes are frequently found in limestone beds and occur in many parts of the world. They average 2 to 6 inches (5 to 15 centimeters) in diameter.

What are **Indian Dollars**?

They are six-sided disk-shaped twin crystals of aragonite ($CaCO_3$), which have altered to calcite but retained their outer form. They occur in large numbers in northern Colorado where they are known as "Indian Dollars." In New Mexico they are called "Aztec Money" and in western Kansas they are called "Pioneer Dollars."

How does a **rock** differ from a **mineral**?

Mineralogists use the term "mineral" for a substance that has all four of the following features: it must be found in nature; it must be made up of substances that were never alive (organic); it has the same chemical makeup wherever it is found; and its atoms are arranged in a regular pattern and form solid crystals.

While "rocks" are sometimes described as an aggregate or combination of one or more minerals, geologists extend the definition to include clay and loose sand and certain limestones.

Who is regarded as the father of modern **mineralogy**?

Georg Bauer (1495-1555) was a German mineralogist, physician, and scholar. Also known by the Latin name Georgius Agricola, he studied mining methods and mining ores, and wrote on the classification of minerals, mineral extraction, and geological phenomena. *De re metallica,* his major work, covered the entire field of mining and metallurgy in a systematic and comprehensive manner. Bauer was one of the first to base his writings on observation and scientific inquiry.

Who was the first person to attempt a **color standardization** scheme?

The German mineralogist Abraham Gottlob Werner (1750-1817) devised a method of describing minerals by their external characteristics, including color. He worked out an arrangement of colors and color names, illustrated by an actual set of minerals.

113

What is the Mohs scale?

The scale is a standard of 10 minerals by which the hardness of a mineral is rated. It was introduced in 1812 by the German mineralogist Friedrich Mohs (1773-1839). The minerals are arranged from softest to hardest.

Hardness	Mineral	Comment
1	Talc	Hardness 1–2 can be scratched by a fingernail
2	Gypsum	Hardness 2–3 can be scratched by a copper coin
3	Calcite	Hardness 3–6 can be scratched by a steel pocket knife
4	Fluorite	
5	Apatite	
6	Orthoclase	Hardness 6–7 will not scratch glass
7	Quartz	
8	Topaz	Hardness 8–10 will scratch glass
9	Corundum	
10	Diamond	

What is meant by the term strategic minerals?

Strategic minerals are minerals essential to the national defense—the supply of which a country uses but cannot produce itself. 33% to 50% of the 80 minerals used by industry could be classed as strategic minerals. Wealthy countries, such as the United States, stockpile these minerals to avoid any crippling effect on their economy or military strength if political circumstances were to cut off their supplies. The United States, for instance, stockpiles bauxite (14½ million tons), manganese (2.2 million tons), chromium (1.8 million tons), tin (185,000 tons), cobalt (19,000 tons), tantalum (635 tons), palladium (1.25 million troy ounces), and platinum (453,000 troy ounces).

What is pitchblende?

Pitchblende is a massive variety of uraninite or uranium oxide found in metallic veins. A radioactive material and the most important ore of uranium, it is the original source of radium.

What is galena?

Galena is a lead sulphide (PbS) and the most common ore of lead, containing 86.6% lead. Lead-gray in color, with a brilliant metallic luster, galena has a specific gravity of 7.5 and a hardness of 2.5 on the Mohs scale, and usually occurs as cubes or a modification of an octahedral form. Mined in Australia, it is also found in Missouri, Kansas, and Oklahoma, as well as Colorado, Montana, and Idaho.

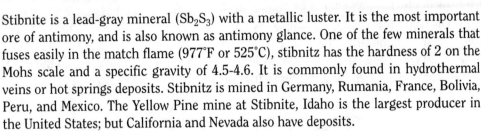

What is **stibnite**?

Stibnite is a lead-gray mineral (Sb_2S_3) with a metallic luster. It is the most important ore of antimony, and is also known as antimony glance. One of the few minerals that fuses easily in the match flame (977°F or 525°C), stibnitz has the hardness of 2 on the Mohs scale and a specific gravity of 4.5-4.6. It is commonly found in hydrothermal veins or hot springs deposits. Stibnitz is mined in Germany, Rumania, France, Bolivia, Peru, and Mexico. The Yellow Pine mine at Stibnite, Idaho is the largest producer in the United States; but California and Nevada also have deposits.

What are **Cape May diamonds**?

They are quartz pebbles, found in the vicinity of the Coast Guard station in Cape May, New Jersey. The pebbles are polished, faceted, and sold to tourists as "Cape May diamonds."

Are there any **diamond mines** in the United States?

The only significant diamond deposit in North America is at Murfreesboro, Arkansas. It is on government-owned land and has never been systematically developed. For a small fee, tourists can dig there and try to find diamonds. The largest crystal found there weighed 40.23 carats and was named the "Uncle Sam" diamond.

Diamonds crystallize directly from rock melts rich in magnesium and saturated with carbon dioxide gas that has been subjected to high pressures and temperatures exceeding 2559°F (1400°C). These rock melts originally came from deep in the Earth's mantle at depths of 93 miles (150 kilometers).

Diamonds are minerals composed entirely of the element carbon, with an isometric crystalline structure. The hardest natural substance, gem diamonds have a density of 3.53, though black diamonds (black carbon cokelike aggregates of microscopic crystals) may have a density as low as 3.15. Diamonds have the highest thermal conductivity of any known substance. This property enables diamonds to be used in cutting tools, because they do not become hot.

How can a **genuine diamond** be identified?

There are several tests that can be performed without the aid of tools. A knowledgeable person can recognize the surface lustre, straightness and flatness of facets, and high light reflectivity. Diamonds become warm in a warm room and cool if the surroundings are cool. A simple test that can be done is exposing the stones to warmth and cold and then touching them to one's lips to determine their appropriate temperature. This is especially effective when the results of this test are compared to the results of the test done on a diamond known to be genuine. Another test is to pick up the stone with a moistened fingertip. If this can be done, then the stone is likely to be a diamond. The majority of other stones cannot be picked up in this way.

The water test is another simple test. A drop of water is placed on a table. A perfectly clean diamond has the ability to almost "magnetize" water and will keep the water from spreading. An instrument called a diamond probe can detect even the most sophisticated fakes. Gemologists always use this as part of their inspection.

What is fool's gold?

Iron pyrite (FeS_2) is a mineral popularly known as "fool's gold." Because of its metallic luster and pale brass yellow color it is often mistaken for gold. Real gold is much heavier, softer, not brittle, and not grooved.

How is the **value of a diamond** determined?

Demand, beauty, durability, rarity, freedom from defects, and perfection of cutting determine the value of a gemstone. But the major factor in establishing the price of gem diamonds is the control over output and price as exercised by the Central Selling Organization's (CSO) Diamond Trading Company Ltd. The CSO is a subsidiary of DeBeers Consolidated Mines Ltd.

How are **diamonds weighed**?

The basic unit is a carat, which is 200 milligrams or $\frac{1}{142}$ of an avoirdupois ounce. A well-cut round diamond of 1 carat measures almost exactly ¼ inch in diameter. Another unit commonly used is the point, which is one hundredth of a carat. A stone of 1 carat weighs 100 points.

Which **diamond** is the **world's largest**?

The Cullinan Diamond, weighing 3,106 carats, was discovered on January 25, 1905, at the Premier Diamond Mine, Transvaal, South Africa. Named for Sir Thomas M. Cullinan, chairman of the Premier Diamond Company, it was cut into 9 major stones and 96 smaller brilliants. The total weight of the cut stones was 1,063 carats, only 35% of the original weight.

Cullinan I, also known as the "Greater Star of Africa" or the "First Star of Africa," is a pear-shaped diamond weighing 530.2 carats. It is 2⅛ inches (5.4 centimeters) long, 1¾ inches (4.4 centimeters) wide, and 1 inch (2.54 centimeters) thick at its deepest point. It was presented to Britain's King Edward VII in 1907, and was set in the British monarch's sceptre with the cross. It is still the largest cut diamond in the world.

Cullinan II, also know as the "Second Star of Africa," is an oblong stone which weighs 317.4 carats. It is set in the British Imperial State Crown.

Besides the Cullinan Diamond, what are the **largest precious stones**?

The largest ruby is a 8,500 carat stone that is 5½ inches (14 centimeters) tall, carved to resemble the Liberty Bell. The largest star ruby is the 2,475 carat "Rajarathna" from India which has a six line star. The largest cut emerald was found in Carnaiba, Brazil, in August 1974. It is 86,136 carats. A 2,302 carat sapphire from at Anakie, Queensland, Australia, was carved into a 1,318 carat head of Abraham Lincoln, making it the largest carved sapphire. "The Lone Star," at 9,719.5 carats, is the largest star sapphire. The largest natural pearl is the "Pearl of Lao-tze," also called the "Pearl of Allah." Found in May 1934 in the shell of a giant clam at Palawan, Philippines, the pearl weighs 14 pounds, 1 ounce (6.38 kilograms).

How does the **emerald** get its color?

Emerald is a variety of green beryl ($Be_3Al_2Si_6O_{18}$) that is colored by a trace of chromium (Cr), which replaces the aluminum (Al) in the beryl structure. Other green beryls exist; but if no chromium is present, they are, technically speaking, not emeralds.

How is the star in **star sapphires** produced?

Sapphires are composed of blue gem-quality corundum (Al_2O_3). When cut in the unfaceted cabochon (dome or convex) form, a sapphire with a trace of titanium dioxide will display a rayed (star) figure usually with six rays. The same effect produces the star in star rubies as well.

METALS

Which **metallic element** is the **most abundant**?

Aluminum is the most abundant metallic element on the surface of the Earth and moon; it composes more than 8% of the Earth's crust. It is never free in nature, combining with oxygen, sand, iron, titanium, etc.; its ores are mainly bauxites (aluminum **117** hydroxide). Nearly all rocks, particularly igneous rocks, contain aluminum as aluminosilicate minerals. Napoleon III (1808–1883) recognized that the physical characteristic of its lightness could revolutionize the arms industry, so he granted a large subsidy to French chemist Sainte-Claire Deville (1818–1881) to develop a method to make its commercial use feasible. In 1854, Deville obtained the first pure aluminum metal through the process of reduction of aluminum chloride. In 1886, the American Charles Martin Hall (1863–1914) and the Frenchman Paul Heroult (1863–1914) independently discovered an electrolytic process to produce aluminum from bauxite. Because of aluminum's resistance to corrosion, low density, and excellent heat-conducting property it is used in cookware manufacturing and can-making industries. It is a good conductor of electricity, and is widely used in overhead cables. Aluminum alloys, such as duralumin, have high tensile strengths and are of considerable industrial importance, especially in the aerospace industry.

Why are alchemical symbols for metals and astrological symbols for planets identical?

The ancient Greeks and Romans knew seven metals and also knew seven "planets" (the five nearer planets plus the sun and the moon). They related each planet to a specific metal. Alchemy, originating in about the 3rd century B.C.E., focused on changing base metals, such as lead, into gold. Although at times alchemy degenerated into mysticism, it contained centuries of chemical experience, which provided the foundation for the development of modern chemistry.

English name	Chemical symbol	Latin name	Alchemical symbol
Gold	Au	*aurum*	☉ (Sun)
Silver	Ag	*argentum*	☽ (Moon)
Copper	Cu	*cuprum*	♀ (Venus)
Iron	Fe	*ferrum*	♂ (Mars)
Mercury	Hg	*hydrargyrum*	☿ (Mercury)
Tin	Sn	*stannum*	♃ (Jupiter)
Lead	Pb	*plumbum*	♄ (Saturn)

What are the noble metals?

The noble metals are gold, silver, mercury, and the platinum group (including palladium, iridium, rhodium, ruthenium, and osmium). The term refers to those metals highly resistant to chemical reaction or oxidation (resistant to corrosion) and is contrasted to "base" metals which are not so resistant. The term has its origins in ancient alchemy whose goals of transformation and perfection were pursued through the different properties of metals and chemicals. The term is not synonomous with "precious metals," although a metal, like platinium, may be both.

What are the **precious metals**?

This is a general term for expensive metals that are used for making coins, jewelry, and ornaments. The name is limited to gold, silver, and platinum. Expense or rarity does not make a metal precious, but rather, it is a value set by law which states that the object made of these metals has a certain intrinsic value. The term is not synonymous with "noble metals," although a metal (such as platinum) may be both noble and precious.

What is **24 karat gold**?

The term "karat" refers to the percentage of gold versus the percentage of an alloy in an object. Gold is too soft to be usable in its purest form; it has to be mixed with other metals.

Karatage	Percentage of fine gold
24	100
22	91.75
18	75
14	58.5
12	50.25
10	42
9	37.8
8	33.75

How thick is **gold leaf**?

Gold leaf is pure gold which is hammered so thin that it takes 300,000 units to make a stack one inch high. The thickness of a single gold leaf is .0000035 inches (3½ millionths of an inch), although this may vary widely according to which manufacturer makes it. Also called gold foil, it is used for architectural coverings and for hot-embossed printing on leather.

What are the chief **gold-producing countries**?

South Africa is by far the leading producer of gold; other principal areas of production are in the western United States (Utah, Nevada, California, South Dakota, Alaska), Australia, Mexico, Canada, China, India, and the former Soviet Union.

1988 Top World Gold Producers

Country	Number of tons	Number of Tonnes (Metric tons)	% of world total gold production
South Africa	684.5	621.0	32.6%
USSR	308.6	280.0	14.7%
United States	226.3	205.3	10.8%
Australia	167.5	152.0	8.0%
Canada	141.1	128.5	6.8%

In the United States, Nevada is the leading gold producer, with California a distant second, followed by South Dakota in third place.

Is **white gold** really gold?

White gold is the name of a class of jeweler's white alloys used as substitutes for platinum. Different grades vary widely in composition, but usual alloys consist of from 20% to 50% nickel, with the balance gold. A superior class of white gold is made of 90% gold and 10% palladium. Other elements used may be copper and zinc. The main use of these alloys is to give the gold a white color.

What is **German silver**?

Nickel silver, sometimes known as German silver, is a silver-white alloy composed of 52% to 80% copper, 10% to 35% zinc, and 5% to 35% nickel. It may also contain a small percent of lead and tin. There are other forms of nickel silver, but the term "German silver" is the name used in the silverware trade.

Which metal is the main component of **pewter**?

Tin. Roman pewter has about 70% tin. The best pewter used for expensive articles today contains 100 parts tin, 8 parts antimony, 2 parts bismuth, and 2 parts copper. This alloy is easy to work with, does not become brittle when repeatedly beaten and can be worked cold. However, it is too soft to use for heavy tools or weapons, so it is confined to domestic utensils.

Where were the **first successful ironworks** in America?

Although iron ore in this country was first discovered in North Carolina in 1585, and the manufacture of iron was first undertaken (but never accomplished) in Virginia in 1619, the first successful ironworks in America was established by Thomas Dexter and Robert Bridges near the Saugus River in Lynn, Massachusetts. As the original promoters of the enterprise, they hired John Winthrop, Jr. from England to begin production. By 1645, a blast furnace had begun operations and in 1648, a forge was working there.

What is the Q-BOP process for **steelmaking**?

A variation of the *basic oxygen process* (BOP) involves blowing high-purity oxygen through a bath of molten pig iron. This is called the Q-BOP process, used in the production of low-alloy steel, which is defined as steel that has no more than 5% total combined alloying elements. It has mainly a surface hardness. The depth of hardness depends on alloy content. There are five types of processes currently employed in the production of low-alloy steels: oxygen top-blowing, known as LD (Linz–Donawitz) process, BOF (*basic oxygen furnace*), or BOP; oxygen and lime bottom blowing, known as OBM or Q- BOP; top and bottom mixed blowing; open-hearth furnace; and electric-arc furnace.

What is **high speed steel**?

High speed steel is a general name for high alloy steels which retain their hardness at very high temperatures and are used for metal-cutting tools. All high speed steels are based on either tungsten or molybdenum (or both) as the primary heat-resisting alloying element. These steels require a special heat so that their unique properties can be fully realized. This procedure consists of heating to a high temperature of 2150 to 2400°F (1175 to 1315°C) to obtain solution of a substantial percentage of the alloy carbides, quenching to room temperature, tempering at 1000 to 1150°F (535 to 620°C), and again cooling to room temperature.

Which countries have **uranium** deposits?

Uranium, a radioactive metallic element is the only natural material capable of sustaining nuclear fission. But only one isotope, uranium-235, which occurs in one molecule out of 40 of natural uranium, can undergo fission under neutron bombardment. Mined in various parts of the world, it must then be converted during purification to uranium dioxide (UO_2). Uranium deposits occur throughout the world. The United States (especially Arizona, Colorado, New Mexico, North Carolina, and Utah), Canada, France, South Africa, Zaire, Australia, and the former Soviet Union have significant uranium resources. Canada and Zaire provide the best sources.

What is **technetium**?

Technetium (Tc, element 43) is a radioactive metallic element that does not occur naturally either in its pure form or as compounds; it is produced during nuclear fission (nuclear reaction during which the nucleus splits almost equally). It was the first element to be made artificially in 1937 when it was isolated and extracted by C. Perrier and Emilio Segre (1905-1989). A fission product of molybdenum (Mo, element 42), Tc can also occur as a fission product of uranium (U, element 92).

Tc has found some application in diagnostic medicine. Ingested soluble technetium compounds tend to concentrate in the liver and are valuable in labeling and in radiological examination of that organ. Also, by technetium labeling of blood serum, components, diseases involving the circulatory system can be explored.

NATURAL SUBSTANCES
See also: Energy

Is **lodestone** a magnet?

Lodestone is a magnetic variety of natural iron oxide. It can be made to swing like a magnet, and was used by early mariners to find the magnetic north. Lodestone is frequently called a natural magnet.

What is **red dog**?

Red dog is the residue from burned coal dumps. The dumps are composed of waste products incidental to coal mining. Under pressure in these waste dumps, the waste frequently ignites from spontaneous combustion, producing a red-colored ash, which is used for driveways, parking lots, and roads.

In coal mining what is meant by **damp**?

A poisonous or explosive gas in a mine. Carbon monoxide is known as white damp, and methane is known as firedamp. Blackdamp is formed by mine fires and explosion of firedamp in mines. It extinguishes light and suffocates its victims. The average blackdamp contains 10% to 15% carbon dioxide and 85% to 90% nitrogen.

What is **diatomite**?

Diatomite (also called diatomaceous Earth) is a white or cream–colored, friable, porous rock composed of the fossil remains of diatoms (small water plants with silica cell walls). These fossils build up on the ocean bottoms to form diatomite, and in some places, these areas have become dry land or diatomacceous Earth. Chemically inert and having a rough texture and other unusual physical properties, it is suitable for many scientific and industrial purposes, such as filtering agent; building material; heat; cold; and sound insulator; catalyst carrier; filler absorbent; abrasive; and ingredient in pharmaceutical preparations. Dynamite is made from it by soaking it in the liquid explosive nitroglycerin.

What is **fuller's Earth**?

It is a naturally occurring white or brown clay containing aluminum magnesium silicate. Once used to clean wool and cloth (fulling), it is currently used for decolorizing oils and fats, as a pigment extender, a filter, and an absorbent, and in floor sweeping compounds.

Which products come from the **tropical forests**?

Products from Tropical Forests

Woods	Houseplants	Spices	Foods
Balsa	*Anthurium*	Allspice	Avocado
Mahogany	Croton	Black pepper	Banana
Rosewood	*Dieffenbachia*	Cardamom	Coconut
Sandalwood	*Dracaena*	Cayenne	Grapefruit
Teak	Fiddle-leaf fig	Chili	Lemon
	Mother-in-law's tongue	Cinnamon	Lime
Fibers	Parlor ivy	Cloves	Mango
Bamboo	*Philodendron*	Ginger	Orange
Jute/Kenaf	Rubber tree plant	Mace	Papaya
Kapok	*Schefflera*	Nutmeg	Passion fruit
Raffia	Silver vase bromeliad	Paprika	Pineapple
Ramie	*Spathiphyllum*	Sesame seeds	Plantain
Rattan	Swiss cheese plant	Turmeric	Tangerine
	Zebra plant	Vanilla bean	Brazil nuts
Gums,resins			Cane sugar
Chicle latex	**Oils, etc.**		Cashew nuts
Copaiba	Camphor oil		Chocolate
Copal	Cascarilla oil		Coffee
Gutta percha	Coconut oil		Cucumber
Rubber latex	Eucalyptus oil		Hearts of palm
Tung oil	Oil of star anise		Macadamia nuts
	Palm oil		Manioc/tapioca
	Patchouli oil		Okra
	Rosewood oil		Peanuts
	Tolu balsam oil		Peppers
	Annatto		Cola beans
	Curare		Tea
	Diosgenin		
	Quinine		
	Reserpine		
	Strophanthus		
	Strychnine		
	Yang-Yang		

How much wood is used to make a ton of **paper**?

In the United States, the wood used for the manufacture of paper is mainly from small diameter bolts and pulpwood. It is usually measured by the cord or by weight. Although the fiber used in making paper is overwhelmingly wood fiber, a large percentage of other ingredients is needed. One ton of a typical paper requires 2 cords of wood, but also requires 55,000 gallons of water, 102 pounds of sulfur, 350 pounds of lime, 289 pounds of clay, 1.2 tons of coal, 112 kilowatt hours of power, 20 pounds of dye and pigments, and 108 pounds of starch, as well as other ingredients.

Which woods are used for **telephone poles**?

The principal woods used for telephone poles are southern pine, Douglas fir, western red cedar, and lodgepole pine. Ponderosa pine, red pine, jack pine, northern white cedar, other cedars, and western larch are also used.

Which woods are used for **railroad ties**?

Many species of wood are used for ties. The more common are oaks, gums, Douglas fir, mixed hardwoods, hemlock, southern pine, and mixed softwoods.

Does any **wood sink** in water?

Ironwood is a name applied to many hard, heavy woods. Some ironwoods are so dense that their specific gravities exceed 1.0 and they are therefore unable to float in water. North American ironwoods include the American hornbeam, the mesquite, the desert ironwood, and leadwood (*Krugiodendron ferreum*) with a specific gravity of 1.34–1.42 to make it the heaviest in the United States.

The heaviest wood is black ironwood (*Olea laurifolia*), also called South Afri ironwood. Found in the West Indies, it has a specific gravity of 1.49 and weighs up to 93 pounds (42.18 kilograms) per foot. The lightest wood is *Aeschynomene hispida,* found in Cuba, with a specific gravity of 0.044 and a weight of 2½ pounds (1.13 kilograms) per foot. Balsa tree (Ochroma pyramidale) varies between 2½ (1.13 kilograms) and 24 pounds (10.88 kilograms) per foot.

How is **petrified wood** formed?

Water containing dissolved minerals such as calcium carbonate ($CaCO_3$) and silicate infiltrates the wood, and the minerals gradually replace the organic matter. The process takes thousands of years. After a time, the wood seems to have turned to stone because the original outer form and structure are retained during this process of petrification. The wood itself does not turn into stone.

What is **rosin**?

Rosin is the resin produced after the distillation of turpentine, obtained from several varieties of pine trees, especially the longleaf pine (*Pinus palustris*) and the slash pine (*Pinus caribaea*). Rosin is used in varnishes, paint driers, soluble oils, paper-sizing, belt dressings, and for producing many chemicals.

Why are **essential oils** called "essential"?

Called essential oils because of their ease of solubility in alcohol to form essences, essential oils are used in flavorings, perfumes, disinfectants, medicine, and other products. They are naturally occurring volatile aromatic oils found in uncombined forms within various parts of plants (leaves, pods, etc.). These oils contain as one of its main ingredients a substance belonging to the terpene group. Examples of essential oils include bergamot, eucalyptus, ginger, pine, spearmint, and wintergreen oils. Extracted by distillation or enfleurage (extraction using fat) and mechanical pressing, these oils can now be made synthetically.

What is **ambergris**?

Ambergris, a highly odorous waxy substance found floating in tropical seas, is a secretion from the sperm whale (*Physeter catodon*). The whale secretes ambergris to protect its stomach from the sharp bone of the cuttlefish, a squid-like sea mollusk, which it ingests. Ambergris is used in perfumery as a fixature to extend the life of a perfume and as a flavoring for food and beverages. Today ambergris is synthesized and used by the perfume trade, which has voluntarily refused to purchase ambergris to protect the sperm whales from exploitation.

Where does **isinglass** come from?

Isinglass is the purest form of animal gelatin. It is manufactured from the swimming bladder of the sturgeon. It is used in the clarification of wine and beer as well as in the making of some cements, jams, jellies, and soups.

Which products are made from **horsehair**?

Over 40 products are listed as being made from horsehair, some of which are baskets, belts, bird nests, hair and industrial brushes, buttons, carpet, curlers, fishing lines, furniture padding, hats, lariats, fishing nets, plumes for military hats or horse bridles, surgical sutures (during the Civil War), upholstery cloth, bows (for violin, cello, and viola) whips, and wigs.

Frankincense is an aromatic resin obtained by tapping the trunks of trees belonging to the genus *Boswellia*. The milky resin hardens when exposed to the air and forms irregular lumps, the form in which it is usually marketed.

Myrrh comes from a tree of the genus *Commiphora*, a native of Arabia and Northeast Africa. The resin is also obtained from the tree trunk.

Where does a **luffa sponge** come from?

Luffas are nonwoody vines of the cucumber family. The interior fibrous skeletons of the fruit are used as sponges. The common name is sometimes spelled loffah. Dishcloth gourd, rag gourd, and vegetable sponge are other popular names.

MAN–MADE PRODUCTS

How is **dry ice** made?

Dry ice is composed of carbon dioxide, which at normal temperatures is a gas. The carbon dioxide is stored and shipped as a liquid in tanks that are pressurized at 1073 pounds per square inch. To make dry ice, the carbon dioxide liquid is withdrawn from the tank and allowed to evaporate at a normal pressure in a porous bag. This rapid evaporation consumes so much heat that part of the liquid CO_2 freezes to a temperature of -109°F (-78°C). The frozen liquid is then compressed by machines into blocks of "dry ice" which will melt into a gas again when set out at room temperature.

It was first made commercially in 1925 by the Prest-Air Devices Company of Long Island City, New York through the efforts of Thomas Benton Slate. Dry ice was used by Schrafft's of New York in July 1925 to keep ice cream from melting. The first large sale of dry ice was made later in that year to Breyer Ice Cream Company of New York.

Why is **sulfuric acid** so important?

Sometimes called "oil of vitriol," sulfuric acid (H_2SO_4) has become one of the most important of all chemicals. It was little used until it became essential for the manufacture of soda in the 18th century. It is prepared industrially by the reaction of water with sulfur trioxide, which in turn is made by chemical combination of sulfur dioxide and oxygen by of two processes (the contact process or the chamber process). Many manufactured articles in common use depend in some way on sulfuric acid for their production. Its greatest use is in the production of fertilizers, but it is also used in the refining of petroleum, and production of automobile batteries, explosives, pigments, iron and other metals, and paper pulp.

What is "aqua regia"?

"Aqua regia," also known as nitrohydrochloric acid, is a mixture of 1 part concentrated nitric acid and 3 parts of concentrated hydrochloric acid. The chemical reaction between the acids makes it possible to dissolve all metals except silver. The reaction of metals with nitrohydrochloric acid typically involves oxidation of the metals to a metallic ion and the reduction of the nitric acid to nitric oxide. The term comes from Latin and means royal water. It was named by the alchemists for its ability to dissolve gold, sometimes called the "royal metal."

Who developed the process for making **ammonia**?

Known since ancient times, ammonia (NH_3) has been commercially important for more than 100 years and has become the second largest chemical in terms of tonnage production and the first chemical in value of production. The first breakthrough in the large-scale synthesis of ammonia resulted from the work of Fritz Haber (1863–1934). In 1913, Haber found that ammonia could be produced by combining nitrogen and hydrogen ($N_2 + 3H_2$ $2NH_3$) with a catalyst (iron oxide with small quantities of cerium and chromium) at 131°F (55°C) under a pressure of about 200 atmospheres. The process was adapted for industrial-quality production by Karl Bosch (1874–1940). Thereafter, many improved ammonia-synthesis systems, based on the Haber-Bosch process, were commercialized using various operating conditions and synthesis loop-designs. One of the five top inorganic chemicals produced in the United States, it is used in refrigerants, detergents and other cleaning preparations, explosives, fabrics, and fertilizers. A little over 75% of ammonia production in the United States is used for fertilizers. It has been shown to produce cancer of the skin in humans, in doses of 1,000 milligrams (or 1 gram) per kilogram (2.2 pounds) of body weight.

What does the symbol H_2O_2 stand for?

Hydrogen peroxide, a syrupy liquid used as a strong bleaching, oxidizing, and disinfecting agent, is the compound. It is made from barium peroxide and diluted phosphoric acid. A 3% solution of hydrogen peroxide is used medicinally as an antiseptic and germicide. Undiluted it can cause burns to human skin and mucous membranes, is a fire and explosion risk, and can be highly toxic.

Who discovered deuterium?

American chemist Harold C. Urey (1893–1981), 1934 winner of the Nobel Prize for Chemistry, discovered deuterium (heavy hydrogen, symbol D) in 1931 with F.G. Brickwedde and G.M. Murphy. This isotope (a form of an element that differs in the number of neutrons and atomic weight) has twice the weight of hydrogen, while all the other isotopes differed slightly in their atomic weights. Deuterium and its oxide makes heavy water (D_2O), which is used to slow down the neutrons in atomic piles of nuclear reactors.

What is the lightest solid material?

The lightest substance is silica aerogels, made of tiny spheres of bonded silicon and oxygen atoms linked together into long strands separated with air pockets. They appear almost like frozen wisps of smoke. In February, 1990, the lightest of these aerosols, having a density of only 5 ounces per cubic foot, was produced at the Lawrence Livermore Laboratory in California. It will be used in window insulation, as traps for sampling cosmic dust in space, and in liquid rocket fuel storage.

What is buckminsterfullerene?

Buckminsterfullerene

It is a large molecule in the shape of a soccer ball, containing 60 carbon atoms, whose structure is the shape of a truncated icosahedron (a hollow, spherical object with 32 faces, 12 of them pentagons and the rest hexagons). This molecule was named buckminsterfullerene because of the structure's resemblance to the geodesic domes designed by American architect R. Buckminster Fuller (1895–1983). The molecule was formed by vaporizing material from a graphite surface with a laser. Large molecules containing only carbon atoms have been known to exist around certain types of cabon-rich stars. Similar molecules are also thought to be present in soot formed during the incomplete combustion of organic materials. Chemist Richard Smalley identified buckminsterfullerene in 1985 and speculated that it may be fairly common throughout the universe. Since that time, other stable, large, even-numbered carbon clusters have been produced. This new class of molecules has been called "fullerenes" since they all seem to have the structure of a geodesic dome. Buckministerfullerene (C_{60}) seems to function as an insulator, conductor, semi-conductor and superconductor in various compounds.

Is **glass** a solid or a liquid?

Even at room temperature, glass appears to be a solid in the ordinary sense of the word. However, it actually is a fluid with an extremely high viscosity. It has been documented that century-old windows show signs of flow. The internal friction of fluids is called viscosity. It is a property of fluids by which the flow motion is gradually damped (slowed) and dissipated by heat. Viscosity is a familiar phenomenon in daily life. An opened bottle of wine can be poured: the wine flows easily under the influence of gravity. Maple syrup, on the other hand, cannot be poured so easily; under the action of gravity, it flows sluggishly. The syrup has a higher viscosity than the wine.

Glass is usually composed of mixed oxides based around the silicon dioxide (SiO_2) unit. A very good electrical insulator, and generally inert to chemicals, commercial glass is manufactured by the fusing of sand (silica, SiO_2), limestone ($CaCO_2$), and soda (sodium carbonate, Na_2CO_3) at temperatures around 2552°F to 2732°F (1400 to 1500°C). On cooling the melt becomes very viscous and at about 932°F (500°C, known as glass transition temperature), the melt "solidifies" to form soda glass. Small amounts of metal oxides are used to color glass, and its physical properties can be changed by the addition of substances like lead oxide (to increase softness, density, and refractive ability for cutglass and lead crystal), and borax (to lower significantly thermal expansion for cookware and laboratory equipment). Other materials can be used to form glasses if cooled sufficiently rapid from the liquid or gaseous phase to prevent an ordered crystalline structure from forming. Glasses such as obsidian occur naturally. Glass objects might have been made as early as 2500 B.C.E. in Egypt and Mesopotamia, and glass blowing developed about 100 B.C.E. in Phoenicia.

What is **crown glass**?

In the early 1800s, window glass was called crown glass. It was made by blowing a bubble, then spinning it until flat. This left a sheet of glass with a bump, or crown, in the center. This blowing method of window-pane making required great skill and was very costly. Still, the finished crown glass produced a distortion through which everything looked curiously wavy, and the glass itself was also faulty and uneven. By the end of the 19th century, flat glass was mass-produced and was a common material. The cylinder method replaced the old method, and used compressed air to produce glass which could be slit lengthwise, reheated, and allowed to flatten on an iron table under its own weight. New furnaces and better polishing machines made the production of plate-glass a real industry. Today, glass is produced by a float-glass process which reheats the newly-formed ribbon and glass and allows it to cool without touching a solid surface. This produces inexpensive glass that is flat and free from distortion.

How is **bulletproof glass** made?

Bulletproof glass is composed of two sheets of plate glass with a sheet of transparent resin in between, molded together under heat and pressure. When subjected to a severe blow, it will crack without shattering. Today's bulletproof glass is a development of laminated or safety glass, invented by the French chemist Edouard Benedictus. It is basically a multiple lamination of glass and plastic layers.

When were **glass blocks** invented?

Glass building bricks were introduced in 1931. They were invented in Europe in the early 1900s as thin blocks of glass supported by a grid. They have been in and out of favor since they were introduced. They are small, the largest being 12 inches by 12 inches (30 centimeters by 30 centimeters). They offer the following advantages: they allow natural light to filter through, but their patterns can afford privacy; they have the insulating value of a 12 inch-thick (30 centimeters) concrete wall; they absorb outside noise; and they are much more secure than ordinary flat glass.

Who invented **thermopane glass**?

Thermopane insulated window glass was invented by C.D. Haven in the United States in 1930. It is two sheets of glass that are bonded together in such a manner that they enclose a captive air space between the two pieces of glass. Often this space is filled with an inert gas which increases the insulating quality of the window. The insulation value of a window is determined by the infiltrated heat loss and by the conducted heat loss. Glass is also one of the best transparent materials because it allows the short wavelengths of solar radiation to pass through it, but prohibits nearly all of the long waves of reflected radiation from passing back through it.

What is the **float glass process**?

Manufacture of high-quality flat glass, needed for large areas and industrial uses, depends on the float glass process, invented by Alistair Pilkington in 1952. The float process departs from all other glass processes where the molten glass flows from the melting chamber into the float chamber which is a molten tin pool approximately 160 feet (49 meters) long and 12 feet (3.5 meters) wide. During its passage over this molten tin, the hot glass assumes the perfect flatness of the tin surface and develops excellent thickness uniformity. The finished product is as flat and smooth as plate glass without having been ground and polished.

How is the **glass used in movie stunts** made?

The "glass" might be made of candy (sugar boiled down to a translucent pane) or plastic. This looks like glass and will shatter like glass, but will not cut a performer.

Who developed **fiberglass**?

Coarse glass fibers were used for decoration by the Egyptians more than two thousand years ago. Other developments were made in Roman times. Parisian craftsman Dubus–Bonnel was granted a patent for the spinning and weaving of drawn glass strands in 1836. In 1893, the Libbey Glass Company exhibited lampshades, at the World's Columbian Exposition in Chicago, that were made of coarse glass thread woven together with silk. However, this was not a true woven glass. Between 1931 and 1939, the Owens Illinois Glass Company and the Corning Glass Works developed practical methods of making fiberglass commercially. Once the technical problem of drawing out the glass threads to a fraction of their original thinness was solved—basically an endless strand of continuous glass filament as thin as 1/5000 of an inch—the industry began to produce glass fiber for thermal insulation and air filters, among other uses. When glass fibers were combined with plastics during World War II, a new material was formed. Glass fibers did for plastics what steel did for concrete—gave strength and flexibility. Glass-fiber-reinforced plastics (GFRP) became very important in modern engineering. Fiberglass combined with epoxy resins and thermosetting polyesters are now used extensively in boat and ship construction, sporting goods, automobile bodies, and circuit boards in electronics.

When was **cement** first used?

Cements are finely ground powders that, when mixed with water, set to a hard mass. The cement used by the Egyptians was calcined gypsum, and both the Greeks and Romans used a cement of calcined limestone. Roman concrete (a mixture of cement, sand, and some other fine aggregate) was made of broken brick embedded in a pozzolanic lime mortar. This mortar consisted of lime putty mixed with brick dust or volcanic ash. Hardening was produced by a prolonged chemical reaction between these components in the presence of moisture. With the decline of the Roman Empire, concrete fell into disuse. The first step toward its reintroduction was in about 1790, when the English engineer, John Smeaton (1724–1792), found that when lime containing a certain amount of clay was burnt, it would set under water. This cement resembled what had been made by the Romans. Further investigations by James Parker in the same decade led to the commercial production of natural hydraulic cement. In 1824, Englishman Joseph Aspdin (1799–1855) took out a patent for what he called "portland cement," a material produced from a synthetic mixture of limestone and clay. He called it "portland" because it resembled a building stone that was quarried on the Isle of Portland off the coast of Dorset. The manufacture of this cement spread rapidly to Europe and the United States by 1870. Today, concrete is often reinforced or prestressed, increasing its load-bearing capabilities.

How was early **macadam** different from modern paved roads?

Macadam roads developed originally in England and France and are named after the Scottish road builder and engineer, John Louden MacAdam (1756-1836). The term "macadam" originally designated road surface or base in which clean, broken, or crushed ledge stone was mechanically locked together by rolling with a heavy weight and bonded together by stone dust screenings which were worked into the spaces and then "set" with water. With the beginning of the use of bituminous material (tar or asphalt), the term "plain macadam," "ordinary macadam," or "waterbound macadam" was used to distinquish the original type from the newer "bituminous macadam. Waterbound macadam surfaces are almost never built now in the United States, mainly because they are expensive and the vacuum effect of vehicles loosens them. Many miles of bituminous macadam roads are still in service, but their principal disadvantages are their high crowns and narrowness. Today's roads that carry heavy traffic are usually surfaced with very durable portland cement.

What is **Belgian block**?

Belgian block is a road-building material, first used in Brussels, Belgium, and introduced into New York about 1850. Its shape is a truncated pyramid with a base of about 5 or 6 inches square (13–15 centimeters) and a depth of 7 to 8 inches (18–20.5 centimeters). The bottom of the block is not more than 1 inch (2.5 centimeters) different from the top. The original blocks were cut from trap-rock from the Palisades of New Jersey.

Belgian blocks replaced cobblestones mainly because their regular, cubical shape allowed them to remain in place better than cobblestones. They were not universally adopted, however, because they would wear round and create joints or openings which would then form ruts and hollows. Although they provided a smooth surface compared to the uneven cobblestones, they still made for a rough and noisy ride.

What is **solder**?

Solder is an alloy of two or more metals used for joining other metals together. The most common solder is called half and half, or "plumber's" solder, and is composed of equal parts of lead and tin. Other metals used in solder are aluminum, cadmium, zinc, nickel, gold, silver, palladium, bismuth, copper, and antimony. Various melting points to suit the work are obtained by varying the proportions of the metals.

Solder is an ancient joining method, mentioned in the Bible (*Isaiah* 41:7). There is evidence of its use in Mesopotamia some 5,000 years ago, and later in Egypt, Greece, and Rome. For the near future, it appears that as long as a combination of conductors, semiconductors, and insulators is used to build circuitry, based on electrical and magnetic impulses, solder will remain indispensable.

What is **creosote**?

Creosote is a yellowish poisonous oily liquid obtained from the distillation of coal tar.

Coal tar constitutes the major part of the liquid condensate obtained from the "dry" distillation or carbonization of coal to coke. Crude creosote oil, also called dead oil or pitch oil, is used as a wood preservative. Railroad ties, poles, fence posts, marine pilings, and lumber for outdoor use are impregnated with creosote in large cylindrical vessels. This treatment can greatly extend the useful life of wood that is exposed to the weather.

What is **carbon black**?

Carbon black is finely divided carbon produced by incomplete combustion of methane or other hydrocarbon gases (by letting the flame impinge on a cool surface). This forms a very fine pigment containing up to 95% carbon, which gives a very intense black color that is widely used in paints, inks, and protective coatings and as colorants for paper and plastics.

How is **sandpaper** made?

Sandpaper is a coated abrasive which consists of a flexible-type backing (paper) upon which a film of adhesive holds and supports a coating of abrasive grains. Various types of resins and hide glues are used as adhesives. The first record of a coated abrasive is in 13th century China, when crushed seashells were bound to parchment using natural gums. The first known article on coated abrasives was published in 1808 and described how calcined, ground pumice was mixed with varnish and spread on paper with a brush. Most abrasive papers are now made with aluminum oxide or silicon carbide, although the term sandpapering is still used. Quartz grains are also used for wood polishing. The paper used is heavy, tough, and flexible, and the grains are bonded with a strong glue.

Why is **titanium dioxide** the most widely used white pigment?

Titanium dioxide has become the predominant white pigment in the world because of its high refractive index, lack of absorption of visible light, ability to be produced in the right size range, and its stability and nontoxicity. It is the whitest known pigment, unrivalled in respect of color, opacity, stain resistance, and durability; it is also non-toxic. The main consuming industries are paint, printing inks, plastics, and ceramics, which together account for 60% to 70% of the total demand.

When and where was **gunpowder** invented?

The explosive mixture of saltpeter (potassium nitrate), sulfur, and charcoal called *gunpowder* was known in China at least by 850 C.E., and probably was discovered by Chinese alchemists searching for components to make artificial gold. Early mixtures had too little saltpeter (50%) to be truly explosive; 75% minimum is needed to get a detonation. So the first use of the mixture was in making fireworks. Later, the Chinese used it in incendiary-like weapons, for warfare. Eventually it is thought that the Chinese found the correct proportions to utilize its explosive effects in rockets and "bamboo bullets." However, some authorities still maintain that the "Chinese gunpowder" really had only pyrotechnic qualities, and "true" gunpowder was an European invention. Roger Bacon (1214–1292) had a formula for it and so might have the German monk Berthold Schwartz (1353). But its first European use depended on the development of firearms in the 14th century. Not until the 17th century was gunpowder used in peacetime in mining and civil engineering applications.

How are colored **fireworks** made?

Fireworks existed in ancient China in the 9th century C.E. where saltpeter (potassium nitrate), sulfur, and charcoal were mixed to produce the dazzling effects. Magnesium burns with a brilliant white light and is widely used in making flares and fireworks. Various other colors can be produced by adding certain substances to the flame. Strontium compounds color the flame scarlet and barium compounds produce a yellowish-green color; borax produces a green color, and lithium a purple color.

What is the chemical formula for **TNT**?

TNT is the abbreviation for (2,4,6-*trinitro*toluene ($C_7H_5N_3O_6$). TNT is a powerful, highly explosive compound widely used in conventional bombs. Discovered by J. Wilbrand in 1863, it is made by treating toluene with nitric acid and sulfuric acids. This yellow crystalline solid with a low melting point, has low shock sensitivity and even burns without exploding. This makes it safe to handle and cast; but once detonated, it explodes violently.

Who invented **dynamite**?

Dynamite was no chance discovery but the result of a methodical search by the Swedish technologist, Alfred Nobel (1833–1896). Nitroglycerine had been discovered in 1849, by the Italian organic chemist, Ascanio Sobriero (1812–1888), but it was so sensitive and difficult to control that it was useless. Nobel sought to turn nitroglycerine into a manageable solid by absorbing it into a porous substance. In 1866–1867, he tried an unusual though by no means rare mineral, kieselguhr, and created a dough-like explosive that was controllable. He also invented a detonating cap incorporating mercury fulminate with which nitroglycerine could be detonated at will. Nobel made a great fortune and bequeathed it to a foundation for awarding prizes for contributions to science, literature, and the promotion of peace.

When was **plastic** first invented?

About 1850 Alexander Parkes (1813–1890) experimented with nitrocellulose (or gun-cotton). Mixed with camphor, it gave a hard but flexible transparent material, which he called "Parkesine." He teamed up with a manufacturer to produce it, but there was no call for it, and the firm went bankrupt. An American, John Wesley Hyatt (1837–1920), acquired the patent in 1868 with the idea of producing artificial ivory for billiard-balls. Improving the formula and with an efficient manufacturing process, he marketed the material, to make a few household articles, under the name "celluloid." Celluloid became the medium for cinematography: celluloid strips coated with a light-sensitive "film" were ideal for shooting and showing movie pictures. Celluloid was the only plastic material until 1904 when a Belgian scientist, Leo Hendrik Baekeland (1863–1944), succeeded in producing a synthetic shellac from formaldehyde and phenol. Called *bakelite,* it was the first of the "thermosetting" plastics (i.e., synthetic materials which, having once been subjected to heat and pressure, became extremely hard and resistant to high temperatures).

How can **plastics** be made **biodegradable**?

Plastic does not rust nor rot. This is an advantage in its usage, but when it comes to disposal of plastic, the advantage turns into a liability. Degradable plastic has starch in it so that it can be attacked by starch-eating bacteria to eventually disintegrate the plastic into bits. Chemically degradable plastic can be broken up with a chemical solution that dissolves it. Biodegradable plastic is also used in surgery where stitching material is the type of plastic that slowly dissolves in the body fluids. Photodegradable plastic contains chemicals that slowly disintegrate when exposed to light over a period of one to three years. 25% of the plastic yokes used to package beverages are made from a plastic called Ecolyte® which is photodegradable.

Who invented **teflon**?

In 1938, the American engineer Roy J. Plunkett (b. 1910), at DuPont de Nemours, discovered the polymer of tetrafluorethylene (PTFE) by accident. This fluorocarbon is marketed under the name of Fluon in Great Britain and Teflon® in the United States. Patented in 1939 and first exploited commercially in 1954, PTFE is resistant to all acids and has exceptional stability and excellent electrical insulating properties. It is used in making piping for corrosive materials, in insulating devices for radio transmitters, in pump gaskets, and in computer microchips. In addition, its non-stick properties make PTFE an ideal material for surface coatings. In 1956, French engineer Marc Gregoire discovered a process whereby he could fix a thin layer of teflon on an aluminum surface. He then patented the process of applying it to cookware, and the no–stick frying pan was created.

Who made the first successful **synthetic gemstone**?

In 1902, Auguste Victor Louis Verneuil (1856–1913), synthesized the first man-made gemstone, a ruby. Verneuil perfected a "flame-fusion" method of producing crystals of ruby and other corundums within a short time period.

ENERGY

NON-NUCLEAR FUELS

What are the three types of **primary energy** that flow continuously on or to the surface of the Earth?

Geothermal energy is heat contained beneath the Earth's crust, and brought to the surface in the form of steam or hot water. The five main sources of this geothermal reservoir are dry, super–heated steam from steam fields below the Earth's surface; mixed hot water, wet steam, etc., from geysers, etc.; dry rocks (into which cold water is pumped to create steam); pressurized water fields of hot water and natural gas beneath ocean beds; and magma (molten rock in or near volcanoes and 5 to 30 miles below the Earth's crust). Most Iceland buildings are heated by geothermal energy, and a few communities in the United States, such as Boise, Idaho, use geothermal home heating. Electric power production, industrial processing, space heating, etc., are fed from geothermal sources. The California Geysers project is the world's largest geothermal electric generating complex with 200 steam wells that provide some 1,300 megawatts of power. The first geothermal power station was built in 1904 at Larderello, Italy.

Solar radiation utilization depends on the weather, number of cloudy days, and the ability to store energy for night use. The process of collecting and storing is diffi-cult and expensive. A solar thermal facility (LUZ International Solar Thermal Plant), in the Mojave Desert, currently produces 274 megawatts and is used to supplement power needs of the Los Angeles utilities companies. Japan has 4 million solar panels on roofs and ⅔ of the houses in Israel have them; 90% of Cyprus homes do as well. Solar photo voltaic cells can generate electric current when exposed to the sun. Virtually every spacecraft and satellite since 1958 utilizes this kind of resource.

Tidal and wave energy contain enormous amounts of energy to be harnessed. The first tidal-powered mill was built in England in 1100; another in Woodbridge, England, built in 1170, has functioned for over 800 years. The Rance River Power Station in France, in operation since 1966, was the first large tidal electric generator plant, producing 160 megawatts. A tidal station works like a hydropower dam, with its turbines spinning as the tide flows through them. Unfortunately, the tidal period of 13½ hours causes problems of integrating the peak use with the peak generation ability. Ocean wave energy can also be made to drive electrical generators.

What is the difference between passive **solar energy systems** and active solar energy systems?

Passive solar energy systems use the architectural design, the natural materials or absorptive structures of the building as an energy saving system. The building itself serves as a solar collector and storage device. An example would be thick-walled stone and adobe dwellings that slowly collect heat during the day and gradually release it at night. Passive systems require little or no investment of external equipment.

Active solar energy systems require a separate collector, a storage device, and controls linked to pumps or fans that draw heat from storage when it is available. Active solar systems generally pump a heat-absorbing fluid medium (air, water, or an antifreeze solution) through a collector. Collectors, such as insulated water tanks, vary in size, depending on the number of sunless days in a locale. Another heat storage system uses eutectic (phase-changing) chemicals to store a large amount of energy in a small volume.

What is **biomass energy**?

The catch-all term *biomass* includes all the living organisms in an area. Wood, crops and crop waste, and wastes of plant, mineral, and animal matter are part of the biomass. Much of it is in garbage, which can be burned for heat energy, or allowed to decay and produce methane gas. It has been estimated that 90% of United States waste products could be burned to provide energy equal to 100 million tons of coal (20% will not burn, but can be recycled). Western Europe, has more than 200 power plants that burn rubbish to produce electricity. Biomass can be converted into biofuels such as biogas or methane, methanol, ethanol, etc. However, the process has been more costly than the conventional fossil fuel processes. Rubbish buried in the ground can provide methane gas through an aerobic decomposition. One ton of refuse can produce eight thousand cubic feet (227 cubic meters) of methane. Worldwide there are 140 such schemes that tap into the underground rubbish "tips."

Which woods have the best heating quality in a wood-burning stove?

The woods with the best heating quality are white ash, beech, yellow birch, shagbark hickory, hophornbeam, black locust, sugar maple, red oak, and white oak. Hardwood will cost more than softwood, but will deliver more heat for the money.

What are some examples of **biomass crops** that could be used for energy production?

Some crops grown specifically for energy include sugar cane, sorghum, ocean kelp, water hyacinth, and various species of trees.

Which **plant** has been investigated as a **source of petroleum**?

The shrub called the gopher plant (*Euphorbia lathyrus*) produces significant quantities of a milk-like sap—called latex—that is an emulsion of hydrocarbons in water.

Why are coal, oil, and natural gas called **fossil fuels**?

They are composed of the remains of organisms that lived as long ago as 500 million years. Organisms such as *phytoplankton* became incorporated into sediments and then were converted, with time, to oil and gas. Coal resulted when the remains of plants and trees were buried and subjected to pressure, temperature, and chemical processes (changing into peat and then lignite) for millions of years.

How and when was **coal formed**?

Coal is formed from the remains of plants that have undergone a series of far-reaching changes, turning into a substance called peat, which subsequently was buried. Through million of years, the Earth's crust buckled and folded, subjecting the peat deposits to very high pressure and changing the deposits into coal. The Carboniferous, or coal-bearing period, occurred about 250 million years ago. Geologists in the United States sometimes divide this period into the Mississippian and the Pennsylvanian periods. Most of the high-grade coal deposits are to be found in the strata of the Pennsylvanian period.

How is **underground coal** mined?

There are two basic types of underground mining methods: 1) room and pillar and 2) longwall. In room and pillar mines, coal is removed by cutting rooms, or large tunnels, in the solid coal, leaving pillars of coal for roof support. Longwall mining takes successive slices over the entire length of a long working face. In the United States, almost all of the coal recovered by underground mining is by room and pillar method. In Great Britain, however, most of the coal is mined by longwall methods.

What is a **"miner's canary"**?

Both mice and birds have been used by miners to test the purity of the air in the mines. A canary or two carried in a small cage will show signs of distress in the presence of carbon monoxide more quickly thans humans will, thus alerting the miner to toxic air. This method of safety was used prior to the more sophisticated equipment used today.

How thick is the coal in a **coal seam**?

Coal seams in the United States range in thickness from a thin film to 50 feet (15 meters) or more. The thickest coalbeds are in the western states, ranging from 10 feet (3 meters) in Utah and New Mexico to 50 feet (15 meters) in Wyoming.

Where are the **largest oil and gas fields** in the world and in the United States?

The Ghawar field, discovered in 1948 in Saudi Arabia, is the largest in the world; it measures 150 x 22 miles (241 x 35 kilometers). In the continental United States, the Hugoton field in Kansas is second in size only to the Prudhoe field in Alaska.

When was the first **oil well** in the United States drilled?

The Drake well at Titusville, Pennsylvania, was completed on August 28, 1859 (some sources list the date as August 27). The driller, William "Uncle Billy" Smith, went down 69.5 feet (21.18 meters) to find oil for Edwin L. Drake (1819–1880), the well's operator. Within 15 years, Pennsylvania oil field production reached over 10 million 360-pound (163.3-kilogram) barrels a year.

When was **offshore drilling for oil** first done?

The first successful offshore oil well was built off the coast at Summerland, Santa Barbara County, California, in 1896.

Why is **Pennsylvania crude oil** so highly valued?

The waxy, sweet paraffinic oils found in Pennsylvania first became prominent because high quality lubricating oils and greases could be made from them.

What is the process known as **hydrocarbon cracking**?

Cracking is a process that uses heat to decompose complex substances. Hydrocarbon cracking is the decomposition by heat, with or without catalysts, of petroleum or heavy petroleum fractions (groupings) to give materials of lower boiling points. Thermal cracking, developed by William Burton in 1913, uses heat and pressure to break some of the large heavy hydrocarbon molecules into smaller gasoline-grade ones. The cracked hydrocarbons are then sent to a flash chamber where the various fractions (groupings) are separated. Thermal cracking not only doubles the gasoline yield, but improves gasoline quality, producing gasoline components with good anti-knock characteristics.

What kinds of **additives** are **in gasoline** and why?

Additive	Function
Antiknock compounds	Increase octane number
Scavengers	Remove combustion products of antiknock compounds
Combustion chamber	Suppress surface ignition and spark plug deposit modifiers fouling
Antioxidants	Provide storage stability
Metal deactivators	Supplement storage stability
Antirust agents	Prevent rusting in gasoline-handling systems
Anti-icing agents	Suppress carburetor and fuel system freezing
Detergents	Control carburetor and induction system cleanliness
Upper cylinder lubricants	Lubricate upper cylinder areas and control intake system deposits
Dyes	Indicate presence of antiknock compounds and identify makes and grades of gasoline

Why is **lead added to gasoline** and why is **lead-free gasoline** used in new cars?

Tetraethyl lead has been used for more than 40 years to improve the combustion characteristics of gasoline. It reduces or eliminates "knocking" (pinging caused by premature ignition) in high-performance large engines and in smaller high-compression engines. It provides lubrication to the extremely close–fitting engine parts where oil has a tendency to wash away or burn off. However, lead will ruin and effectively destroy the catalyst presently used in emission control devices installed in new cars, so lead-free gasoline must be used.

141

What is a **reformulated gasoline?**

Oil companies are being required to offer new gasolines that burn more cleanly and have less impact on the environment. Typically, reformulated gasolines contain lower concentrations of benzene, aromatics, and olefins; less sulfur; a lower *Reid vapor pressure* (RVP); and some percentage of an oxygenate (non-aromatic component) such as *methyl tertiary butyl ether* (MTBE). MTBE is a high-octane gasoline blending components produced by the reaction of isobutylene and methanol.

What do the **octane numbers** of gasoline mean?

The octane number is a measure of the gasoline's ability to resist engine knock (pinging caused by premature ignition). Two test fuels, normal heptane and isooctane, are blended for test results to determine octane number. Unblended heptane has an octane number of zero and isooctane a value of 100. Gasolines are compared with the test blends to find one that makes the same knock as the test fuel. The octane rating of the gasoline under testing is the percentage by volume of isooctane required to produce the same knock. For example, if the test blend has 85% isooctane, the gasoline has an octane rating of 85. The octane rating that appears on gasoline pumps is an average of research octane determined in laboratory tests with engines running at low speeds, and motor octane, determined at higher speeds.

When did gasoline stations open?

The first service station (or garage) was opened in Bordeaux, France, in December 1895 by A. Barol. It provided overnight parking, repair service, and refills of oil and "motor spirit." In April, 1897, a parking and refueling establishment—Brighton Cycle and Motor Co.—opened in Brighton, England.

The pump that eventually came to be used to dispense gasoline was devised by Sylanus Bowser of Fort Wayne, Indiana, but in September, 1885, it dispensed kerosene. Twenty years later Bowser manufactured the first self-regulating gasoline pump. In 1912, a Standard Oil of Louisiana superstation opened in Memphis, Tennessee, featuring 13 pumps, a ladies' rest room, and a maid who served ice water to waiting customers. On December 1, 1913, in Pittsburgh, Pennsylvania, the Gulf Refining Company opened the first drive-in station as a 24 hour-a-day operation. Only 30 gallons (114 liters) of gasoline were sold the first day.

How is **gasohol** made?

Gasohol, a mixture of 90% unleaded gasoline and 10% ethyl alcohol (ethanol), has gained some acceptance as a fuel for motor vehicles. It is comparable in performance to 100% unleaded gasoline with the added benefit of superior antiknock properties (no premature fuel ignition). No engine modifications are needed for the use of gasohol.

Since corn is the most abundant United States grain crop, it is predominately used in producing ethanol. However, the fuel can be made from other organic raw materials, such as oats, barley, wheat, milo, sugar beets, or sugar cane. Potatoes, and cassava (a starchy plant) and cellulose (if broken up into fermentable sugars) are possible other sources. The corn starch is processed through grinding and cooking. The process requires the conversion of a starch into a sugar, which in turn is converted into alcohol by reaction with yeast. The alcohol is distilled and any water is removed until it is 200 proof (100% alcohol).

One acre of corn yields 250 gallons (946 liters) of ethanol; an acre of sugar beets yields 350 gallons (1,325 liters), while an acre of sugar can produce 630 gallons (2,385 liters). In the future, motor fuel could be produced almost exclusively from garbage, but currently its conversion remains an expensive process.

What are the advantages and disadvantages of the **alternatives to gasoline** to power automobiles?

Because the emissions of gasoline is a major air pollution problem in most U.S. urban areas, alternatives are being worked on. Currently none of the alternatives deliver as much energy content as gasoline so more of each of these fuels must be consumed to equal the distance that the energy of gasoline propels the automobile. The most viable alternative is flexible fuel, a combination of methanol and gasoline, which would add at least 300 dollars to car prices for an expensive fuel sensor and a longer fuel tank.

Alternative	Advantages	Disadvantages
Electricity from batteries	No vehicle emissions, good for stop–and–go driving	Short-lived bulky batteries; limited trip range
Ethanol from corn, biomass, etc.	Relatively clean fuel	Costs, corrosive damage
Hydrogen from electrolysis; etc.	Plentiful supply; non-toxic emissions	High cost; highly flammable
Methanol from methanol gas, coal, biomass, wood	Cleaner combustion; less volatile	Corrosive; some irritant emissions
Natural gas from hydrocarbons and petroleum deposits	Cheaper on energy basis; relatively clean	Cost to adapt vehicle bulky storage; sluggish performance

What is **cogeneration**?

Cogeneration is an energy production process involving the simultaneous generation of thermal (steam or hot water) and electric energy by using a single primary heat source. By producing two kinds of useful fuels in the same facility the net energy yield from the primary fuel increases from 30–35% to 80–90%.

NUCLEAR POWER

How many **nuclear power plants** are there **worldwide**?

As of 1990 there were 423 reactors operational with 83 power plants under construction and three planned reactors cancelled.

Country	Number of units
Argentina	2
Belgium	7
Brazil	1
Bulgaria	5
Canada	20
Czechoslovakia	8
Finland	4
France	56
Germany	26
Hungary	4
India	7
Japan	41
Korea, South	9
Mexico	1
Netherlands	2
Pakistan	1
South Africa	2
Spain	9
Sweden	12
Switzerland	5
United Kingdom	37
United States	112
U.S.S.R.	45
Yugoslavia	1
Total	423

As of August, 1991, the United States had 111 commercial plants in operation, generating 576,784 million net kilowatt hours or 20.6% of domestic electricity generation, with eight more under construction.

What is the **life of a nuclear power plant**?

The working life of a nuclear power plant is approximately 40 years, which is about the same as that of other types of power stations.

Where is the **oldest operational nuclear power plant** in the United States?

The Yankee Plant at Rowe, Massachusetts, which was constructed in 1960, is the oldest United States power plant.

What is the **Rasmussen report**?

Professor Rasmussen of the Massachusetts Institute of Technology (MIT) conducted a study of nuclear reactor safety for the United States Atomic Energy Commission. The study cost 4 million dollars and took 3 years to complete. It concluded that the odds against a worst-case nuclear accident occurring were astronomically large—ten million to one. The worse–case accident projected about 3,000 early deaths and 14 billion dollars in property damage due to contamination. Cancers occurring later due to the event might number 1,500 per year. The study concluded that the safety features engineered into a plant are very likely to prevent serious consequences from a meltdown. Other groups criticized the Rasmussen report and in particular declared that the estimates of risk were too low. After the Chernobyl disaster in 1986, some scientists estimated that a major nuclear accident might in fact happen every decade.

Which **nuclear reactors** have had **accidents**?

Incidents with core damage in nuclear reactors

Description of incident	Site	Date	Adult thyroid dose (in rems)
Minor core damage (no release of radiologic material)	Chalk River, Ontario, Canada	1952	not applicable
	Breeder Reactor Idaho	1955	not applicable
	Westinghouse Test Reactor	1960	not applicable
	Detroit Edison Fermi, Michigan	1966	not applicable

Description of incident	Site	Date	Adult thyroid dose (in rems)
Major core damage (radioiodine released)			
Noncommercial	Windscale, England	1957	16
	Idaho Falls SL-1, Idaho	1961	0.035
Commercial	Three Mile Island, Pennsylvania	1979	0.005
	Chernobyl, Soviet Union	1986	100 (estimated)

What is a **meltdown**?

A meltdown is a type of accident in a nuclear reactor in which the fuel core melts, resulting in the release of dangerous amounts of radiation. In most cases the large containment structure that houses a reactor would prevent the radioactivity from escaping. However, there is a small possibility that the molten core could become hot enough to burn through the floor of the containment structure and go deep into the Earth. Nuclear engineers call this type of situation the "China Syndrome." All reactors are equipped with emergency systems to prevent such an accident from occurring.

Where did the expression **"China Syndrome"** originate?

In a discussion on the theoretical problems of a nuclear reactor if it were to undergo a meltdown, a scientist commented that the molten core could bore a hole through the Earth, coming out—if the reactor happened to be melting down in America—in China. Ever since then, this scenario has been nicknamed the "China Syndrome." Although the scientist was grossly exaggerating, some people took him seriously. In fact, the core would bore a hole about 30 feet into the Earth, but this would have grave repercussions.

What actually happened at **Three Mile Island**?

The Three Mile Island nuclear power plant in Pennsylvania experienced a partial meltdown of its reactor core and radiation leakage. On March 28, 1979, just after 4:00 a.m., a water pump in the secondary cooling system of the Unit 2 pressurized water reactor failed. A relief valve jammed open, flooding the containment vessel with radioactive water. A backup system for pumping water was down for maintenance. Temperatures inside the reactor core rose, fuel rods ruptured, and a partial (52%) meltdown occurred, because the radioactive uranium core was almost entirely uncovered by coolant for 40 minutes. The thick steel-reinforced containment building prevented nearly all the radiation from escaping—the amount of radiation released into the atmosphere was one-millionth of that at Chernobyl. However, if the coolant had not been replaced, the molten fuel would have penetrated the reactor containment vessel, where it would have come into contact with the water, causing a steam explosion, breaching the reactor dome, and leading to radioactive contamination of the area similar to the Chernobyl accident.

What caused the **Chernobyl** accident?

Site of the worst nuclear power accident in history, the Chernobyl nuclear power plant in Byelorussia will affect, in one form or another, 20% of the republic's population (2.2 million people). On April 26, 1986, at 1:23:40 a.m., during unauthorized experiments by the operators, in which safety systems were deliberately circumvented in order to learn more about the plant's operation, one of the four reactors rapidly overheated and its water coolant "flashed" into steam. The hydrogen formed from the steam reacted with the graphite moderator to cause two major explosions and a fire. The explosions blew apart the thousand-ton (907 metric ton) lid of the reactor, and released radioactive debris high into the atmosphere. It is estimated that 3.5% of the reactor's fuel and 10% of the graphite reactor itself was emitted into the atmosphere. Human error and design features (positive void coefficient type of reactor, use of graphite in construction, and lack of a containment building) are generally cited as the causes of the accident. Thirty-one people initially died from trying to stop the fires. More than 240 others sustained severe radiation sickness. Eventually 150,000 people living near the reactor were relocated; some of them may never be allowed to return home. Fallout from the explosions, containing radioactive isotope cesium–137, was carried by the winds westward across Europe.

MEASURES AND MEASUREMENT

What is the **weight per gallon** of common fuels?

One gallon of fuel	Weight in pounds
Butane	4.86
Propane	4.23
Kerosene	6.75
Gasoline	6.00
Aviation gasoline	6.46–6.99

How much does a **barrel of oil weigh**?

A barrel of oil weighs about 306 pounds (140 kilograms).

How many **gallons** are contained in a **barrel of oil**?

The barrel, a common measure of crude oil, contains 42 U.S. gallons and 34.97 imperial gallons.

How do **various energy sources compare**?

Below are listed some comparisons (approximate equivalents) for energy sources as of 1990:

Energy unit	Equivalent
1 BTU of energy	1 match tip 250 calories (International Steam Table) 0.25 kilocalories (food calories)
1,000 BTU of energy	2 5-ounce glasses of wine 250 kilocalories (food calories) 0.8 peanut butter and jelly sandwiches
1 million BTU of energy	90 pounds of coal 120 pounds of oven-dried hardwood 8 gallons of motor gasoline 10 therms of dry natural gas 11 gallons of propane 2 months of the dietary intake of a laborer
1 quadrillion BTU of energy	45 million short tons of coal 60 million short tons of oven-dried hardwood 1 trillion cubic feet of dry natural gas 170 million barrels of crude oil 470 thousand barrels of crude oil per day for 1 year 28 days of U.S. petroleum imports 26 days of U.S. motor gasoline 26 hours of world energy use (1989)
1 barrel of crude oil	5.6 thousand cubic feet of dry natural gas 0.26 short tons (520 pounds) of coal 1,700 kilowatthours of electricity
1 short ton of coal	3.8 barrels of crude oil 21 thousand cubic feet of dry natural gas 6,500 kilowatt hours of electricity
1,000 cubic feet of natural gas	0.18 barrels (7.4 gallons) of crude oil 0.05 short tons (93 pounds) of coal 300 kilowatthours of electricity
1,000 kilowatthours of electricity	0.59 barrels of crude oil 0.15 short tons (310 pounds) of coal 3,300 cubic feet of dry natural gas

Notes: One quadrillion equals 1,000,000,000,000,000.

Because of energy losses associated with the generation of electricity, about three times as much fossil fuel is required to generate 1,000 kilowatthours: 1.8 barrels of oil, 0.47 short tons of coal, or 10,000 cubic feet of dry natural gas.

What are the **fuel equivalents** to produce one quad of energy?

One quad (meaning 1 quadrillion) is equivalent to:

1×10^{15} BTU

252×10^{15} calories or 252×10^{12} K calories.

In fossil fuels, one quad is equivalent to

180 million gallons of crude oil

0.98 trillion cubic feet of natural gas

37.88 million tons of anthracite coal

38.46 million tons of bituminous coal

In nuclear fuels, one quad is equivalent to 2500 tons of U_{308} if only U_{235} is used.

In electrical output, one quad is equivalent to 2.93×10^{11} kilowatt-hours electric.

What are the **approximate heating values** of fuels?

Fuel	BTU	Unit of measure
Oil	141,000	gallon
Coal	31,000	pound
Natural gas	1,000	cubic feet
Steam	1,000	cubic feet
Electricity	3,413	kilowatt hour
Gasoline	124,000	gallon

How much heat will 100 cubic feet of **natural gas** provide?

One hundred cubic feet of natural gas can provide about 75,000 BTUs (*British thermal units*) of heat. A British thermal unit, a common energy measurement, is defined as the amount of energy required to raise the temperature of one pound of water by one Fahrenheit degree.

149

How is a **heating degree day** defined?

Early this century engineers developed the concept of heating degree days as a useful index of heating fuel requirements. They found that when the daily mean temperature is lower than 65°F, most buildings require heat to maintain a 70°F temperature. Each degree of mean temperature below 65°F is counted as "one heating degree day." For every additional heating degree day, more fuel is needed to maintain a 70°F indoor temperature. For example, a day with a mean temperature of 35°F would be rated as 30 heating degree days and would require twice as much fuel as a day with a mean temperature of 50°F (15 heating degree days). The heating degree concept has become a valuable tool for fuel companies for evaluation of fuel use rates and efficient scheduling of deliveries. Detailed daily, monthly, and seasonal totals are routinely computed for the stations of the National Weather Service.

What does the term **cooling degree day** mean?

It is a unit for estimating the energy needed for cooling a building. One unit is given for each degree Fahrenheit above the daily mean temperature when the mean temperature exceeds 75°F.

How many BTUs are equivalent to one ton of **cooling capacity**?

1 ton = 288,000 BTUs/24 hours or 12,000 BTUs/hour.

How are **utility meters** read?

On older electric and gas meters, a series of four or five dials indicate the amount of energy being consumed. The dials are read from left to right, and if the pointer falls between two numbers, the lower number is recorded. In the electric meter depicted below, the dial registers 13021 kilowatt hours. Gas meters are set to read hundreds of cubic feet. New meter models have digital displays.

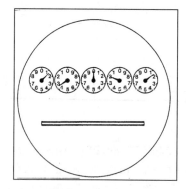

Electric meter

Gas meter

How much **wood** is in a **cord**?

A cord of wood is a pile of logs 4 feet (1.2 meters) wide and 4 feet (1.2 meters) high and 8 feet (2.4 meters) long. It may contain from 77 to 96 cubic feet of wood. The larger the unsplit logs the larger the gaps, with fewer cubic feet of wood actually in the cord.

CONSUMPTION AND CONSERVATION

Which countries **consume the most energy**?

Top Energy-Consuming Countries (1987)

Country	World total %	Oil %	Coal %	Gas %
United States	24.6	41.1	23.4	22.4
U.S.S.R.	18.0	32.3	22.6	38.3
China	7.8	18.4	75.5	2.3
Japan	4.9	55.9	18.0	9.9
West Germany	3.6	42.2	27.5	17.1
Canada	3.2	31.5	10.7	19.8
United Kingdom	2.7	36.2	32.9	23.8
France	2.7	42.6	9.6	12.4
India	2.0	32.1	55.8	4.2
Italy	2.0	59.3	9.9	21.9

What is the current **per capita energy consumption** in the United States?

In 1990 in the United States, it was about 260 million/BTUs (*B*ritish *T*hermal *U*nits) per year. In France and Japan it is about 100 million BTUs per year. A BTU, a common energy measurement, is defined as the amount of energy required to raise the temperature of 1 pound of water by one Fahrenheit degree.

End-use is the total energy consumption minus losses from generation, transmission, and distribution of power. Below is listed per capita consumption for representative years:

Year	Total (Per capita) (million BTU)	End-use (Per capita) (million BTU)
1950	219	194
1955	235	206

Year	Total (Per capita) (million BTU)	End-use (Per capita) (million BTU)
1960	244	212
1965	272	232
1970	327	270
1975	327	261
1980	335	259
1985	310	232
1989	328	246
1990	327	260

Does the United States currently produce **enough energy** to meet its consumption needs?

No. From 1958 forward, the United States consumed more energy than it produced, and the difference was met by energy imports. In 1990, 67.59 quadrillion BTUs was the total United States energy production with coal supplying 22.6 quadrillion BTUs; natural gas supplying 18 quadrillion BTUs; crude oil and natural plant liquids, 17.6; hydroelectric, 2.9; nuclear, 6.2; and geothermal, biomass, photovoltaic, wind, and solar energy, 0.2. In that year the United States imported 13.83 quadrillion BTUs in petroleum to make up the deficit, since its 1990 energy consumption was 81.44 quadrillion BTUs (29.2 in residential and commercial; 30.2 in industrial; and 22 in transportation).

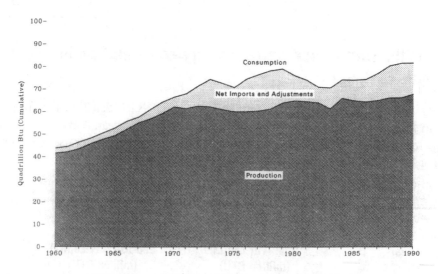

U.S. energy consumption, production, and imports 1960-1990

How long, at the present rate of consumption, will **U.S. current major energy reserves** last?

Assuming 1986 production rates, U.S. estimated energy reserves amount to 32.5 years for oil, 58.7 years for gas, and 226 years for coal.

How much money can be saved by lowering the setting on a **home furnace thermostat**?

Tests have shown that a 5°F reduction in the home thermostat setting for approximately eight hours will save up to 10% in fuel costs.

How much energy is saved by raising the setting for a **house air conditioner**?

For every 1°F the inside temperature is increased the energy needed for air conditioning is reduced by 3%. If all consumers raised the settings on their air conditioners by 6°F, for example, 190,000 barrels of oil could be saved each day.

How are **fireplace logs** made from **newspaper**?

In a large tub, make a neat stack of old newspapers. Cover the pile with water and let them soak until all the layers are saturated. Pick up the layers of the paper and roll them tightly into logs. Make thin ones for kindling. Stand on end to dry.

How much energy is required to use various **electrical appliances**?

The table below indicates the annual estimated energy consumption for various household electrical products.

Appliance	Estimated kilowatt-hours
Air conditioner (room)	1,389
Blender	15
Broiler	100
Clock	17
Clothes dryer	993
Clothes washer	103
Coffee maker	106

Appliance	Estimated kilowatt-hours
Computer	25–400
Dehumidifier	377
Dishwasher	165–363
Fan (circulating)	43
Fan (attic)	291
Food mixer	13
Freezer (frost-free)	1,820
Frying pan	186
Garbage disposal	30
Hair dryer	14
Iron	144
Microwave oven	300
Radio	86
Range (self-cleaning oven)	1,205
Refrigerator-freezer (frost-free)	1,591–1,829
Television (black-and-white)	362
Television (color)	502
Toaster	39
Vacuum cleaner	46
Video cassette recorder (VCR)	10–70
Water heater (standard)	4,219

What is the advantage of switching from **incandescent** to **fluorescent light bulbs**?

One 18-watt fluorescent bulb provides the light of a 75-watt incandescent bulb and lasts ten times as long. Even though the purchase price is higher, over its useful life, an 18-watt fluorescent light bulb saves 80 pounds of coal used to produce electricity. This translates into 250 fewer pounds of carbon dioxide released into the Earth's atmosphere.

How much energy is saved by recycling one aluminum can?

Some sources indicate that one recycled aluminum can saves as much energy as it takes to run a TV set for four hours or the energy equivalent of ½ gallon (1.9 liters) of gasoline.

When should a **fluorescent light** be turned off to save energy?

Fluorescent lights use much electric current getting started, and frequent switching the light on and off will shorten the lamp's life and efficiency. It is energy-efficient to turn off a fluorescent light only if it will not be used again within an hour or more.

How does **driving speed** affect **gas mileage** for most automobiles?

Most automobiles get about 28% more miles per gallon of fuel at 50 miles (80.5 kilometers) per hour than at 70 miles (112.6 kilometers) per hour, and about 21% more at 55 miles (88.5 kilometers) per hour than at 70 miles per hour.

When is it more economical to **restart an automobile** rather than let it idle?

Tests by the Environmental Protection Agency have shown that is it more economical to turn the engine off rather than let it idle if the idle time would exceed 60 seconds.

How much gasoline do **underinflated tires** waste?

Underinflated tires waste as much as 1 gallon out of every 20 gallons of gasoline consumed by an automobile.

What is a **pedicar**?

As a response to the concern of energy conservation, the pedicar was introduced in 1973. It was a pedal-powered, all-weather one passenger vehicle with straight-line pedal action, disc brakes, five forward speeds plus neutral and reverse. Costing about $550 in 1973, the vehicle was conceived as an alternative to the automobile. It had a speed of 8 to 15 miles (12.9 to 24.1 kilometers) per hour and was developed mainly for fun use around parks, resorts, college campuses, country clubs, and similar protected areas.

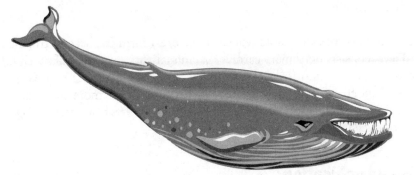

ENVIRONMENT

ECOLOGY, RESOURCES, ETC.

What is a **biome**?

A biome is a plant and animal community that covers a large geographical area. Complex interactions of climate, geology, soil types, water resources, and latitude all determine the kinds of plants and animals that thrive in different places. Fourteen major ecological zones, called "biomes," exist over five major climatic regions and eight zoo-geographical regions. Three types of biomes are desert, tropical rain forest, and tundra.

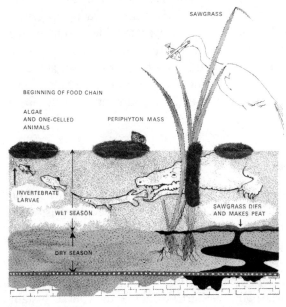

SAWGRASS

BEGINNING OF FOOD CHAIN

ALGAE
AND ONE-CELLED
ANIMALS

PERIPHYTON MASS

INVERTEBRATE
LARVAE

WET SEASON

SAWGRASS DIES
AND MAKES PEAT

DRY SEASON

Marsh food chain

How does the process work in a **food chain**?

A food chain is the transfer of food energy from the source in plants through a series of organisms with repeated eating and being eaten. The number of steps or "links" in a sequence is usually four to five. The first trophic (group of organisms that get their energy the same way) is **157** plants; the animals that eat plants (called herbivores) form the second trophic level. The third level consists of primary carnivores (animal-eating animals like wolves) who eat herbivores, and the fourth level are animals (like killer whales) that eat primary carnivores. Food chains overlap because many organisms eat more than one type of food, so that these chains can look more like food webs. In 1891 German zoologist Karl Semper introduced the food chain concept.

What is eutrophication?

A process in which the supply of plant nutrients in a lake or pond is increased. In time, the result of natural eutrophication may be dry land where water once flowed, caused by plant overgrowth.

Natural fertilizers, washed from the soil, result in an accelerated growth of plants, producing overcrowding. As the plants die off, the dead and decaying vegetation depletes the lake's oxygen supply causing fish to die. The accumulated dead plant and animal material eventually changes a deep lake to a shallow one, then to a swamp, and finally it becomes dry land.

While the process of eutrophication is a natural one, it has been accelerated enormously by human activities. Fertilizers from farms, sewage, and industrial wastes and some detergents all contribute to the problem.

How does ozone benefit life on Earth?

Ozone in the upper atmosphere (stratosphere) is a major factor in making life on Earth possible. The ozone belt shields the Earth from excessive ultraviolet radiation generated by the sun. Scientists predict that depletion of this layer could lead to increased health problems for humans and disruption of sensitive terrestrial and aquatic ecosystems. Ozone, a form of oxygen with three atoms instead of the normal two, is highly toxic; less than one part per million of this blue-tinged gas is poisonous to humans. While beneficial in the stratosphere, near ground level it is a pollutant that helps form photochemical smog and acid rain.

What is the greenhouse effect?

The greenhouse effect is a warming near the Earth's surface that results when the Earth's atmosphere traps the sun's heat. The atmosphere acts much like the glass walls and roof of a greenhouse. The effect was described by John Tyndall (1820–1893) in 1861. It was given the greenhouse analogy much later in 1896 by the Swedish chemist Svante Arrhenius (1859–1927). The greenhouse effect is what makes the Earth habitable. Without the presence of water vapor, carbon dioxide, and other gases in the atmosphere, too much heat would escape and the Earth would be too cold to sustain life. Carbon dioxide, methane, nitrous oxide, and other "greenhouse gases" absorb the infrared radiation rising from the Earth and hold this heat in the atmosphere instead of reflecting it back into space.

In the twentieth century, the increased build-up of carbon dioxide, caused by the burning of fossil fuels has been a matter of concern. There is some controversy concerning whether the increase noted in the Earth's average temperature is due to the increased amount of carbon dioxide and other gases, or is due to other causes. Volcanic activity, destruction of the rain forests, use of aerosols, and increased agricultural activity may also be contributing factors.

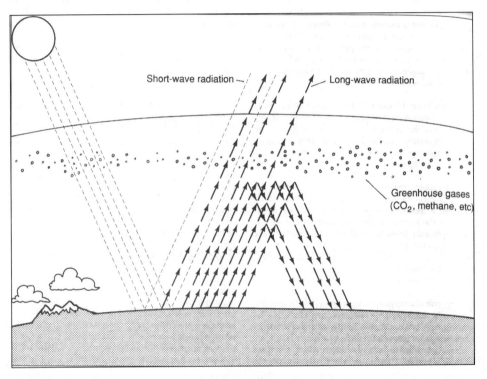

Short-wave radiation

Long-wave radiation

Greenhouse gases
(CO_2, methane, etc)

Greenhouse effect

Why is **El Niño** harmful?

Along the west coast of South America, near the end of each calendar year, a warm current of nutrient-poor tropical water moves southward, replacing the cold, nutrient-rich surface water. Because this condition frequently occurs around Christmas, local residents call it *El Niño* (Spanish for child), referring to the Christ child. In most years the warming lasts for only a few weeks. However, when *El Niño* conditions last for many months, the economic results can be catastrophic. It is this extended episode of extremely warm water that scientists now refer to as *El Niño*. During a severe *El Niño*, **159** large numbers of fish and marine plants may die. Decomposition of the dead material depletes the water's oxygen supply which leads to the bacterial production of huge amounts of smelly hydrogen sulfide. A greatly reduced fish (especially anchovy) harvest affects the world's fishmeal supply, leading to higher prices for poultry and other animals which normally are fed fishmeal.

Studies reveal that *El Niño* is not an isolated occurrence, but is instead part of a pattern of change in the global circulation of the oceans and atmosphere. The 1982–1983 *El Niño* was one of the most severe climate events of the twentieth century in both its geographical extent as well as in the degree of warming (14°F, or 8°C). The 1986–1987 *El Niño* might have been responsible in part for record global warmth in 1987—the warmest year in the last 100 years.

What is **red tide** and what causes it?

Red tide is a term used for a brownish or reddish discoloration occurring in ocean, river, or lake water. It is caused by the rapid reproduction of a variety of toxic organisms, especially the toxic red dinoflagellates that are members of two genera, *Gymnodidium* and *Gonyaulax*. Some red tides are harmless, but millions of fish may be killed during a "bloom," as the build-up is called. Other red tides can poison shellfish and the birds or humans who eat the contaminated food. Scientists do not fully understand why the "bloom" occurs.

How many **acres of tropical forest** does the world lose annually?

Using data from satellite observations, it is estimated that between 16.4 to 20.4 million hectares are being destroyed each year (a hectare equals 107,639.1 square feet or 10,000 square meters). Only 50% of the mature tropical forests remain with 750 to 800 million hectares of the original 1.5 to 1.6 billion hectares destroyed. There are two types of tropical forests: wet and dry; the wet ("rain") forests have been incurring the most losses. Latin America has lost 37% of them; Asia, 42%; and Africa, 52%. Logging, fuelwood gathering, and conversion of forests to agriculture are the main causes. Yet tropical forests have been called "green deserts," because their soils are poor in nutrients. The forest vegetation, seemingly lush, has survived through ingenious life–support systems. When the trees are stripped away, the exposed soil deteriorates rapidly, eroded by torrential rains. After the rain ceases, the sun bakes the Earth into a hard mass, rendering the soil incapable of vegetative growth.

How rapidly is **deforestation** occurring?

In the 1990 figures given below, the annual deforestation rate is given in percent of forests eradication as well as amount in square kilometers.

Country	Deforestation (square kilometers)	Percent
Brazil	13,820	0.4
Colombia	6,000	1.3
Mexico	7,000	1.5
Indonesia	10,000	0.9
Peru	2,700	0.4
Malaysia	3,100	1.5
Ecuador	3,400	2.4
India	10,000	2.7
Zaire	4,000	0.4
Madagascar	1,500	1.5

How long will it be before all **tropical forests** have been destroyed, if present **rates of destruction** continue?

If present deforestation rates continue, all tropical forests will be cleared in 177 years. These forests contain 155,000 of the 250,000 known plant species and innumerable insect and animal species. Half of all medicines prescribed worldwide are originally derived from wild products; the United States National Cancer Institute has identified more than 2,000 tropical rainforest plants with the potential to fight cancer. Rubber, timber, gums, resins and waxes, pesticides, lubricants, nuts and fruits, flavorings and dyestuffs, steroids, latexes, essential and edible oils, and bamboo are among the forest's products that would be drastically affected by the depletion of the tropical forests.

Who owns America's **forest lands**?

More than 282 million acres, or 39% of the total amount of United States forest land, are owned by private individuals. Government owns 137 million acres, or 19%, and the forest industry owns 68 million acres—just 9% of the total United States forest land. The remaining 33% is the 242 million acres of park lands, wilderness reserves, and non-working forest lands.

What causes the most **forest fires** in the western United States?

Lightning is the single largest cause of forest fires in the western states.

161

When did the symbol of **Smokey the Bear** begin to be used for forest fire prevention?

A nationwide effort in wildfire prevention was begun in 1942. A poster featuring Smokey the Bear was circulated in 1945. The symbol soon became very popular. The slogan "Only you can prevent forest fires" was coined in 1947.

How many acres in the United States are infested by the **gypsy moth?**

In 1990, the gypsy–moth infested area in the United States was 7,374,816 acres, of which 4,357,700 were in Pennsylvania. More defoliation has occurred in Pennsylvania than in any other state.

How much of the Earth is protected as national parks and similar sites?

Below is a table of protected areas by country for the year 1990.

Country	Percent of total land area
Venezuela	22.2
Bhutan	19.8
Chile	18.2
Botswana	17.4
Panama	16.9
Czechoslovakia	15.4
Namibia	12.7
United States	10.5
Indonesia	9.3
Australia	5.9
Canada	5.0
Mexico	4.8
Brazil	2.4
Madagascar	1.8
Former Soviet Union	1.1
WORLD	4.9

How large is **Dinosaur National Monument?**

Dinosaur National Monument consists of 211,272 acres (85,499 hectares) on the border of northeast Utah and northwest Colorado. It contains the largest known concentration of fossilized bones. Most of the monument is a scenic wilderness area of canyons formed by the Green River and its tributary, the Yampa. For a comparison of its size with other parks, the ten largest national parks in the 48 contiguous states are listed below:

Park	Location	Acreage (Federal and Non-Federal)
Yellowstone National Park	Idaho, Montana, Wyoming	2,219,785
Death Valley National Monument	California, Nevada	2,067,628
Lake Mead National Recreation Area	Arizona, Nevada	1,496,600
Everglades National Park	Florida	1,398,938
Grand Canyon National Park	Arizona	1,218,375
Glacier National Park	Montana	1,013,572
Olympic National Park	Washington	914,818
Yosemite National Park	California	761,170
Big Bend National Park	Texas	735,416
Isle Royale National Park	Michigan	571,790
Big Cypress National Preserve	Florida	570,000
Joshua Tree National Monument	California	559,954

Where is **Hawk Mountain Sanctuary?**

Hawk Mountain Sanctuary, founded in 1934 as the first sanctuary in the world to offer protection to migrating hawks and eagles, is near Harrisburg, Pennsylvania, on the Kittatinny Ridge. Each year between the months of August and December, over 15,000 migrating birds pass by. Rare species such as Golden Eagles may be seen there.

Who was the father of **conservation?**

American naturalist John Muir (1838–1914) was the father of conservation and the founder of the Sierra Club. He fought for the preservation of the Sierra Nevada Mountains in California, and the creation of Yosemite National Park. He directed most of the Sierra Club's conservation efforts and was a lobbyist for the Antiquities Act.

Who coined the term **"Spaceship Earth"?**

American inventor and environmentalist, Buckminster Fuller (1895–1983), coined the term "Spaceship Earth" as an analogy of the need for technology to be self-contained and to avoid waste.

Who started **Earth Day**?

The first Earth Day was April 22, 1970. It was coordinated by Denis Hayes at the request of United States Senator from Wisconsin, Gaylord Nelson, who is sometimes referred to as the "father" of Earth Day. His main objective was to organize a nation-wide public demonstration so large it would get the attention of politicians and force the environmental issue into the political dialogue of the nation.

Which **environmental landmarks** soon followed the celebration of the first **Earth Day**?

Important official environmental actions that began in 1970 or shortly thereafter were: the birth of the EPA (Environmental Protection Agency), the enactment of the National Environmental Policy Act, the creation of the President's Council on Environmental Quality, and the passage of the Clean Air Act establishing national air quality standards for the first time.

What is a **"green product"**?

Green products are environmentally safe products that contain no chlorofluorocar-bons, are degradable (can decompose), and are made from recycled materials. Deep-green products are those from small suppliers who build their identities around their claimed environmental virtues. Greened-up products come from the industry giants and are environmentally improved versions of established brands.

EXTINCT AND ENDANGERED PLANTS AND ANIMALS

Did **dinosaurs and humans** ever coexist?

No. Dinosaurs first appeared in the Triassic period (about 220 million years ago) and dis-appeared at the end of the Cretaceous period (about 65 million years ago). Modern humans (homo sapiens) appeared only about 25,000 years ago. Movies that show humans and dinosaurs existing together are Hollywood fantasies.

What were the **smallest and largest dinosaurs**?

Compsognathus, a carnivore from the late Jurassic period (131 million years ago) was about the size of a chicken and measured, at most, 35 inches (89 centimeters) from the tip of its snout to the tip of its tail. It probably weighed up to 15 pounds (6.8 kilograms).

The largest species for which a whole skeleton is known is *Brachiosaurus*. A specimen in the Humboldt Museum in Berlin measures 72.75 feet (22.2 meters) long and 46 feet (14 meters) high. It weighed an estimated 34.7 tons (31,480 kilograms). *Brachiosaurus* was a four-footed plant-eating dinosaur with a long neck and a long tail and lived from about 155 to 121 million years ago.

How **long** did **dinosaurs live?**

The lifespan has been estimated at 75 to 300 years. Such estimates are educated guesses. From examination of the microstructure of dinosaur bones, scientists have inferred that they matured slowly and probably had proportionately long lifespans.

How does a **mastodon** differ from a **mammoth?**

Although the words are sometimes used interchangeably, the mammoth and the mastodon were two different animals. The mastodon seems to have appeared first and a side branch may have led to the mammoth.

The *mastodon* lived in Africa, Europe, Asia, and North and South America. It appears in the Oligocene (25 to 38 million years ago) and survived until less than 1 million years ago. It stood a maximum of 10 feet (3 meters) tall and was covered with dense woolly hair. Its tusks were straight forward and nearly parallel to each other.

The *mammoth* evolved less than 2 million years ago and died out about 10,000 years ago. It lived in North America, Europe, and Asia. Like the mastodon, the mammoth was covered with dense, woolly hair, with a long, coarse layer of outer hair to protect it from the cold. It was somewhat larger than the mastodon, standing 9 to 15 feet (2.7 to 4.5 meters). The mammoth's tusks tended to spiral outward, then up.

The gradual warming of the Earth's climate and the change in environment were probably primary factors in the animals' extinction. But early man killed many of them as well, perhaps hastening the process.

Why did **dinosaurs** become extinct?

Many theories exist as to why dinosaurs disappeared from the Earth about 65 million years ago. Scientists argue over whether the dinosaurs became extinct gradually or all at once. The gradualists believe that the dinosaur population steadily declined at the end of Cretaceous period. Numerous reasons have been proposed for this. Some claim the dinosaurs' extinction was caused by biological changes which made them less competitive with other organisms, especially the mammals who were just beginning to appear. Overpopulation has been argued, as has the theory that mammals ate too many dinosaur eggs for the animals to reproduce themselves. Others believe that disease—everything from rickets to constipation—wiped them out. Changes in climate, continental drift, volcanic eruptions, and shifts in the Earth's axis, orbit, and/or magnetic field have also been held responsible.

The catastrophists argue that a single disasterous event caused the extinction, not only of the dinosaurs, but also of a large number of other species that coexisted with them. In 1980, American physicist Luis Alvarez (1911–1988) and his geologist son, Walter Alvarez (b. 1940), proposed that a large comet or meteoroid struck the Earth 65 million years ago. They pointed out that there is a high concentration of the element iridium in the sediments at the boundary between the Cretaceous and Tertiary Periods. Iridium is rare on Earth, so the only source of such a large amount of it had to be outer space. This iridium anamoly has since been discovered at over 50 sites around the world. In 1990, tiny glass fragments, which could have been caused by the extreme heat of an impact, were identified in Haiti. A 110-mile-wide (177-kilometer) crater in the Yucatan Peninsula, long covered by sediments, has been dated to 64.98 million years ago, making it a leading candidate for the site of this impact.

A hit by a large extraterrestrial object, perhaps as much as 6 miles (9.3 kilometers) wide, would have had a catastrophic effect upon the world's climate. Huge amounts of dust and debris would have been thrown into the atmosphere, reducing the amount of sunlight reaching the surface. Heat from the blast may also have caused large forest fires which would have added smoke and ash to the air. Lack of sunlight would kill off plants and have a domino-like effect on other organisms in the food chain, including the dinosaurs.

It is possible that the reason for the dinosaurs' extinction may have been a combination of both theories. The dinosaurs may have been gradually declining when the impact of a large object from space delivered the coup de grâce.

The fact that dinosaurs became extinct has been cited as proof of their inferiority and that they were evolutionary failures. However, these animals flourished for 150 million years. By comparison, the earliest ancestors of humanity appeared only about 3 million years ago. Humans have a long way to go before they can claim the same sort of success as the dinosaurs.

How did the dodo become extinct?

The dodo became extinct around 1800. Although thousands were slaughtered for meat, pigs and monkeys, which destroyed dodo eggs, were probably most responsible for the dodo's extinction. They were native to the Mascarene Islands in the Central Indian Ocean. They became extinct on Mauritius soon after 1680 and on Réunion about 1750. They remained on Rodriguez until 1800.

What is a **quagga**?

The quagga, a native of South Africa, was basically a brown, rather than striped, zebra with white legs and tail. In the early 19th century, it lived in the wild in great herds and was tamed to become a harness animal or was killed for its skin. The species became extinct in 1883.

When did the last **passenger pigeon** die?

At one time, 200 years ago, the passenger pigeon (*Ectopistes migratrius*) was the world's most abundant bird. Although the species was found only in eastern North America, it had a population of 3 to 5 billion birds (25% of the North American land bird population). Overhunting caused a chain of events that reduced their numbers below minimum threshold for viability. In the 1890s several states passed laws to protect the pigeon, but it was too late. The last known wild bird was shot in 1900. The last passenger pigeon, named Martha, died on September 1, 1914, in the Cincinnati Zoo.

Under what conditions is a **species considered "endangered"**?

This determination is a complex process that has no set of fixed criteria that can be applied consistently to all species. The known number of living members in a species is not the sole factor. A species with a million members known to be alive but living in only one small area could be considered endangered, whereas another species having a smaller number of members, but spread out in a broad area, would not be considered so threatened. Reproduction data—the frequency of reproduction, the average number of offspring born, the survival rate, etc.—enter into such determinations. In the United States, the director of the Fish and Wildlife Service (within the Department of the Interior) determines which species are to be considered endangered, based on research and field data from specialists, biologists, botanists, and naturalists.

According to the Endangered Species Act of 1973, a species can be listed if it is threatened by any of the following:

1. The present or threatened destruction, modification, or curtailment of its habitat or range.
2. Utilization for commercial, sporting, scientific, or educational purposes at levels that detrimentally affect it.
3. Disease or predation.
4. Absence of regulatory mechanisms adequate to prevent the decline of a species or degradation of its habitat.
5. Other natural or man–made factors affecting its continued existence.

If the species is so threatened, the director then determines the "critical habitat," that is, the species' inhabitation areas containing the essential physical or biological **167** features necessary for the species' preservation. The critical habitat can include non-habitation areas, which are deemed necessary for the protection of the species.

What was the **first animal** on the United States **endangered species list**?

The peregrine falcon was the first to be listed in the late 1970s.

Which animal species has become **extinct since 1973** when the Endangered Species Act was signed?

Seven domestic species have been declared extinct: Florida's dusky seaside sparrow, the Santa Barbara song sparrow, the blue pike, the Tecopa pupfish, the Sampson's pearly mussel, and the fishes longjaw cisco and the Amistad gambusic.

How many species have been **removed from the** United States' **endangered and threatened species list**, since 1973?

Six species have been removed from the federal endangered and threatened species list since 1973 because they have recovered. Seven species have been removed from the federal list because they have become extinct.

How many species of **plants and animals** are **threatened** in the United States?

There are 157 threatened species (96 animals and 61 plants) in the United States; and 504 endangered species (275 animals and 229 plants).

Category	Threatened	Endangered	Threatened & Endangered
Mammals	8	55	63
Birds	12	73	85
Reptiles	18	16	34
Amphibians	5	6	11
Fishes	34	54	88
Snails	6	7	13
Clams	2	40	42
Crustaceans	2	8	10
Insects	9	13	22

Category	Threatened	Endangered	Threatened & Endangered	
Arachnids	0	3	3	
Plants	61	229	290	
TOTAL	157	504	661	

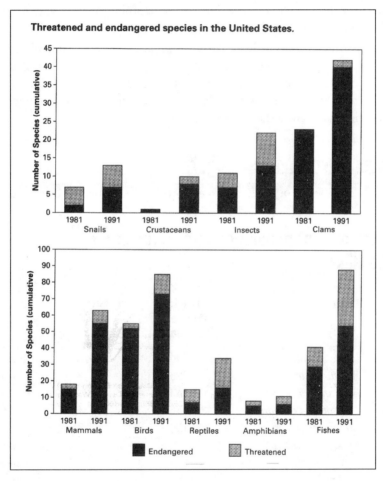

U.S. threatened and endangered species

Are there any **endangered species** in the United States that can be **legally hunted**?

The grizzly bear is the only one of more than 600 threatened species in the United States that can be legally hunted. However, this situation is currently being challenged in the courts under the terms of the Endangered Species Act.

What is the status of an **elephant** in Africa?

From 1979 to 1989, Africa lost half of its elephants from poaching and illegal ivory trade, with the population decreasing from an estimated 1.3 million to 600,000. This led to the transfer of the African elephant from threatened to endangered status in October, 1989, by CITES (the Convention in International Trade in Endangered Species). An ivory ban took effect on January 18, 1990; but six African countries (South Africa, Zimbabwe, Botswana, Namibia, Malawi, and Zambia) are trying to downlist this mammal to a threatened status to allow them to trade in ivory again.

Altogether there are 35 African nations that have elephants, and all want to conserve this resource for their countries' benefit. Kenya now values a living elephant at $14,375 in tourism income for every year of its life giving it a potential life-time value of $900,000. The ivory from an average elephant killed for its tusk would only be worth $1,000 (the price paid before the ban on ivory as of October, 1989).

Are **turtles** an endangered group?

Of the world's 270 species of turtles, 42% are rare or threatened with extinction.

What is the current population and status of the great whales?

Species	Latin Name	Original Population (thousands)	Current Population (thousands)	Status
Sperm	*Physeter macro cephalas*	2400	1974	Insufficient
Blue	*Balenoptera*	226	13	Endangered
Finback	*Balenoptera physalus*	543	123	Vulnerable
Humpback	*Megaptera novaeangliae*	146	4	Vulnerable
Right	*Eubalaena glacialis*	120	3	Endangered
	Eubalaena australis			Vulnerable
Sei	*Balaenoptera*	254	51	Vulnerable
Gray	*Eschrichtius*	20	11	Unlisted
Bowhead	*Balaena mysticetus*	20	2	Vulnerable
Bryde's	*Balaenoptera edeni*	92	92	Insufficient
Minke	*Balaenoptera acutorostrata*	295	280	Insufficient

What is a **dolphin-safe tuna**?

The order Cetacea, composed of whales, dolphins, and porpoises, were spared from the extinction of large mammals at the end of the Pleistocene about 10,000 years ago. But from 1,000 c.e. on they, especially the whale, have been relentlessly hunted by man for their valuable products. The 20th century, with its many technological improvements, has become the most destructive period for the cetacea. In 1972, the United States Congress passed the Marine Mammal Protection Act; one of its goals was to reduce the number of small cetaceans (notably *Stenalla* and *Delphinus*) killed and injured during commercial fishing operations, such as the incidental catch of dolphins in tuna purse-seines (nets that close up to form a huge ball that is hoisted aboard ship). Dolphins are often found swimming with schools of yellowfin tuna and are caught along with the tuna by fisherman who use purse-seine nets. The dolphins are drowned by this fishing method because they must be able to breathe air to survive. The number of incidental deaths and injuries in 1972 was estimated at 368,000 for United States fishing vessels and 55,078 for non-United States vessels. In 1979 the figures were reduced to 17,938 and 6,837 respectively. But in the 1980s, the dolphins killed by foreign vessels rose dramatically to over 100,000 a year. Most of the slaughter occurs in the eastern Pacific Ocean from Chile to Southern California.

To further reduce the numbers of dolphins killed during tuna catches, the three largest sellers of canned tuna in the United States, spearheaded by the Starkist company, decided that they would not sell tuna that has been caught by these methods harmful to dolphins. Their products are thus called dolphin-safe tuna.

POLLUTION

See also: Health and Medicine—Health Hazards, Risks, etc.

What is the **Toxic Release Inventory (TRI)?**

TRI is a government-mandated, publicly available compilation of information on the release of over 300 individual toxic chemicals and 20 categories of chemical compounds by manufacturing facilities in the United States. The law requires manufacturers to state the amounts of chemicals they release directly to air, land, or water, or that they transfer to off-site facilities that treat or dispose of wastes. The U.S. Environmental Protection Agency compiles these reports into an annual inventory and makes the information available in a computerized database. In 1989, 22,560 facilities released 5.7 billion pounds of toxic chemicals into the environment. Over 189 million pounds of this total were released into the water; 2.4 billion pounds emitted into the air; 445 million pounds into landfills; 1.2 billion pounds into underground wells; 551 million pounds to municipal wastewater treatment plants; and 916 million pounds to treatment and disposal facilities (offsite).

What are **PCBs?**

Polychlorinated biphenyls (PCBs) are a group of chemicals that were widely used before 1970 in the electrical industry, as a coolant for transformers and in capacitors and other **171** electrical devices. They caused environmental problems because they do not break down, and can spread through the water, soil, and air. They have been linked by some scientists to cancer and reproductive disorders and have been shown to cause liver function abnormalities. Government action has resulted in the control of the use, disposal, and production of PCBs in nearly all areas of the world including the United States.

What was the distribution of **radioactive fallout** after the Chernobyl accident in Byelorussia?

Radioactive fallout, containing the isotope cesium 137, and nuclear contamination covered an enormous area including Byelorussia, Latvia, Lithuania, the central portion of the then Soviet Union, the Scandinavian countries, the Ukraine, Poland, Austria, Czechoslovakia, Germany, Switzerland, northern Italy, eastern France, Romania, Bulgaria, Greece, Yugoslavia, the Netherlands, and the United Kingdom. The fallout, extremely uneven because of the shifting wind patterns, extended 1,200 to 1,300 miles (1,930 to 2,090 kilometers) from the point of the accident. Roughly 5% of the reactor fuel or seven tons of fuel containing 50 to 100 million curies were released. Estimates of the effects of this fallout range from 28,000 to 100,000 deaths from cancer and genetic defects within the next 50 years. In particular, livestock in high rainfall areas received unacceptable dosages of radiation.

Why are the **"greenhouse gases"** important?

These gases influence how much of the sun's energy is absorbed on Earth and how much is radiated back into space. The popular conception is that these gases act like a giant greenhouse, trapping energy emitted from ground, etc., to make Earth inhabitable. But too much of these gases in the atmosphere has caused global warming. Greenhouse gases include a variety of man-made air pollutants. Carbon dioxide is the most troublesome gas. Others are chlorofluorocarbons (CFCs), methane, and nitrogen oxides.

How do **spray cans** of deodorant, paint, air fresheners, whipped cream, etc., affect the atmosphere?

Every time a button is pushed on a spray can using fluorocarbon propellant, a tiny bit of the propellant is released into the air. The fluorocarbon drifts hundreds of miles into the Earth's upper atmosphere where it changes some ozone into oxygen creating a "hole" that lets through more ultraviolet light to the Earth's surface. This creates health problems for humans such as cataracts and skin cancer, and disturbs delicate ecosystems (for example, plants produce less seed). Currently propellants have been changed to hydrocarbons, such as butane. Chlorofluorocarbons (organic compounds containing chlorine and fluorine) have been banned in aerosols and are being phased out elsewhere.

How do **chlorofluorocarbons** affect the Earth's ozone layer?

Chlorofluorocarbons (CFCs) are hydrocarbons, such as freon, in which part or all of the hydrogen atoms have been replaced by fluorine atoms. These can be liquids or gases, are non-flammable and heat-stable, and are used as refrigerants, aerosol propellants, and solvents. When released into the air, they slowly rise into the Earth's atmosphere where they are broken apart by ultraviolet rays from the sun. Some of the resultant molecular fragments react with the ozone in the atmosphere reducing the amount of ozone. The CFC molecules' chlorine atoms act as catalysts in a complex set of reactions that convert two molecules of ozone into three molecules of ordinary hydrogen. This is depleting the beneficial ozone layer faster than it can be recharged by natural processes. In 1978, the United States government banned the use of fluorocarbon aerosols. Chemical industries must cut CFC manufacture to 50% by the year 2000; they have mounted a massive research effort to find a safer chemical substitute.

What are the components of **smog**?

Photochemical air pollution, commonly known as smog, is the result of a number of complex chemical reactions. The hydrocarbons, hydrocarbon derivations, and nitric oxides emitted from such sources as automobiles are the raw materials for photochemical reactions. In the presence of oxygen and sunlight, the nitric oxides combine with organic compounds, such as the hydrocarbons from unburned gasoline, to produce a whitish haze, sometimes tinged with a yellow-brown color. In the process, a large number of new hydrocarbons and oxyhydrocarbons are produced. These secondary hydrocarbon products may compose as much as 95% of the total organics in a severe smog episode.

What are **flue gas "scrubbers"**?

The scrubbing of flue gases refers to the removal of sulfur dioxide (SO_2) and nitric oxide (NO), which are major components of air pollution. Wet scrubbers use a chemical solvent or lime, limestone, sodium alkali, or diluted sulfuric acid to remove the SO_2 formed during combustion. Dry scrubbing uses either a lime/limestone slurry or ammonia sprayed into the flue gases.

Does the United States government or any of the states regulate the **pollutant emissions** of gasoline powered landscaping equipment?

The United States government does not regulate these everyday tools of American living. The State of California, through its Air Resources Board, wants these emissions reduced in two stages in 1994 and 1999.

What is **acid rain**?

The term "acid rain" was coined by British chemist Robert Angus Smith (1817–1884) who, in 1872, published *Air & Rain; The Beginnings of a Chemical Climatology*. Since then, acid rain has unfortunately become an increasingly used term for rain, snow, sleet, or other precipitation that has been polluted by acids such as sulfuric and nitric acids.

When gasoline, coal, or oil are burned, their waste products of sulfur dioxide and nitrogen dioxide combine in complex chemical reactions with water vapor in clouds to form acids. The United States alone discharges 40 million metric tons of sulfur and nitrogen oxides into the atmosphere. This, combined with natural emissions of sulfur and nitrogen compounds, has resulted in severe ecological damage. Hundreds of lakes in North America (especially northeastern Canada and United States) and in Scandinavia are so acidic that they cannot support fish life. Crops, forests, and building materials, such as marble, limestone, sandstone, and bronze, have been affected as well, but the extent is not as well documented. However, in Europe, where so many living trees are stunted or killed, a new word *Waldsterben* (forest death) has been coined to describe this new phenomenon.

In 1990, amendments to the (United States) Clean Air Act contained provisions to control emissions that cause acid rain. It included the reductions of sulfur dioxide emissions from 19 million tons to 9.1 million tons annually and the reduction of industrial nitrogen oxide emissions from six to four million tons annually, both by the year 2000. Also the elimination of 90% of industrial benzine, mercury, and dioxin emissions, the reduction of automotive nitrogen oxide by 60%, and hydrocarbons by 40% by year 1997 were specified.

How **acidic** is acid rain?

Acidity or alkalinity is measured by a scale known as the pH (potential for Hydrogen) scale. It runs from zero to 14. Zero is extremely acid, 7 is neutral, and 14 is very alkaline. Since the scale is logarithmic, a change in one unit equals a tenfold increase or decrease. So a solution at pH 2 is 10 times more acidic than one at pH 3 and 100 times as acidic as a solution at pH 4. Any rain below 5.0 is considered acid rain; some scientists use the value of 5.6 or less. Normal rain and snow containing dissolved carbon dioxide (a weak acid) measure about pH 5.6. Actual values range according to geographical area. Eastern Europe and parts of Scandinavia have 4.3 to 4.5; the rest of Europe is 4.5 to 5.1; eastern United States and Canada ranges from 4.2 to 4.6, and Mississippi Valley has a range of 4.6 to 4.8. The worst North American area, having 4.2, is centered around Lake Erie and Lake Ontario.

For comparison, some common items and their pH values are listed below:

Concentrated sulfuric acid	1.0
Lemon juice	2.3
Vinegar	3.3
Acid rain	4.3
Normal rain	5.0 to 5.6
Normal lakes and rivers	5.6 to 8.0
Distilled water	7.0
Human blood	7.35 to 7.45
Seawater	7.6 to 8.4

How much oil is dumped into the oceans?

Every year well over three million metric tons of oil contaminate the sea. One half comes from ships, but the rest comes from land–based pollution, with only 33% of it spilled by accident. More than 1.1 million metric tons of oil are deliberately discharged from tankers washing out their tanks.

Source of oil	% of total
Tankers operational discharge	22%
Municipal wastes	22%
Tanker accidents	12.5%
Atmospheric rainout (oil released by industry and cars)	9.5%
Bilge and fuel oils	9%
Natural seeps	7.5%
Non-refining industrial waste	6%
Urban runoff	3.5%
Coastal oil refineries	3%
Offshore production	1.5%
River runoff	1%
Others	2.5%

Where did the first major oil spill occur?

The first major commercial oil spill occurred on March 18, 1967, when the tanker *Torrey Canyon* grounded on the Seven Stones Shoal off the coast of Cornwall, **175**

England, spilling 830,000 barrels (119,000 tons) of Kuwaiti oil into the sea. This was the first major tanker accident. However during World War II, German U–boat attacks on tankers, between January and June of 1942, off the United States East Coast, spilled 590,000 tons of oil. Although the *Exxon Valdez* was widely publicized as a major spill of 35,000 tons in 1989, it is dwarfed by the deliberate dumping of oil from Sea Island into the Persian Gulf on January 25, 1991. It is estimated that that spill equaled almost 1.5 million tons of oil.

Date	Cause	Thousands tons spilled
1/42-6/42	German U-boat attacks on tankers off the East Coast of U.S. during World War II	590
3/18/67	Tanker Torrey Canyon grounds off Land's End in the English Channel	119
3/20/70	Tanker Othello collides with another ship in Tralhavet Bay, Sweden	60–100
12/19/72	Tanker Sea Star collides with another ship in Gulf of Oman	115
5/12/76	Urquiola grounds at La Coruna, Spain	100
3/16/78	Tanker Amoco Cadiz grounds off Northwest France	223
6/3/79	Itox I oil well blows in Southern Gulf of Mexico	600
7/79	Tankers Atlantic Express and Aegean Captain collide off Trinidad and Tobago	300
2/19/83	Blowout in Norwuz oil field in the Persian Gulf	600
8/6/83	Fire aboard Castillo de Beliver off Cape Town, South Africa	250
1/25/91	Iraq begins deliberately dumping oil into Persian Gulf from Sea Island, Kuwait	1,450

How harmful are balloon releases?

Both latex and metallic balloons can be harmful. A latex balloon can land in water, lose its color, and resemble a jellyfish, which if eaten by sea animals can cause their death because they cannot digest it. A metal balloon can get caught in electric wires and cause power outages.

What are **Operation Ranch Hand** and **Agent Orange?**

Operation Ranch Hand was the tactical military project for the aerial spraying of herbicides in South Vietnam during the Vietnam Conflict (1961–1975). In these operations Agent Orange, the collective name for the herbicides 2,4-D and 2,4,5-T, was used for the defoliation. The name derives from the color-coded drums in which the herbicides were stored.

Which pollutants lead to **indoor air pollution?**

Indoor air pollution, also known as "tight building syndrome," results from conditions in modern, high energy efficiency buildings, which have reduced outside air exchange, or have inadequate ventilation, chemical contamination, and microbial contamination. Indoor air pollution can produce various symptoms, such as headache, nausea, and eye, nose, and throat irritation. In addition houses are affected by indoor air pollution emanating from consumer and building products and from tobacco smoke. Below are listed some pollutants found in houses:

Pollutant	Sources	Effects
Asbestos	Old or damaged insulation, fireproofing, or acoustical tiles	Many years later, chest and abdominal cancers and lung diseases
Biological pollutants	Bacteria, mold and mildew, viruses, animal dander and cat saliva, mites, cockroaches, and pollen	Eye, nose, and throat irritation; shortness of breath; dizziness; lethargy; fever; digestive problems; asthma; influenza and other infectious diseases
Carbon monoxide	Unvented kerosene and gas heaters; leaking chimneys and furnaces; wood stoves and fireplaces; gas stoves; automobile exhaust from attached garages; tobacco smoke	At low levels, fatigue at higher levels, impaired vision and coordination; headaches; dizziness; confusion; nausea Fatal at very high concentrations
Formaldehyde	Plywood, wall paneling, particle board, fiber-board; foam insulation; fire and tobacco smoke; textiles, and glues	Eye, nose, and throat irritations; wheezing and coughing; fatigue; skin rash; severe allergic reactions; may cause cancer

177

Pollutant	Sources	Effects
Lead	Automobile exhaust; sanding or burning of lead paint; soldering	Impaired mental and physical development in children; decreased coordination and mental abilities; kidneys, nervous system, and red blood cells damage
Mercury	Some latex paints	Vapors can cause kidney damage; long–term exposure can cause brain damage
Nitrogen dioxide	Kerosene heaters, unvented gas stoves and heaters; tobacco smoke	Eye, nose, and throat irritation; may impair lung function and increase respiratory infections in young children

What causes **formaldehyde contamination** in homes?

Formaldehyde contamination is related to the widespread construction use of wood products bonded with urea-formaldehyde resins and products containing formaldehyde. Major formaldehyde sources include subflooring of particle board; wall paneling made from hardwood plywood or particle board; and cabinets and furniture made from particle board, medium density fiberboard, hardwood plywood, or solid wood. *Urea-formaldehyde foam insulation* (UFFI) has received the most media notoriety and regulatory attention. Formaldehyde is also used in drapes, upholstery, carpeting, and wallpaper adhesives, milk cartons, car bodies, household disinfectants, permanent-press clothing, and paper towels. In particular, mobile homes seem to have higher formaldehyde levels than do houses. Six billion pounds of formaldehyde are used in the United States each year.

The release of formaldehyde into the air by these products (called outgassing) can develop poisoning symptoms in humans. The EPA classifies formaldehyde as a potential human carcinogen (cancer-causing agent).

How is **asbestos** removed from existing sites?

Three acceptable procedures are used to remove asbestos from existing sites: removal, encapsulation, and enclosure. In the removal process, asbestos is stripped from the underlying surface, collected, and placed in containers for burial in an approved site. Encapsulation, the quickest and least expensive method, coats the asbestos material with a sealant. This is often used when the asbestos is not readily accessible. In the enclosure process, a barrier is placed between the asbestos and the building environment. This is rarely used because the asbestos remains in place and has the potential for exposure during routine maintenance.

RECYCLING, CONSERVATION, AND WASTE

See also: Energy—Consumption and Conservation

What is the **NIMBY** syndrome?

NIMBY is the acronym for *"Not In My Back Yard,"* referring to major community resistance to new incinerator sitings, landfills, prisons, roads, etc. NIMFY is "Not In My Front Yard."

Which states have the greatest number of **hazardous waste sites**?

The top five are New Jersey, with 109 hazardous waste sites, Pennsylvania, with 95, New York, 83, California, 88, and Michigan, 77.

Where are the six **nuclear dump sites** in the United States?

In 1991, the six nuclear waste dump sites are in West Valley, New York; Sheffield, Illinois; Maxey Flats, Kentucky; Barnwell, South Carolina; Beatty, Nevada; and Richland, Washington.

How is **nuclear waste** stored?

Nuclear wastes consist either of fission products formed from atom splitting of uranium, cesium, strontium, or krypton, or from transuranic elements formed when uranium atoms absorb free neutrons. Wastes from transuranic elements are less radioactive than fission products; however, these elements remain radioactive far longer—hundreds of thousands of years. The types of waste are irradiated fuel (spent fuel) in the form of 12–foot–long rods, high–level radioactive waste in the form of liquid or sludge, and low–level waste (non-transuranic or legally high-level) in the form of reactor hardware, piping, toxic resins, water from fuel pool, etc.

In the United States most of the spent fuel has been left for 10 years or more in water-filled pools at the individual plant sites. They are waiting permanent disposal that is set for 1998 by a mandate in the Nuclear Waster Policy Act (1982). Most low-level radioactive waste has been stored in steel drums in shallow landfills at the six nuclear dump sites and at the Hanford Nuclear Reservation in the state of Washington. Most high-level nuclear waste has been stored in double–walled stainless-steel tanks surrounded by 3 feet (1 meter) of concrete. The current best storage method, developed by the French in 1978, is to incorporate the waste into a special molten glass mixture, then enclosing it with a steel container and burying it in special pits.

How much **garbage** does the average American generate?

According to one study, Americans produce about 230 million tons of refuse a year—5.1 pounds per person a day or about 1,900 pounds per year. Another survey reported that a typical suburban family of three generated 40 pounds of garbage weekly with the components of the content given in percentages below:

Type of solid waste	Percent
Aluminum	1%
Polystyrene meat trays, cups, egg cartons, and packing	3%
Disposable diapers	3%
Wood, textiles, old clothing	5%
Metal cans and nails	5%
Plastic soda bottles and bags	5%
Glass	8%
Miscellaneous	10%
Food	11%
Paper	21%
Yard waste and grass clippings	23%

An analysis by the U.S. Environmental Protection Agency published in 1990, indicated not only an increase in the amount of garbage per person, but also a change in the content of it as well:

Waste materials	Garbage In pounds per person per day			
	1960	1970	1980	1988
Total nonfood product wastes	1.65	2.26	2.57	2.94
Paper and paperboard	0.91	1.19	1.32	1.60
Glass	0.20	0.34	0.36	0.28
Metals	0.32	0.38	0.35	0.34
Plastics	0.01	0.08	0.19	0.32

Waste materials	1960	1970	1980	1988
Rubber and leather	0.06	0.09	0.10	0.10
Textiles	0.05	0.05	0.06	0.09
Wood	0.09	0.11	0.12	0.14
Other	0.00	0.02	0.07	0.07
Other Wastes				
Food wastes	0.37	0.34	0.32	0.29
Yard wastes	0.61	0.62	0.66	0.70
Miscellaneous inorganic waste	0.04	0.05	0.05	0.06
Total waste generated	2.66	3.27	3.61	4.00

How much **solid waste** is generated annually in the United States?

The United States produces 4.543 million tons of solid waste annually.

Type of waste	Million tons	Percent
Agriculture	2,340	52
Mineral industries (mining and milling waste)	1,620	36
Industrial (nonhazardous)	225	5
Municipal (domestic)	180	4
Utility	90	2
Hazardous	45	2
Low-level radioactive	3	0.0007

Of the 179.6 million tons of *municipal solid wastes* (MSW) in 1988 in the above table, paper waste ranked the highest.

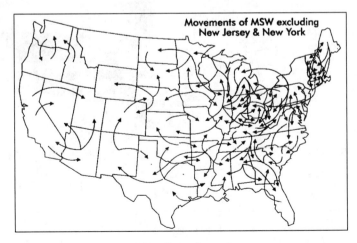

Movements of MSW excluding New Jersey & New York

Where does garbage go?

Municipal Solid Waste

Type of waste	Million tons
Paper	71.8
Yard wastes	31.6
Rubber, Textile, Wood, etc.	20.8
Metals	15.3
Plastics	14.4
Food wastes	13.2
Glass	12.5

How critical is the problem of **landfilling** in the United States?

According to one source, landfilling is currently over used in the United States, but will continue to be an essential component of waste management. The U.S. has more than 9,000 landfills. In 1960, 62% of all garbage was sent to landfills; in 1980, the figure was 81%; and in 1988 it decreased to 73%. After as much waste as possible has been reduced, recovered, reused, or converted to energy, there will still be some waste that cannot be disposed of in any other way except in landfills. However, landfill capacity is already dwindling severely in the most populous regions of the country, as shown in the accompanying illustration.

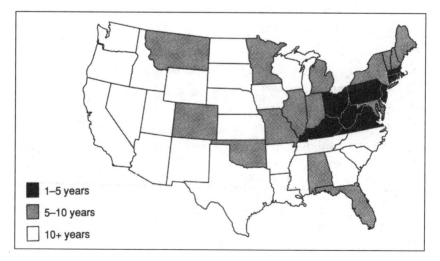

1–5 years
5–10 years
10+ years

Landfill capacity left

Is the United States a leader in the percentage of **municipal solid waste** that is **incinerated**?

Incinerating 15% of its total municipal solid waste, the United States ranks fifth. First place Switzerland incinerates 74% of its waste; Japan incinerated 66%; Sweden, 50%; and France, 15%.

How much does **packaging** contribute to municipal solid waste?

Packaging accounts for 30.3% of municipal solid waste in 1990. The material components of this figure are paper (47.7%), glass (24.5%), plastic (14.5%), steel (6.5%), wood (4.5%), and aluminum (2.3%).

How much space does a **recycled ton of paper** save in a landfill?

Each ton saves more than three cubic yards of landfill space.

How many trees are saved by **recycling paper**?

One ton (907 kilograms) of recycled waste paper spares 17 trees.

How much **newspaper** must be **recycled to save one tree**?

One 35- to 40-foot (10.6- to 12-meter) tree produces a stack of newspapers 4 feet (1.2 meters) thick; this much newspaper must be recycled to save a tree.

How much **waste paper** does a **newspaper** generate?

An average yearly newspaper subscription (for example *The San Francisco Chronicle*) received every day, produces 550 pounds (250 kilograms) of waste paper (per subscription per year). The average *New York Times* Sunday edition produces 8 million pounds (3.6 million kilograms) of waste paper.

What is the major use for **recycled telephone books**?

They can be recycled to make record album and CD covers.

How many **paper mills** in the United States use waste paper to produce new products?

Of the approximately 600 mills in the United States producing pulp, paper, paperboard, or building products, 200 depend almost entirely on waste paper for their raw material. Another 300 mills use 10 to 50% waste paper in their manufacturing processes.

What problems may be encountered when **polyvinyl chloride (PVC) plastics** are burned?

Chlorinated plastics, such as PVC, contribute to the formation of hydrochloric acid gases. They also may be a part of a mix of substances containing chlorine that form a precursor to dioxin in the burning process. Polystyrene, polyethylene and polyethylene terephthalate (PET) do not produce these pollutants.

What do the **numbers inside the recycling symbol** on plastic containers mean?

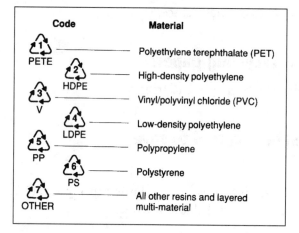

Code	Material
1 PETE	Polyethylene terephthalate (PET)
2 HDPE	High-density polyethylene
3 V	Vinyl/polyvinyl chloride (PVC)
4 LDPE	Low-density polyethylene
5 PP	Polypropylene
6 PS	Polystyrene
7 OTHER	All other resins and layered multi-material

Codes for plastic containers

The Society of the Plastics Industry has developed a voluntary coding system for plastic containers to assist recyclers in sorting plastic containers. The symbol is designed to be imprinted on the bottom of the plastic containers. The numerical code appears inside a three-sided triangular arrow. A guide to what the numbers mean is listed below.

Code	Material	Examples
1	Polyethylene terephthalate (PET)	Soft drink bottles
2	High-density polyethylene (HDPE)	Milk and water jugs
3	Vinyl	Shampoo bottles
4	Low-density polyethylene (LDPE)	Ketchup bottles
5	Polypropylene	Squeeze bottles
6	Polystyrene	Fast-food packaging
7	Other	

Are **cloth diapers** or **disposable diapers** better for the environment?

This is a complex matter with both alternatives having an environmental impact. Disposable diapers make up 2% of the total solid waste while cloth diapers account for only 1% of the solid waste. In 1990, 16 to 17 billion disposable diapers were sold. However, cleaning cloth diapers requires the use of detergents (polluting-agent) and hot water (energy user). In addition, if a professional diaper service is used, then there is extra gasoline for delivery and the exhaust from trucks contributes to air pollution.

How many **automobile tires** are scrapped each year and what can be done with them?

Some 279 million tires are scrapped each year. The aim of the Tire Recycling Incentives Act of 1990 is to motivate people to find suitable uses for the scrap tires. Some firms are turning waste tires into energy while others are turning them into powder, pellets, and other compound forms for many new uses.

What is a **WOBO**?

A WOBO (world bottle) is the first mass-produced container designed for secondary use as a building product. It was conceived by Albert Heineken of the Heineken beer family. The beer bottles were designed in a special shape to be used, when empty, as glass bricks for building houses. The actual building carried out with WOBOs was only a small shed and a double garage built on the Heineken estate at Noordwijk, near Amsterdam. Although not implemented, WOBO was a sophisticated and intelligent design solution to what has emerged as a major environmental issue in recent years.

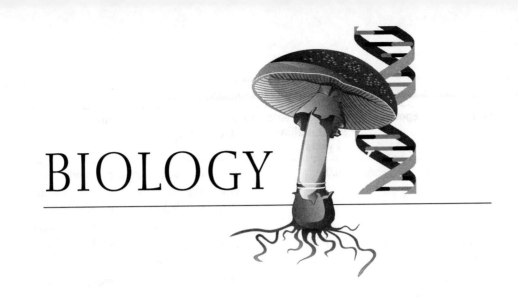

BIOLOGY

EVOLUTION AND GENETICS

Which **biological events** occurred during the **geologic time divisions?**

		Cenozoic Era (Age of Mammals)		
Period	Epoch	Beginning date in est. millions of years	Plants and microorganisms	Animals
Quaternary	Holocene (Recent)	10,000 years ago	Decline of woody plants and rise of herbaceous plants	Age of *Homo sapiens*; humans dominate
	Pleistocene	1.9	Extinction of many species (from 4 ice ages)	Extinction of many large mammals (from 4 ice ages)
Tertiary	Pliocene	6.0	Development of grass lands; decline of forests; flowering plants	Large carnivores; many grazing mammals; first known human-like primates
	Miocene	25.0		Many modern mammals evolve

Cenozoic Era
(Age of Mammals)

Period	Epoch	Beginning date in est. millions of years	Plants and microorganisms	Animals
	Oligocene	38.0	Spread of forests; flowering plants, rise of monocotyledons	Apes evolve; all present mammal families evolve; saber-toothed cats
Tertiary	Eocene	55.0	Gymnosperms and angiosperms dominant	Beginning of age of mammals; modern birds
	Paleocene	65.0		Evolution of primate mammals

Mesozoic Era
(Age of Reptiles)

Period	Epoch	Beginning date in est. millions of years	Plants and microorganisms	Animals
Cretaceous		135.0	Rise of angiosperms; gymnosperms decline	Dinosaurs reach peak and then become extinct; toothed birds become extinct; first modern birds; primitive mammals
Jurassic		200.0	Ferns and gymnosperms common	Large, specialized dinosaurs; first modern birds; insectivorous marsupials
Triassic		250.0	Gymnosperms and ferns dominate	First dinosaurs; egg-laying mammals.

Paleozic Era
(Age of Ancient Life)

Period	Epoch	Beginning date in est. millions of years	Plants and microorganisms	Animals
Permian		285.0	Conifers evolve.	Modern insects appear; mammal-

Period	Epoch	Beginning date in est. millions of years	Plants and microorganisms	Animals
				like reptiles; extinction of many Paleozic invertebrates
Carboniferous (divided into Mississippian and Pennsylvanian periods by some in the U.S.)		350.0	Forests of ferns and gymnosperms; swamps; club mosses and horsetails	Ancient sharks abundant; many echinoderms, mollusks and insect forms; first reptiles; spread of ancient amphibians
Devonian		410.0	Terrestrial plants established; first forests; gymnosperms appear	Age of fish; amphibians; wingless insects and millipedes appear
Silurian		425.0	Vascular plants appear; algae dominant	Fish evolve; marine arachnids dominant; first insects; crustaceans
Ordovician		500.0	Marine algae dominant; terrestrial plants first appear	Invertebrates dominant; first fish appear
Cambrian		570.0	Algae dominant	Age of marine invertebrates

Precambrian Era

Period	Epoch	Beginning date in est. millions of years	Plants and microorganisms	Animals
Archeozoic and Protozoic Eras		3800.0	Bacterial cells; then primitive algae and fungi; marine protozoans	Marine invertebrates at end of period
Azoic		4600.0	Origin of the Earth.	

How did **humans evolve**?

Evolution of the *Homo* lineage of modern humans (*Homo sapiens*) began with the hunter of nearly five feet tall, *Homo habilis*, who is widely presumed to have evolved from an australopithecine ancestor. Near the beginning of the Pleistocene (2 million years ago), *Homo habilis* was transformed into *Homo erectus* (Java Man), who used fire and possessed culture. Middle Pleistocene populations of Homo erectus are said to show steady evolution toward the anatomy of *Homo sapiens* (Neanderthals, Cro-Magnons, and modern humans), 120,000 to 40,000 years ago. Pre-modern *Homo sapiens* built huts and made clothing.

What is meant by **Mendelian inheritance**?

Mendelian inheritance refers to genetic traits carried through heredity; the process was studied and described by the Austrian monk Gregor Mendel (1822–1889). Mendel was the first to deduce correctly the basic principles of heredity. Mendelian traits are also called single gene or monogenic traits, because they are controlled by the action of a single gene or gene pair. More than 4,300 human disorders are known or suspected to be inherited as Mendelian traits, encompassing autosomal dominant (e.g., neurofibromatosis), autosomal recessive (e.g., cystic fibrosis), sex-linked dominant, and recessive conditions (e.g., color-blindness and hemophilia).

Overall, incidence of Mendelian disorders in the human population is about 1%. Many non-anomalous characteristics that make up human variation are also inherited in Mendelian fashion.

Who is generally known as the **father of genetics**?

The English biologist William Bateson (1861–1926) coined the term *genetics* to indicate the study of the causes and effects of heritable characteristics, and was a leading proponent of Mendelian views. Although Gregor Mendel (1822–1884) is considered the founder of genetics, other contenders, such as William Bateson, share the honor with Mendel.

What is the significance of **The Origin of Species**?

Charles Darwin (1809–1882) first proposed a theory of evolution based on natural selection in his treatise on *The Origin of Species*. The publication of *The Origin of Species* ushered in a new era in our thinking about the nature of man. The intellectual revolution it caused and the impact it had on man's concept of himself and the world were greater than those caused by the works of Newton and others. The effect was immediate, the first edition being sold out on the day of publication (November 24, 1859). The *Origin* has been referred to as "the book that shook the world." Every modern discussion of man's future, the population explosion, the struggle for existence, the purpose of man and the universe, and man's place in nature rests on Darwin.

The work was a product of his analyses and interpretations of his findings from his voyages on the H.M.S. *Beagle,* as a naturalist. In Darwin's day, the prevailing explanation for organic diversity was the story of creation in the Book of Genesis in the Bible. The *Origin* was the first publication to present scientifically sound, well-organized evidence for evolution. Darwin's theory of evolution was based on natural selection in which the best, the fittest, survive, and if there is a difference in genetic endowment among individuals, the race will, by necessity, steadily improve. It is a two-step process: the first consists of the production of variation, and the second, of the sorting of this variability by natural selection in which the favorable variations tend to be preserved.

Who coined the phrase "survival of the fittest"?

Although frequently associated with Darwinism, this phrase was coined by Herbert Spencer (1820–1903), an English sociologist. It is the process by which organisms that are less well-adapted to their environment tend to perish and better-adapted organisms tend to survive.

What is **Batesian mimicry**?

In 1861, Henry Walter Bates (1825–1892), a British naturalist, proposed that a non-toxic species can evolve (especially in color and color pattern) to look like a toxic or unpalatable species, or to act like a toxic species, to avoid being eaten by a predator. The classic example is the viceroy butterfly, which resembles the unpalatable monarch butterfly. This is called Batesian mimicry. Subsequently, Fritz Müller (1821–1897), a German-born zoologist, discovered that all the species of similar appearance become distasteful to predators. This phenomenon is called Müllerian mimicry.

When was the **Scopes (monkey) trial**?

John T. Scopes (1900–1970), a high-school biology teacher, was brought to trial by the State of Tennessee in 1925 for teaching the theory of evolution. He challenged a law passed by the Tennessee legislature that made it unlawful to teach in any public school any theory that denies the divine creation of man. He was convicted and sentenced, but the decision was reversed later and the law repealed in 1967.

What is **genetic engineering**?

Genetic engineering is the deliberate alteration of the genetic make-up (genome) of an organism by manipulation of its DNA (*d*eoxyribo*n*ucleic *a*cid) molecule (a double helix **191** chemical structure containing genetic information) to effect a change in heredity traits.

Genetic engineering techniques include cell fusion, and the use of recombinant DNA (RNA) or gene-splicing. In cell fusion, the tough outer membranes of sperm and egg cells are stripped off by enzymes, and then the fragile cells are mixed and combined with the aid of chemicals or viruses. The result may be the creation of a new life form from two species. Recombinant DNA techniques transfer a specific genetic activity from one organism to the next through the use of bacterial plasmids (small circular pieces of DNA lying outside the main bacterial chromosome) and enzymes, such as restriction endonucleases (which cut the DNA strands); reverse transcriptase (which makes a DNA strand from an RNA strand); DNA ligase (which joins DNA strands together); and tag polymerase (which can make a double-stranded DNA molecule from a single stranded "primer" molecule). The process begins with the isolation of suitable DNA strands and fragmenting them. After these fragments are combined with vectors, they are carried into bacterial cells where the DNA fragments are "spliced" on to plasmid DNA that has been opened up. These hybrid plasmids are now mixed with host cells to form transformed cells. Since only some of the transformed cells will exhibit the desired characteristic or gene activity, the transformed cells are separated and grown individually in cultures. This methodology has been successful in producing large quantities of hormones (such as insulin) for the biotechnology industry. However, it is more difficult to transform animal and plant cells. Yet the technique exists to make plants resistant to diseases and to make animals grow larger. Because genetic engineering interferes with the processes of heredity and can alter the genetic structure of our own species, there is much concern over the ethical ramifications of such power, as well as the possible health and ecological consequences of the creation of these bacterial forms. Some applications of genetic engineering in the various fields are listed below:

Agriculture—Crops having larger yields, disease- and drought-resistancy; bacterial sprays to prevent crop damage from freezing temperatures (1987); and livestock improvement through changes in animal traits (1982).

Industry—Use of bacteria to convert old newspaper and wood chips into sugar; oil- and toxin-absorbing bacteria for oil spill or toxic waste clean-ups (1991); and yeasts to accelerate wine fermentation.

Medicine—Alteration of human genes to eliminate disease (1990, experimental stage); faster and more economical production of vital human substances to alleviate deficiency and disease symptoms (but not to cure them) such as insulin (1982), interferon (cancer therapy), vitamins, human growth hormone ADA (1990), antibodies, vaccines, and antibiotics.

Research—Modification of gene structure in medical research (1989), especially cancer research (1990).

Food processing—Rennin (enzyme) in cheese aging (1989).

What is the **Genome Project**?

In late 1990, the biomedical community began work on a 15-year, $3 billion (government-financed) program to map the entire human *genome* (the complete genetic structure of a species), which involves locating 100 genes. So far they have plotted geographic locations on the chromosomes of about 4 human genes (4% of the total). The goal is not only to pinpoint these genes, but also to decode the biochemical information down to the so-called "letters" of inheritance, the four basic constituents of all genes, called *nucleotides:* (A (adenine), C (cytosine), G (guanine), and T (thymine)). Since these letters are linked in pairs of sequences in the double helix of DNA, this means that three billion pairs are involved in this process, of which so far 35 million have been deciphered. The justification of the project is that since genetics is a powerful way to study human disease and other aspects of biology, Genome will accelerate biomedical research.

Can **human beings** be **cloned**?

In theory, yes. There are, however, many technical obstacles to human cloning, as well as moral, ethical, philosophical, religious, and economic issues to be resolved before a human being could be cloned.

A clone is a group of cells derived from the original cell by fission (one cell dividing into two cells) or by mitosis (cell nucleus division with each chromosome splitting into two). It perpetuates an existing organism's genetic make-up. Gardeners have been making clones (copies) of plants for centuries by taking cuttings of plants to make genetically identical copies. For plants that refuse to grow from cuttings or for the animal world, modern scientific techniques have greatly extended the range of cloning. The technique for plants starts with taking a cutting of a plant, usually the "best" one in terms of reproductivity or decorativeness or other standard. Since all the plant's cells contain the genetic information from which the entire plant can be reconstructed, the cutting can be taken from any part of the plant. Placed in a culture medium having nutritious chemicals and a growth hormone, the cells in the cutting divide, doubling in size every six weeks until the mass of cells produces small white globular points called embryoids. These embryoids develop roots, or shoots, and begin to look like tiny plants. Transplanted into compost, these plants grow into exact copies of the parent plant. The whole process takes 18 months. This process, called tissue culture, has been used to make clones of oil palm, asparagus, pineapples, strawberries, brussels sprouts, cauliflower, bananas, carnations, ferns, etc. Besides making high productive copies of the best plant available, this method controls viral diseases that are passed through seed generations.

For animals, a technique called nuclear transfer enables up to 32 clones to be produced at one time. An embryo at the 32-cell stage of development is split up using tiny surgical tools. Each of the 32 cells then are combined with single cell embryos (from the same species) from which the nucleus has been removed. This method has **193** been used on mice, frogs, sheep, and cattle. Ultimately, there seems to be no biological reason why human beings could not be cloned, sometime in the future.

Who originated the idea called **panspermia**?

Panspermia is the idea that microorganisms, spores, or bacteria attached to tiny particles of matter have travelled through space, eventually landing on a suitable planet and initiating the rise of life there. The word itself means "all-seeding." British scientist Lord Kelvin (1824–1907) suggested, in the 19th century, that life may have arrived here from outer space, perhaps carried by meteorites. In 1903, Swedish chemist Svante Arrhenius (1859–1927) put forward the more complex panspermia idea that life on Earth was "seeded" by means of extraterrestrial spores, bacteria, and microorganisms landing here on tiny bits of cosmic matter.

What is the difference between **DNA** and **RNA**?

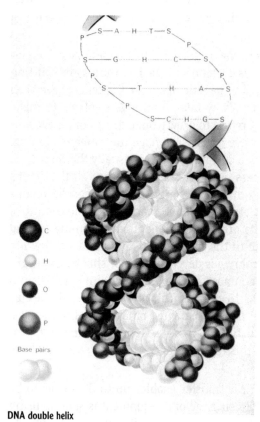

DNA double helix

DNA (*de*oxyribo*nucleic a*cid) is a nucleic acid formed from a repetition of simple building blocks called nucleotides. The nucleotides consist of phosphate (PO_4), sugar (deoxyribose), and a base which is either adenine (A), thymine (T), guanine (G), or cytosine (C). In a DNA molecule, this basic unit is repeated in a double helix structure made from two chains of nucleotides linked between the bases. The links are either between A and T or between G and C. The structure of the bases does not allow other kinds of links. The famous double helix structure resembles a twisted ladder. The 1962 Nobel Prize in physiology or medicine was awarded to James Watson (b. 1928), Francis Crick (b. 1916), and Maurice Wilkins (b. 1916) for determining the molecular structure of DNA.

RNA (*ribo*nucleic *a*cid) is also a nucleic acid, but it consists of a single chain and the sugar is ribose rather than deoxyribose. The bases are the same except that the thymine (T) which appears in DNA is replaced by another base called uracil (U), which links only to adenine (A).

What are the stages in the type of cell division called **mitosis**?

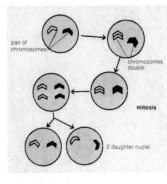

Mitosis process

Cell division in eukaryotes (higher organisms) consists of two stages, *mitosis,* the division of the nucleus, and *cytokinesis,* the division of the whole cell.

The first process in the actual division of the cell is mitosis. In mitosis, the replicated chromosomes are maneuvered so that each new cell gets a full complement of chromosomes—one of each. The process is divided into four phases: prophase, metaphase, anaphase, and telophase.

Nuclear division of sex cells is called meiosis. Sexual reproduction generally requires two parents and it always involves two events (meiosis and fertilization).

How many **mitochondria** are there in a cell?

The number of mitochondria varies according to the type of cell, but each cell in the human liver has over 1,000 mitochondria. A mitochondrion (singular form) is a self-replicating double-membraned body found in the cytoplasm of all eukaryotic (having a nucleus) cells. The number of mitochondria per cell varies between one and 10,000, and averages about 200. The mitochondria are the sites for much of the metabolism necessary for the production of ATP, lipids, and protein synthesis.

Who is considered the "father" of **embryology**?

Kaspar Friedrich Wolff (1733–1794), a German surgeon, is regarded as the "father" of embryology. Wolff produced his revolutionary work *Theoria generationis* in 1759. Until that time it was generally believed that each living organism developed from an exact miniature of the adult within the seed or sperm. Wolff introduced the idea that cells in a plant or animal embryo are initially unspecified (undefined) but later differentiate to produce the separate organs and systems with distinct types of tissues.

What does **"ontogeny recapitulates phylogeny"** mean?

Ontogeny is the course of development of an organism from fertilized egg to adult; phylogeny is the evolutionary history of a group of organisms. The phrase, originating in nineteenth century biology, means that, as an embryo of an advanced organism grows, it will pass through stages that look very much like the adult phase of less **195** advanced organisms. For example, at one point the human embryo has gills and resembles a tadpole.

LIFE PROCESSES, STRUCTURES, ETC.

What is a biological clock?

First recognized by the Chinese in the 3rd century B.C.E., the biological clock is an intrinsic mechanism that controls the rhythm of various metabolic activities of plants and animals. Some, such as mating, hibernation, and migration have a yearly cycle; others, such as ovulation and menstrual cycles of women, follow a lunar month. The majority, however, have a 24-hour, day-night cycle called circadian rhythm. This day-night cycle, first recognized in plants over 250 years ago and existing in virtually all species of plants and animals, regulates these organisms' metabolic functions: plants opening and closing their petals or leaves, germination and flowering functions, changes in human body temperature, hormone secretion, blood sugar and blood pressure levels, and sleep cycles.

Research in chronobiology—the study of these daily rhythms—reveals that many accidents occur between 1 and 6 a.m., that most babies are born in the morning hours, that heart attacks tend to occur between 6 and 9 a.m., and that most Olympic records are broken in the late afternoon. The clock regulator in animals may be the pineal gland, an appendage of the brain.

Who was regarded as the father of **biochemistry**?

Jan Baptista van Helmont (1577–1644) is called the father of biochemistry because he studied and expressed vital phenomena in chemical terms. The term "biochemistry," coined by F. Hoppe-Seyler in 1877, is the science dealing with the dynamics of living chemical processes or metabolism. It was formed from both the chemists' animal and vegetable chemistry, and the biologists' and doctors' physiological, zoological, or biological chemistry.

Helmont devoted his life to the study of chemistry as the true key to medicine and is considered as one of the founders of modern pathology because he studied the external agents of diseases as well as the anatomical changes caused by diseases.

What is the "Spiegelman monster"?

Sol Spiegelman, an American microbiologist, conducted an experiment to identify the smallest molecule capable of replicating itself. He began with a virus called Q_B, which consisted of a single molecule of ribonucleic acid (RNA) composed of 4,500 nucleotides (units of nucleic acid).

Usually this virus is able to make copies of itself only by invading a living cell, because it requires a cellular enzyme called replicase. Spiegelman added the replicase along with a supply of free nucleotides to the virus in a test tube, and the virus replicated itself for several generations when a mutant appeared having fewer than 4,500 nucleotides. Being smaller, this mutant replicated faster than the original virus. Then another, still smaller, mutant appeared and displaced the first, and so it went. Finally, the virus degenerated into a little piece of ribonucleic acid with only 220 nucleotides, the smallest bit necessary for recognizing the replicase. This little test tube monster could continue to replicate at high speed, providing it was supplied with a source of building blocks.

Who is Melvin Calvin?

Calvin (b. 1911) is an American chemist who won the 1961 Nobel Prize for Chemistry for his achievement of working out the chemical reaction cycles called biosynthetic pathways in photosynthesis. Photosynthesis is the process by which green plants use the energy of the sunlight to convert water and carbon dioxide into carbohydrates and oxygen. This cycle of reactions is now called the Calvin cycle. He also conducted research in organometallic chemistry, the chemical origin of life, and taught chemistry at the University of California at Berkeley.

CLASSIFICATION, MEASUREMENT, AND TERMS

What is radiocarbon dating?

Radiocarbon dating is a process for determining the age of a prehistoric object by measuring its radiocarbon content. The technique was developed by an American chemist, **197**

Dr. Willard F. Libby (1908–1980), in the late 1940s. All living things contain radiocarbon (carbon 14), an isotope that occurs in a small percentage of atmospheric carbon dioxide as a result of cosmic ray bombardment. After an animal or plant dies, it no longer absorbs radiocarbon and the radiocarbon present begins to decay (break down by releasing particles) at an exact and uniform rate. Its half-life of 5,730 years made it useful for measuring prehistory and events occurring within the past 35 to 50 thousand years. The remaining radiocarbon can be measured and compared to that of a living sample. In this way, the age of an animal or plant 50,000 years old or less (or, more precisely, the elapsed time since its death) can be determined. A recent development, called the Accelerated Mass Spectrometer, which separates and detects atomic particles of different mass, can establish more accurate dates with a smaller sample.

Since Libby's work, other isotopes having longer half-lives have been used as "geologic clocks" to date very old rocks. The isotope Uranium-238 (decaying to Lead-206) has a half-life of 4.5 billion years, Uranium-235 (decaying to Lead-207) has a value of 704 million years, Thorium-232 (decaying to Lead-278) has a half-life of 14 billion years, Rubidium-87 (decaying to Strontium-87) has a half-life value of 48.8 billion years, Potassium-40 (decaying to Argon-40) has a value of 1.25 billion years and Samarium-147 (decaying to Neodymium-143) has a value of 106 billion years. These isotopes are used in dating techniques of gas formation light emission (called thermoluminescence). Other ways to date the past are dating by tree rings (counting its annual growth rings), and dating by thermoremanent magnetism (the magnetic field of the rock is compared to a date chart of changes in the Earth's magnetic field).

Who coined the term **biology**?

Biology as a term was first used by Karl Burdach (1776–1847) to denote the study of man. Jean Baptiste Pierre Antoine de Monet Lamarck (1744–1829) gave the term a broader meaning in 1812. He believed in the integral character of science. For the special sciences, chemistry, meteorology, geology, and botany-zoology, he coined the term *biology*.

Lamarckism epitomizes the belief that changes acquired during an individual's lifetime as the result of active, quasi-purposive, functional adaptations can somehow be imprinted upon the genes, thereby becoming part of the heritage of succeeding generations. Today, very few professional biologists believe that anything of the kind occurs—or can occur.

Biology is the science that deals with living things (Greek *bios,* "life"). Formerly broadly divided into two areas, zoology (Greek *zoon,* "animal"), the study of animals, and botany (Greek *botanes,* "plant"), the study of plants, biology is now divided and sub-divided into hundreds of special fields involving the structure, function, and classification of the forms of life. These include anatomy, ecology, embryology, evolution, genetics, paleontology, and physiology.

Who devised the current **animal and plant classification** system?

The naming and organizing of the millions of species of plants and animals is frequently called taxonomy; such classifications provide a basis for comparisons and generalizations. A common classification format is an hierarchical arrangement in which a group is classified within a group, and the level of the group is denoted in a ranking.

Carolus Linnaeus (1707–1778) composed a hierarchical classification system for plants (1753) and animals (1758) using a system of nomenclature (naming) that continues to be used today. Every plant and animal was given two scientific names (binomial method) in Latin, one for the group or genus and the other for the species within the group or genus. He categorized the organisms by perceived physical differences and similarities. Although Linnaeus started with only two kingdoms, contemporary classifiers have expanded them into five kingdoms. Each kingdom is divided into two or more phyla (major groupings; phylum is the singular form). Members within one phylum are more closely related to one another than they are to members of another phylum. These phyla are subdivided into parts again and again; members of each descending level have a closer relationship to each other than those of the level above. Generally the ranking of the system, going from general to specific, are Kingdom, Phylum (plant world uses the term Division), Class, Order, Family, Genus, and Species. In addition, intermediate taxonomic levels can be created by adding the prefixes "sub" or "super" to the name of the level, for example, "subphylum" or "superfamily." Zoologists working on parts of the animal classification are not always uniform in their groupings. The system is still evolving and changing as new information emerges and new interpretations develop. Below is listed a comparison of hierarchy for four of the five kingdoms.

Taxonomic level	Human	Grasshopper	White Pine	Typhoid Bacterium
Kingdom	Animalia	Animalia	Plantae	Protista
Phylum	Chordata	Anthropoda	Tracheophyta	Schizomyco-phyta
Class	Mammalia	Insecta	Gymnospermae	Schizomycetes
Order	Primates	Orthoptera	Coniferales	Eubacteriales
Family	Hominidae	Arcridiidae	Pinaceae	Bacteriaceae
Genus	*Homo*	*Schistocerca*	*Pinus*	*Eberthella*
Species	*sapiens*	*americana*	*strobus*	*typhosa*

What are the **five kingdoms** presently used to categorize living things?

Carolus Linnaeus in 1735 divided all living things into two kingdoms in his classification system that was based on similarities and differences of organisms. However, since **199** then, fungi seemed not to fit nicely into either kingdom. Although fungi were generally considered plants, they have no chlorophyll, roots, stems, or leaves, and hardly resemble any true plant. They also have several features found in the animal kingdom, as well as unique features characteristic of themselves alone. So fungi was considered the third kingdom. In 1959, R.H. Whittaker proposed the current five- kingdom system, based on new evidence from biochemical techniques and electron microscope observations that revealed fundamental differences among organisms. Each kingdom is listed below.

Monera—One-celled organisms lacking a membrane around the cell's genetic matter. Prokaryote (or procaryote) is the term used for this condition where the genetic material lies free in the cytoplasm of the cell without a membrane to form the nucleus of the cell. The kingdom consists of bacteria and blue-green algae (also called blue-green bacteria or cyanobacteria). Bacteria do not produce their own food but blue-green algae do. Blue-green algae, the primary form of life 3.5 to 1.5 billion years ago, produce most of the world's oxygen through photosynthesis.

Protista—Mostly single-celled organisms with a membrane around the cell's genetic material called eukaryotes (or eucaryotes) because of the nucleus membrane and other organelles found in the cytoplasm of the cell. Protista consists of true algae, diatoms, slime molds, protozoa, and euglena. Protistans are diverse in their modes of nutrition, etc. They may be living examples of the kinds of ancient single cells that gave rise to the kingdoms of multi–celled eukaryotes (fungi, plants, and animals).

Fungi—One-celled or multicelled eukaryotes (having a nucleus membrane or a membrane around the genetic material). The nuclei stream between cells giving the appearance that cells have multiple nuclei. This unique cellular structure, along with the unique sexual reproduction pattern, distinguish the fungi from all other organisms. Consisting of mushrooms, yeasts, molds, etc., the fungi do not produce their own food.

Plantae—Multicellular organisms having cell nuclei and cell walls, which directly or indirectly nourish all other forms of life. Most use photosynthesis (a process by which green plants, containing chlorophyll, utilize sunlight as an energy source to synthesize complex organic material, especially carbohydrates from carbon dioxide, water, and inorganic salts) and most are autotrophs (produce their own food from inorganic matter).

Animalia—Multicellular organisms whose eukaryotic cells (without cell walls) form tissues (and from tissues form organs). Most get their food by ingestion of other organisms; they are heterotrophs (cannot produce their food from inorganic elements). Most are able to move from place to place (mobile) at least during part of the life cycle.

FUNGI, BACTERIA, ALGAE, ETC.

What are **diatoms**?

Diatoms are microscopic algae in the phylum *bacillarrophyte* of the protista kingdom. Yellow or brown in color, almost all diatoms are single-celled algae, dwelling in fresh and salt water, especially in the cold waters of the North Pacific Ocean and the Antarctic. Diatoms are an important food source for marine plankton (floating animal and plant life) and many small animals.

Diatoms have hard cell walls; these "shells" are made from silica that they extract from the water. It is unclear how they accomplish this. When they die, their glassy shells, called frustules, sink to the bottom of the sea, which hardens into rock called diatomite. One of the most famous and accessible diatomites is the Monterrey Formation along the coast of central and southern California.

What is the scientific study of **fungi** called?

Mycology is the science concerning fungi. In the past, fungi have been classified in other kingdoms, but currently they are recognized as a separate kingdom based on their unique cellular structure and their unique pattern of sexual reproduction.

Fungi are heterotrophs (cannot produce their own food from inorganic matter). They secrete enzymes that digest food outside their bodies and their fungal cells absorb the products. Their activities are essential in the decomposition of organic material and cycling of nutrients in nature.

Some fungi, called saprobes, obtain nutrients from non-living organic matter. Other fungi are parasites; they obtain nutrients from the tissues of living host organisms. The great majority of fungi are multicelled and filamentous. A mushroom is a modified reproductive structure in or upon which spores develop. Each spore dispersed from it may grow into a new mushroom.

What is a **lichen**?

Lichens are organisms that grow on rocks, tree branches, or bare ground. They are composed of a green algae and a colorless fungus living together symbiotically. They do not have roots, stems, flowers, or leaves. The fungus, having no chlorophyll, cannot manufacture its own food, but can absorb food from the algae which it enwraps completely, providing protection from the sun and moisture.

This relationship between the fungus and algae is called symbiosis (a close association of two organisms not necessarily to both their benefits). Lichens were the first recognized and are still the best examples of this phenomenon. A unique feature of **201** lichen symbiosis is that it is so perfectly developed and balanced as to behave as a single organism.

What are **petite yeasts**?

In certain yeasts, abnormally small colonies occur. The cells that form petite colonies are mutants that lack certain components for respiration; therefore they exhibit a reduced growth rate. Mutations occur in nuclear DNA or in mitochondrial DNA. Mitochondria are "organelles" found in most eukaryotic (complex) cells. They are the sites for cellular respiration and other cellular processes. Occurrence is one in 500.

Are there **stone-eating bacteria**?

Stone-eating bacteria belong to several families in the genus *Thiobacillus*. They can cause damage to monuments, tombs, buildings, and sculptures by converting marble into plaster. The principal danger seems to come from *Thiobacillus thioparus*. This microbe's metabolic system converts sulfur dioxide gas (found in the air) into sulfuric acid and uses it to transform calcium carbonate (marble) into calcium sulfate (plaster). The bacilli draw their nutrition from carbon dioxide formed in the transformation. *Nitrobacter* and *Nitrosomonas* are other "stone-eating bacteria" that use ammonia from the air to generate nitric and nitrous acid. Still other kinds of bacteria and fungi can produce organic acids (formic, acetic, and oxalic acids) which attack the stone as well. The presence of these microbes was first observed by French scientist Henri Pochon at Angkor Wat, Cambodia, during the 1950s. The increase of these bacteria and other biological-damaging organisms that threaten tombs and buildings of antiquity are due to the sharp climb in the level of free sulfur dioxide gas in the atmosphere from automotive and industrial emissions.

How is a fairy ring formed?

A *fairy ring,* or *fungus ring,* is found frequently in a grassy area. There are three types: those that do not affect the surrounding vegetation, those that cause increased growth, and those that damage the surrounding environment. The ring is started from a mycelium (the underground, food-absorbing part of a fungus). The fungus growth is on the outer edge of the grassy area because the inner band of decaying mycelium depletes the resources in the soil at the center. This creates a ring effect. Each succeeding generation grows farther out from the center.

Who was first to coin the word **virus**?

English physician Edward Jenner (1749–1823), founder of virology and a pioneer in vaccination, first coined the word *virus*. A virus is a minute parasitic organism that reproduces only inside the cell of its host. Viruses replicate by invading host cells and taking over the cell's "machinery" for DNA replication. Viral particles then can break out of the cells, causing disease. Using one virus to immunize against another one was precisely the strategy Jenner used when he inoculated someone with cowpox (a disease that attacks cows) to make them immune to smallpox. This procedure is called vaccination, from the word *vaccine* (the Latin name for cowpox). Vaccines are usually a very mild dose of the disease–causing bacteria or virus (weakened or dead). These vaccines stimulate the creation of antibodies in the body that recognize and attack a particular infection.

Who were the founders of **modern bacteriology**?

German bacteriologist Robert Koch (1843–1910) and French chemist Louis Pasteur (1822–1895) are considered bacteriology's founders. Pasteur devised a way to heat food or beverages at a temperature slow enough not to ruin them, but high enough to kill most of the microorganisms that would cause spoilage and disease. This process is called pasteurization. By demonstrating that tuberculosis was an infectious disease caused by a specific bacillus and not by bad heredity, Koch laid the groundwork for public health measures that significantly reduced such diseases. His working methodologies for isolating microorganisms, his laboratory procedures, and his four postulates for determination of disease agents gave medical investigators valuable insights into the control of bacterial infections.

Why are **Koch's postulates** significant?

German bacteriologist Robert Koch (1843–1910) developed four rules in his study of disease-producing organisms, which later investigators found useful. The following conditions must be met to prove that a bacterium causes a particular disease:

1. The microorganism must be found in large numbers in all diseased animals but not in healthy ones.
2. The organism must be isolated from a diseased animal and grown outside the body in a pure culture.
3. When the isolated microorganism is injected into other healthy animals, it must produce the same disease.
4. The suspected microorganism must be recovered from the experimental hosts (#3), isolated, compared to the first microorganism, and found to be identical.

PLANT WORLD

PHYSICAL CHARACTERISTICS, FUNCTIONS, ETC.

How many **chloroplasts** are there in plant cells?

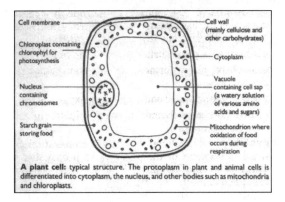

Cell membrane

Chloroplast containing chlorophyl for photosynthesis

Nucleus containing chromosomes

Starch grain storing food

Cell wall (mainly cellulose and other carbohydrates)

Cytoplasm

Vacuole containing cell sap (a watery solution of various amino acids and sugars)

Mitochondrion where oxidation of food occurs during respiration

A plant cell: typical structure. The protoplasm in plant and animal cells is differentiated into cytoplasm, the nucleus, and other bodies such as mitochondria and chloroplasts.

Chloroplasts are the functional units where photosynthesis takes place. They contain the green pigments chlorophyll a and b, which trap light energy for photosynthesis (the process whereby green plants use light energy for the synthesis of sugar from carbon dioxide and water, with oxygen released as a by-product). A unicellular plant may have only a single large chloroplast, whereas a plant leaf cell may have as many as 20 to 100.

What is the best type of **pollination**?

Effective pollination occurs when viable pollen is transferred to plant's stigmas, ovule-bearing organs, or ovules (seed precursors). Without pollination, there would be no fertilization. Since plants are immobile organisms, they usually need external agents to transport their pollen from where it is produced in the plant to where fertilization can occur. This situation produces cross-pollination, wherein one plant's pollen is moved by an agent to another plant's stigma. Some plants are able to self-pollinate—transfer their own pollen to their own stigmas. But of the two methods, cross-pollination seems the better, for it allows new genetic material to be introduced.

The cross-pollination agents are insects, wind, birds, mammals, or water. Many times flowers offer one or more "rewards" to attract these agents—sugary nectar, oil, solid food bodies, perfume, a place to sleep, or sometimes, the pollen itself. Other times the plant can "trap" the agent into transporting the pollen. Generally plants use color and fragrances as attractants to lure these agents. For example, a few orchids use a combination of smell and color to mimic the female of certain bee and wasp species so successfully that the corresponding males will attempt to mate with them. Through this process (pseudocopulation) the orchids achieve pollination. While some plants cater to a variety of agents, other plants are very selective and are pollinated by a single species of insect only. This extreme pollinator specificity tends to maintain the purity of a plant species.

Plant structure can accommodate the type of agent used. For example, plants (such as grasses and conifers) whose pollen is carried by the wind, tend to have a simple structure lacking petals, with freely exposed and branched stigmas (to catch airborne pollen) and dangling anthers (pollen-producing parts) on long filaments. This type of anther allows the light, round pollen to be easily caught by the wind. These plants are found in areas (such as prairies and mountains) where insect agents are rare. In contrast, semi-enclosed, nonsymmetrical, long-lived flowers (such as iris, rose, and snapdragon) offer some "landing platform" and nectar in the flower base to accommodate insect agents such as the bee. The sticky abundant pollen can become attached easily to the insect to be borne away to another flower.

What is **tropism**?

Tropism is the movement of a plant in response to a stimulus. The categories include

Chemotropism—a response to chemicals by plants in which incurling of leaves may occur.

Geotropism or *gravitropism*—a response to gravity in which the plant moves in relation to gravity. Shoots of a plant are negatively geotropic (growing upward), while roots are positively geotropic (growing downward).

Hydrotropism—a response to water or moisture in which roots grow toward the water source.

Paraheliotropism—a response by the plant leaves to avoid exposure to the sun.

Phototropism—a response to light in which the plant may be positively phototropic (moving toward the light) or negatively phototropic (moving away from the light source). Main axes of shoots are usually positively phototropic, whereas roots are generally insensitive to light.

206 *Thermotropism*—a response to temperature by plants.

Thigmotropism or *haptotropism*—a response to touch by the climbing organs of a plant, for example the plant's tendrils curl around a support in a "spring-like manner."

How do different kinds of **music affect plant growth**?

In experiments done in the 1960s and 1970s, plants responded best to classical and Indian devotional music. In a controlled environment, plants exposed to these kinds of music had lush and abundant growth and good root development. Exposure to country music or silence brought about no abnormal growth reaction, while jazz produced more abundant growth. With rock music, plants did poorly. Their roots were scrawny and sparse and they seemed to be in a dying stage. Plants exposed solely to white noise died quickly.

TREES AND SHRUBS

See also: The Environment—Ecology, Resources, etc.; Minerals and Other Materials

What are the **longest-lived tree species** in the United States?

Of the 850 different species of tree in the United States, the oldest species is the bristlecone pine (*Pinus longaeva*), which grows in the deserts of Nevada and southern California (especially in the White Mountains). Some of these trees are believed to be over 4,600 years old. The *potential* life span of these pines is estimated to be 5,500 years. The giant sequoia (*Sequoiadendron giganteum*) has a potential life span of 6,000 years. The oldest tree in the United States was a bristlecone pine (designated as PNW-114) on Mount Wheeler (Nevada) whose age was evaluated to be 5,000 years old. The oldest known living tree is a bristlecone pine named "Methuselah" in the White Mountains of southern California, whose life span was confirmed to be 4,700 years. But these ages are very young when compared to the oldest surviving species in the world, which is the maiden-hair tree (*Ginkgo bibloba*) of Zhexiang, China. This tree first appeared during the Jurassic era some 160 million years ago. Also called icho, or the ginkyo ("silver apricot"), this species has been cultivated in Japan since 1100 C.E.

The Longest Lived Tree Species in the United States

Name of tree	Number of years
Bristlecone pine (*Pinus longaeva*)	3,000–4,700
Giant sequoia (*Sequoiadendron giganteum*)	2,500
Redwood (*Sequoia sempervirens*)	1,000–3,500
Douglas fir (*Pseudotsuga menziesii*)	750
Bald cypress (*Taxodium distichum*)	600

How are **tree rings** used to date historical events?

Tree fragments of unknown age and the rings of living trees can be compared in order to establish the date when the fragment was part of a living tree. Thus, tree rings can be used to establish the year in which an event took place as long as the event involved the maiming or killing of a tree. Precise dates can be established for the building of a medieval cathedral or an American Indian pueblo; the occurrence of an earthquake, landslide, volcanic eruption, or a fire; and even the date when a panel of wood was cut for a Dutch painting.

Every year, a tree produces an annular ring composed of one wide, light ring and one narrow, dark ring. During spring and early summer, tree stem cells grow rapidly and are larger; this is the wide, light ring. In winter growth is greatly reduced and cells are much smaller; this is the narrow, dark ring. In the cold of winter or the dry heat of summer, no cells are produced.

Why do tree **leaves turn color** in the fall?

The carotenoids (pigments in the photosynthesizing cells), which are responsible for the fall colors, are present in the leaves during the growing season. However, the colors are eclipsed by the green chlorophyll. Toward the end of summer, when the chlorophyll production ceases, the other colors of the carotenoids (such as yellow, orange, red, or purple) become visible. Listed below are the autumn leaf colors of some common trees.

Tree	Color
Sugar maple and sumac	Flame red and orange
Red maple, dogwood, sassafras, and scarlet oak	Dark red
Poplar, birch, tulip tree, willow	Yellow
Ash	Plum purple
Oak, beech, larch, elm, hickory, and sycamore	Tan or brown
Locust	Stays green until leaves drop
Black walnut and butternut	Drops leaves before they turn color

What is the **tallest tree**?

The tallest tree ever measured was an Australian eucalyptus (*Eucalyptus regnans*) at Watts River, Victoria, Australia, reported in 1872 by forester William Ferguson. It was 435 feet (132 meters) tall and almost certainly measured over 500 feet (150 meters) originally. Another *Eucalyptus regnans* at Mt. Baw Baw, Victoria, Australia, is believed to have measured 470 feet (143 meters) in 1885. The tallest tree currently standing is the "National Geographic Society" coast redwood (*Sequoia sempervirens*) in Redwood National Park, Humboldt County, California. It stood at 373 feet (113 meters) in October, 1990, according to Ron Hildebrant of California. The 365-foot (111-meter)

"Dyerville" giant redwood (*Sequoia giganteum*) in the same park fell during a storm on March 27, 1991.

The tallest known pine tree is a sugar pine (*Pinus lambertiana*) located in California, which measured 215 feet (65.5 meters) in 1967. The sugar pine gets its name from its sweet resin, which naturalist John Muir found preferable to maple sugar. The cones may weigh as much as four pounds (1.8 kilograms) and can reach a length of 26 inches (65 centimeters). The tallest known fir is the 415-foot (125-meter) Douglas fir (*Pseudotsuga menziesii*) in Lynn Valley, British Columbia, measured in 1902.

In the United States the tallest Douglas fir, with a height of 298 feet (90 meters), is located in the Olympic National Park in the state of Washington. In the same park another Douglas fir measures 212 feet (64 meters) high. The tallest living giant sequoia in Sequoia National Park measured 275 feet (83 meters) in 1975.

Other United States trees measuring more than 100 feet (30 meters) in height are given below:

Species	Height Feet	Meters	When and where measured
Ash (Black) (*Fraximus nigra*)	155	47.2	Adrian, Michigan
Beech (American) (*Fagus grandifolia*)	130	39.6	Ashtabula County, Ohio, 1984
Buckeye (Yellow) Tennessee (*Aesculus octandra*)	145	44.2	Great Smoky Mountains National Park,
Cedar (Port-Orford) (*Chaemaecyparis lawsoniana*)	219	66.8	Siskiyou National Forest, Oregon
Cherry (Black) (*Prunus serotina*)	138	42.1	Washtenaw County, Michigan
Elm (American) (*Ulmus americana*)	125	38.1	Southampton County, Virginia, 1985
Eucalyptus (Bluegum) (*Eucalyptus globulus*)	165	50.3	Fort Ross State Park, Sonoma County, California
Fir (Grand)	251	76.5	Olympic National Park, Washington, 1987
Fir (Noble) (*Abies procera*)	238	70.7	Gifford Pinchot National Forest, Washington
Hackleberry (Common)	111	33.8	Rock County, Wisconsin, 1989
Hemlock (Western) (*Tsuga heterophylla*)	241	73.5	Olympic National Park, Washington
Hickory (Pignut) (*Carya glabra*)	190	57.9	Robbinsville, North Carolina

Species	Height		When and where measured
	Feet	Meters	
Honeylocust	115	35.1	Wayne County, Michigan. 1972
Maple (Red) (*Acer rubrum*)	179	54.6	St. Clair County, Michigan
Maple (Black)	118	40	Allegan County, Michigan, 1976
Oak (White) (*Quercus alba*)	107	32.6	Wye Mills State Park, Maryland
Oak (Swamp) Chestnut (*Quercus michauxii*)	200	61	Layette County, Alabama
Pecan (*Carya illinoensis*)	143	43.6	Cocke County, Texas, 1980
Pine (Bishop)	112	34.1	Mendocino County, California, 1986
Pine (Ponderosa) (*Pinus ponderosa*)	223	68	Plumas, California
Pine (Western) (*Pinus monticola*)	157	47.9	El Dorado National Park, California
Pine (Virginia)	120	36.3	Chambers County, Alabama, 1989
Red cedar (Western) (*Juniperus virginiana*)	178	54.3	Forks, Washington
Spruce (Sitka) (*Picea sitchensis*)	206	62.8	Olympic National Forest, Washington
Spruce (Blue) (*Picea pungens*)	126	38.4	Gunnison National Forest, Colorado
Sweetgum (*Athoxanthum odoratum*)	136	41.5	Craven County, North Carolina, 1986
Sycamore (*Platanus occidentalis*)	129	39.3	Jeromesville, Ohio
Willow (Black) (*Salix nigra*)	109	33.2	Grand Traverse County, Michigan
Willow (Weeping) (*Salix tristis*)	114	34.7	Asheville, North Carolina, 1982

Which U.S. city has the greatest **number of trees**?

According to a survey of 20 cities, Houston, Texas, has the most trees with 956,700, while Hartford, Connecticut, has only 20,000.

What is a **banyan tree**?

The banyan tree (*Ficus benghalensis*), a native of tropical Asia, is a member of the *Ficus* or fig genus. It is a magnificent evergreen, sometimes 100 feet (30 meters) in height. As the massive limbs spread horizontally, the tree sends down roots that develop into secondary, pillar-like supporting trunks. Over a period of years a single tree may spread to occupy a tremendous area, as much as 2,000 feet (600 meters) around the periphery.

How can a **spruce tree** be distinguished from a **fir tree**?

The best way to tell the difference between the two trees is by the cones and leaves. On a fir tree the cones are upright (erect) and the needles flat, while spruces have pendulous (hanging) cones and very angular, four-sided needles.

Does the **rose family** produce any trees?

The apple, pear, peach, cherry, plum, mountain ash, and hawthorn trees are members of the rose family (Rosaceae).

What is a **monkey ball tree**?

The osage orange (*Maclura pomifera*) produces large green orange-like fruits. These are roughly spherical, 3½ to 5 inches (8 to 12 centimeters) in diameter, and have a coarse, pebbly surface.

How much of a cut tree is sold as **timber**?

For the average hardwood tree cut down to make lumber, half the total wood volume is left in the woods as tops, limbs, and logging residue; about a quarter is lost as sawdust, slabs, and edgings in the sawmill; one-eighth disappears as shavings and machining residues, leaving about one-eighth of the original volume to be sold as timber.

How is **glycerine** used to preserve leaves?

Glycerine and water are generally used to preserve coarser leaves such as magnolia, rhodendron, beech, holly, heather, or Japanese maple. The leaves should be fresh. For autumn colors, the leaves should be picked just as they are turning color. The preserving solution is made by adding two parts boiling water to one part glycerine. Place leaf stems, split if necessary, in the warm solution so they are covered about 3 to 4 inches (7.5 to 10 centimeters). When drops of glycerine form on the leaves, enough has been absorbed. Wipe off any excess oil. Entire branches may also be treated by using a container large enough to allow the solution to completely cover the leaves. The solution for this method is made with equal quantities of glycerine and water. The leaves are drained on thick sections of newspapers for a few days, then washed with a little soap and water and pegged on a line to dry.

Are there any **natural predators of gypsy moth caterpillars**?

About 45 kinds of birds, squirrels, chipmunks, and white-footed mice eat this serious insect pest. Among the 13 imported natural enemies of the moth, two flies, *Compislura concinnata* (a tachnid fly) and *Sturnia scutellata,* parasitize the caterpillar. Other parasites and various wasps have also been tried as controls, as well as spraying and male sterilization. Originally from Europe, this large moth (*Porthetria dispar*) lays its eggs on the leaves of oaks, birches, maples, and other hardwood trees. When the yellow hairy caterpillars hatch from the eggs, they devour the leaves in such quantities that the tree becomes temporarily defoliated. Sometimes this can even cause the tree to die. The caterpillars grow from ⅛ inch (3 millimeters) to about 2 inches (5 centimeters) before they spin a pupa in which they will metamorphose (change physically) into adult moths.

Who released the **gypsy moth**?

In 1869 Professor Leopold Trouvelot brought gypsy moth egg masses from France to Medford, Massachusetts. His intention was to breed the gypsy moth with the silkworm to overcome a wilt disease of the silkworm. He placed the egg masses on a window ledge, and evidently the wind blew them away. About 10 years later these caterpillars were numerous on trees in that vicinity, and in 20 years, trees in eastern Massachusetts were being defoliated. In 1911, a contaminated plant shipment from Holland also introduced the gypsy moth to that area. These pests have now spread to 25 states, especially in the northeastern United States. Scattered locations in Michigan and Oregon have also reported occurrences of gypsy moth infestations.

FLOWERS AND OTHER PLANTS

See also: Environment—Extinct and Endangered Plants and Animals

Who is known as the father of **botany**?

The ancient Greek thinker, Theophrastus (ca.372–ca.287 B.C.E.), is known as the father of botany. His two large botanical works, *On the History of Plants* and *On the Causes of Plants* were so comprehensive that 1,800 years passed before any new discovery in botany was made. He integrated the practice of agriculture into botany and established a theory of plant growth and the analysis of plant structure. He related plants to their natural environment and identified, classified, and described 550 different plants.

What are the **parts of a flower**?

Sepals—are found on the outside of the bud or on the underside of the open flower. They serve to protect the flower bud from drying out. Some sepals ward off predators by their spines or chemicals.

Petals—serve to attract pollinators and are usually dropped shortly after pollination occurs.

Nectar—contains varying amounts of sugar and proteins that can be secreted by any of the floral organs. It usually collects inside the flower cup near the base of the cup formed by the flower parts.

Stamen—is the male part of a flower and consists of a filament and an anther, where pollen is produced.

Pistil—is the female part, which consists of the stigma, the style, and an ovary containing ovules. After fertilization the ovules mature into seeds.

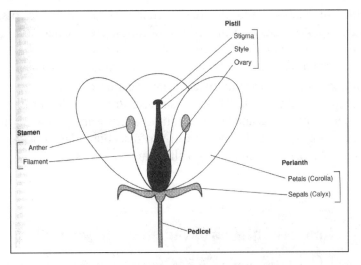

Pistil
Stigma
Style
Ovary

Stamen
Anther
Filament

Perianth
Petals (Corolla)
Sepals (Calyx)

Pedicel

Parts of a flower

What is meant by an **"imperfect" flower**?

An imperfect flower is one that is unisexual, having either stamens (female parts) or pistils (male parts) but not both.

How does a **bulb** differ from a **corm, a tuber,** and a **rhizome**?

Many times the term "bulb" is applied to any underground storage organ in which a plant stores energy for its dormant period. Dormancy is one of nature's devices a plant utilizes to get through difficult weather conditions (winter cold or summer drought). **213**

A true bulb—consists of fleshy scales containing a small basal plate (a modified stem from which the roots emerge) and a shoot. The scales that surround the embryo are modified leaves that contain the nutrients for the bulb during dormancy and early growth. Some bulbs have a tunic (a paper-thin covering) around the scales. The basal plate can also hold the scales together. New bulbs form from the lateral buds on the basal plate. Tulips, daffodils, lilies, and hyacinths are examples of bulb flowers.

A corm—is actually a stem that has been modified into a mass of storage tissue. The eye(s) at the top of the corm is a growing point. The corm is covered by dry leaf bases similar to the tunic-covering of the bulb. Roots grow from the basal plate on the underside of the corm. New corms form on top of or beside the old one. Corm-type flowers include gladiolus, freesia, and crocus.

A tuber—is a solid underground mass of stem like a corm but it lacks both a basal plate and a tunic. Roots and shoots grow from "eyes" (growth buds) out of its sides, bottom, and sometimes its top. Some tubers are roundish. Others are flattened and lumpy. Some examples of tubers are gloxinia, caladium, ranunculuses, and anenome.

A tuberous root—is a swollen root that has taken in moisture and nutrients. It resembles a tuber. New growth occurs on the base of the old stem, where it joins the root. A tuberous root can be divided by cutting off a section with an eye-bearing portion from where the old stem was attached. Dahlias have tuberous roots.

A rhizome or a rootstock—is a thickened, branching storage stem that usually grows laterally along or slightly below the soil surface. Roots develop downward on the bottom surface, while buds and leaves sprout upwards from the top of the rhizome. It is propagated by cutting the parent plant into sections. Japanese, Siberian, and bearded irises; cannas; calla lilies; and trillium are rhizomes.

How are carnivorous plants categorized?

Carnivorous plants, numbering between 450–500 species and 12 genera, have been classified according to the nature of their trapping mechanisms. Active traps display rapid motion in their capture of prey. The Venus fly trap (*Dionaea muscipula*) and the bladderwort (*Utricularia vulgaris*) are active traps. Semi-active traps employ a two-stage trap in which the prey is caught in the trap's adhesive fluid. As it struggles, the plant is triggered to slowly tighten its grip. The sundew (species *Drosera*) and butterwort (*Pinguicula vulgaris*) are semi-active traps. Passive traps entice insects by nectar. The insects fall into a reservoir of fluid and drown. An example of the passive trap is the pitcher plant (5 genera).

What are the **symbolic** meanings of herbs and plants?

Aloe—Healing, protection, affection
Angelica—Inspiration
Arbor vitae—Unchanging friendship
Bachelor's buttons—Single blessedness
Basil—Good wishes, love
Bay—Glory
Carnation—Alas for my poor heart
Chamomile—Patience
Chives—Usefulnessx
Clover, White—Think of me
Coriander—Hidden worth
Cumin—Fidelity
Fennel—Flattery
Fern—Sincerity
Geranium, Oak-leaved—True friendship
Goldenrod—Encouragement
Heliotrope—Eternal love
Holly—Hope
Hollyhock—Ambition
Honeysuckle—Bonds of love
Horehound—Health
Hyssop—Sacrifice, cleanliness
Ivy—Friendship, continuity
Lady's mantle—Comforting
Lavender—Devotion, virtue
Lemon balm—Sympathy
Marjoram—Joy, Happiness
Mints—Eternal refreshment
Morning glory—Affectation
Nasturtium—Patriotism
Oak—Strength
Oregano—Substance
Pansy—Thoughts
Parsley—Festivity
Pine—Humility
Poppy, Red—Consolation
Rose—Love
Rosemary—Remembrance
Rudbeckia—Justice
Rue—Grace, clear vision
Sage—Wisdom, immortality

Salvia, Blue—I think of you
Salvia, Red—Forever mine
Savory—Spice, interest
Sorrel—Affection
Southernwood—Constancy, jest
Sweet-pea—Pleasures
Sweet woodruff—Humility
Tansy—Hostile thoughts
Tarragon—Lasting interest
Thyme—Courage, strength
Valerian—Readiness
Violet—Loyalty, devotion
Violet, Blue—Faithfulness
Violet, Yellow—Rural happiness
Willow—Sadness
Zinnia—Thoughts of absent friends

What special significance does the **passionflower** have?

Spanish friars of the 16th century first gave the name to this flower. They saw in the form of the passionflower (*Passiflora*) a representation of the passion in Christ. The flowers have five petals and five sepals, which symbolize the 10 faithful apostles present at the crucifixion. The corona of five filaments was thought to resemble Christ's crown of thorns. The five stamens represented the five wounds in Christ's body, and the three stigmas stood for the nails driven into his hands and feet. Most species of passionflower are native to the tropical areas of the Western Hemisphere.

What is the **first wildflower to bloom in the spring** in the northern United States?

The first flower of the northern spring is little known and rarely seen, for its habitat is the swamp. Commonly called skunk cabbage (*Spathyema foetidus*), its time of first bloom is February. The first spring flower that is generally known in New England and the Middle West is the *Hepatica*, or liverleaf, which blooms in March or early April.

Is there a **national flower** for the United States?

The national flower of the United States is the rose, adopted on October 7, 1986.

What is the greatest number of leaves a **clover** can have?

A fourteen-leafed white clover (*Trifolium repens*) and a fourteen-leafed red clover (*Trifolium pratense*) have been found in the United States.

Where are **living stones** found?

Various succulent plants native to the stony deserts of South Africa, are called living stones because they mimic their surroundings. Each shoot or plant is made up of two grossly swollen leaves virtually fused together and colored to resemble a pebble. Large daisy-like flowers are borne from between the leaf pair.

What is **wormwood**?

Artemisia absinthium, known as wormwood, is a hardy, spreading, fragrant perennial, 2 to 4 feet (61 to 122 centimeters) tall. It is native to Europe but widely naturalized in North America. The liqueur absinthe is flavored from this plant. Absinthe was formerly used as a bitter tonic.

What is unique about the **water-lily** Victoria amazonica?

It is very big! Found only on the Amazon River, this water-lily has leaves that are up to six feet in diameter. The 12-inch flowers open at dusk on two successive nights.

Why do cats like **catnip**?

Catnip (*Nepeta cataria*) is a hardy perennial herb attractive to cats. Also known as catmint or catnep, it belongs to the mint family. The whole cat family (*Felidae*) reacts to catnip. Mountain lions, lynx, tigers, and lions roll over, rub their face, extend their claws, and twist their bodies when they smell catnip's pungent odor. The scent of catnip probably excites cats because it contains a chemical called trans-neptalactone, which closely resembles an excretion in the dominant female's urine.

GARDENING, FARMING, ETC.

What is the best **soil pH** for growing plants?

Nutrients such as phosphorous, calcium, potassium, and magnesium are most available to plants when the soil pH is between 6.0 and 7.5. Under highly acid (low pH) conditions, these nutrients become insoluble and relatively unavailable for uptake by plants. High soil pH can also decrease the availability of nutrients. If the soil is more alkaline than pH 8, phosphorous, iron, and many trace elements become insoluble and unavailable for plant uptake.

When is the **best time to work the soil?**

Although it is possible to prepare soil at any time of year, fall digging is the best time. Dig in the fall and leave the ground rough. Freezing and thawing during winter breaks up clods and aerates the soil. Insects that otherwise overwinter are mostly turned out. The soil settling during winter will lessen the likelihood of air pockets in the soil when planting the following spring. Fall preparation provides time for soil additives such as manure and compost to break down before planting time.

What is meant by the term **double-digging?**

Double-digging produces an excellent deep planting bed for perennials, especially if the area is composed of heavy clay. It involves removing the top 10 inches of soil and moving it to a holding area, then spading the next 10 inches, and amending this layer with organic matter and/or fertilizer. Then the soil from the "first" digging is replaced after it, too, has been amended.

What is meant by **xeriscaping?**

A xeriscape, a landscape of low water-use plants, is the modern approach to gardening where water shortages are prevalent. Taken from the Greek word *xeros*, meaning dry, this type of gardening utilizes drought-resistant plants and low maintenance grasses, which require water only every 2 to 3 weeks. Drip irrigation, heavy mulching of plant beds, and organic soil improvements are other xeriscape techniques that allow better water absorption and retention, which in turn decrease garden watering time.

What does the term **hydroponics** mean?

This term refers to growing plants in some medium other than soil; inorganic plant nutrients (such as potassium, sulphur, magnesium, and nitrogen) are continuously supplied to the plants in solution. Hydroponics is mostly used in areas where soil is scarce or unsuitable. It is a much-used method of growing plants in research, but for the amateur, it has many limitations and may prove frustrating. Hydroponics was developed in 1959 by two Soviet research scientists at the University of Leningrad.

Which guidelines should be followed **to garden by the moon?**

Guidelines for lunar gardening are simple. The waxing moon occurs between the new moon and full moon, and includes the first and second quarter phases. The waning moon occurs between the full moon and new moon, and includes the third and fourth

quarters. Generally all activities for growth and increase, especially plants that produce above the ground, should take place during the waxing moon. Vegetables and fruits intended to be eaten immediately should be gathered at the waxing moon. Cutting, controlling, and harvesting for food to be conserved or preserved, as well as planting crops that yield below the ground, should take place during the waning moon.

How are **seedlings hardened off** before planting?

Hardening-off is a gardening term for gradually acclimatizing seedlings raised indoors to the outdoor environment. Place the tray of seedlings outdoors for a few hours each day in a semi-protected spot. Lengthen the amount of time they stay out by an hour or so each day; at the end of the week, they will be ready for planting outdoors.

What is the difference between **container-grown, balled-and-burlapped**, and **bare-rooted** plants?

Container-grown plants have been grown in some kind of pot—usually peat, plastic, or clay—for most or all of their lives. Balled-and-burlapped plants have been dug up with the soil carefully maintained around their roots in burlap. Bare-rooted plants have also been dug from their growing place but without retaining the root ball. Typically plants from a mail-order nursery come bare-rooted with their roots protected with damp sphagnum moss. Bare-rooted plants are the most susceptible to damage.

What is meant by the **chilling requirement** for fruit trees?

When a fruit tree's fruiting period has ended, a dormant period must follow during which the plant rests and regains strength for another fruit set the following year. The length of this set is measured in hours of temperatures between 32°F and 45°F (0°F and 7.2°C). A cherry tree requires about 700 hours of chilling time.

What was a **victory garden**?

During World War I, patriots grew "liberty gardens." In World War II, the U.S. Secretary of Agriculture Claude R. Wickard encouraged householders to plant vegetable gardens wherever they could find space. By 1945 there were said to be 20 million victory gardens producing about 40% of all American vegetables in many unused scraps of land. Such sites as the strip between a sidewalk and the street, town squares, and the land around Chicago's Cook County jail were used. The term "victory garden" dates back to Elizabethan England, in a book by that title written in 1603 by Richard Gardner.

What is a recommended size **vegetable garden** for the beginner?

Garden size depends on the amount of space available, how much produce is desired from a garden, and how much work a person is willing to put into the garden. A modest-sized, 10-by-20 foot (3-by-6 meter) plot, laid out in traditional rows, is quite manageable in terms of weeding, cultivating, planting, and harvesting. Even a 10-by-10 foot (3-by-3 meter) plot will suffice for a salad or "kitchen" garden, with plenty of greens and herbs for salads and seasonings on a daily basis. "Intensive" gardening methods, where plants are arranged in blocks rather than rows, allow for increased yield in an even smaller space. One 4-by-4 foot (1-by-1 meter) block, with a vertical frame at one end, can provide salad vegetables for one person throughout the growing season, though two blocks would provide a wider variety of vegetables.

Is there a "best" **time to weed** in the vegetable garden?

Weeding is usually the most unpopular and the most time-consuming garden chore. Some studies (using weeding with peas and beans as examples) have shown that weeding done during the first 3 to 4 weeks of vegetable growth produced the best crops and that unabated weed growth after that time did not significantly reduce the vegetable yields.

Why should **lawn clippings** be left on the grass after mowing?

The clippings are a valuable source of nutrients for the lawn. They provide nitrogen, potassium, and phosphorous to feed the new grass and reduce the need for fertilizer. Young, tender, short clippings decompose quickly. When clippings are left on the lawn instead of being added to the trash collection, the amount of waste added to landfills is decreased.

What are the best annual and perennial **plants** to grow to attract butterflies?

Ageratum, cosmos, globe candytuft, heliotrope, lantana, marigold, mexican sunflower, torch flower, nasturtium, sweet alyssum, and zinnia attract butterflies.

Which **flowers** should be planted in the garden **to attract hummingbirds?**

Scarlet trumpet honeysuckle, weigela, butterfly bush, beardtongue, coralbells, red-hot poker, foxglove, beebalm, nicotiana, petunia, summer phlox, and scarlet sage provide brightly colored (reds and orange shades), nectar-bearing attractants for the hummingbirds.

What prevents **daffodils** from blooming the next year, although they were fertilized?

Try fertilizing daffodils (*Narcissus pseudonarcissus*) as soon as the new shoots appear. This helps the roots renew themselves and will also aid in leaf and flower development. Overcrowding hinders flower production, too. Try digging the bulbs up every third or fifth year, separating them, and respacing them.

How can **geraniums** be carried over from one year to the next?

While they must be kept from freezing, geraniums can survive the winter happily in a cool sunny spot, such as a cool greenhouse, a bay window, or a sunny unheated basement. They need only occasional watering while in this semi-dormant state. Cuttings from these plants can be rooted in late winter or early spring for a new crop of geraniums (some sources suggest rooting in the fall). In homes without a suitable cool and sunny spot, the plants can be forced into dormancy by allowing the soil to dry completely, then gently knocking the soil off of the roots. While the plants can simply be hung in a cool (45° to 50°F; 7° to 10°C), slightly humid room, they will do better if put into individual paper bags, with the openings tied shut. The plants should be checked regularly. The leaves will dry and shrivel, but if the stems shrivel, the plants should be lightly misted with water. If any show mold or rot, cut off the affected sections, move the plants to a drier area, and leave the bags open for a day or two. In the early spring, prune the stems back to healthy green tissue and pot in fresh soil.

What is a **five-in-one-tree**?

These very curious trees consist of a rootstock with five different varieties of the same fruit—usually apples—grafted to it. The blooming period is usually magnificent, with various colors of blooms appearing on the same tree.

What is meant by **espaliering** a fruit tree and why is it done?

To espalier a fruit tree means to train it to grow flat against a surface. It can be grown in small places such as against a wall, and it will thrive even if its roots are underneath sidewalks or driveways. Since many fruit trees must be planted in pairs, espaliered fruit trees can be planted close together, providing pollen for each other, yet taking up little space. **221**

What is **pleaching**?

Pleaching is a method of shearing closely planted trees or shrubs into a high wall of foliage; the wall frequently defines a walkway, or allée. Many kinds of trees can be used—maples, sycamores, lindens, etc. Because of the time needed in caring for pleached allées, they are infrequently seen in American gardens, but are frequently observed in Europe.

What is the secret of **bonsai**—the Japanese art of growing dwarf trees?

These miniature trees with tiny leaves and twisted trunks can be centuries old. To inhibit growth of the plants, they have been carefully deprived of nutrients, pruned of their fastest-growing shoots and buds, and kept in small pots to reduce the root systems. Selective pruning, pinching out terminal buds, and wiring techniques are devices used to control the shape of the trees. Bonsai possibly started during the Chou dynasty (900–250 B.C.E.) in China, when emperors made miniature gardens that were dwarf representations of the provincial lands that they ruled.

What is a **dwarf conifer**?

Conifers are evergreen shrubs and trees with needle-shaped leaves, cones, and resinous wood; examples are pines, spruces, firs, and junipers. After 20 years, dwarf or slow-growing forms of these otherwise tall trees are typically about three feet (90 centimeters) tall.

How are seedless grapes grown?

Seedless grapes are grown from small cuttings of other seedless plants. Because seedless grapes cannot reproduce in the conventional way that grapes usually do (i.e., dropping seeds), growers have to take cuttings from other seedless grape plants, and root them. Although the exact origin of seedless grapes is unknown, they might have first been cultivated in present-day Iran or Afghanistan thousands of years ago. Initially the first seedless grape was a genetic mutation in which the hard seed casing failed to develop—the mutation is called stenospermoscarpy. One modern seedless grape commonly bought today is the green Thompson seedless grape, from which 90% of all raisins are made.

How did the **navel orange** originate?

Every navel orange is derived from a mutant tree that appeared on a plantation in Brazil in the early nineteenth century. A bud from the mutant tree was grafted onto another tree, whose branches were then grafted onto another, and so forth.

How are **chia seeds** grown?

Minuscule black chia seeds are gathered from a type of wild sage that is found in the southwest United States and Mexico. Chia seeds have a very high protein content. Chia seeds cannot be sprouted in the conventional manner because of their highly mucilaginous nature. However, if spread on hollow earthenware vessels that are made especially for the purpose, the seeds will soon produce a green blanket of protein-rich chia sprouts.

How can **fruit trees** be protected from being eaten by field mice?

Valuable trees, especially newly planted fruit trees, can be protected by wrappings or guards of wire, wood veneer, or plastic. Other controls such as pieces of lava rocks soaked in garlic are effective as repellents, and garlic sprays will repel most rodents.

What is the **railroad worm**?

The apple maggot (*Rhagoletis pomonella*), which becomes the apple fruit fly, is frequently called the railroad worm. Inhabiting orchards in eastern United States and Canada, the larvae feed on the fruit pulp of apples, plums, cherries, etc., and cause damage to fruit crops.

Which insects are common problems in **strawberry** cultivation?

Earwigs, slugs, and snails are big problems in some areas. Strawberries (*Fragaria*) are also bothered by Japanese beetles, aphids, thrips, weevils, nematodes, and mites.

Which type of **fence** protects a garden from deer?

A post-and-wire-mesh fence with a sharp-angled, narrow gate, which people and small animals can navigate, but deer cannot, plus the installation of a standard motion-sensing security light with a beeper help keep deer away.

223

How can **squirrels** be kept away from vegetable and flower gardens?

Squirrels like to dig up bulbs, ruin anything colorful in the flower garden; and take a bite out of tomatoes, cucumbers, and melons. The traditional recommendation of spreading mothballs around is not very successful. A new method is to lay down 1- to 2-inch (2.5- to 5-centimeter) mesh sheets of chicken wire. Squirrels will avoid the mesh, apparently because they fear getting their toes stuck in it. Another method is to sprinkle hot pepper around the plants, renewing it after it rains.

How can **poison ivy/oak/sumac** be eliminated from the garden?

Poison oak

First, choose a moist, cool day in early spring or late fall when the plant is less armed with irritant. Wear thick leather or cotton gloves (but not rubber); cut the vines near the base and apply an herbicide to kill the roots. Handle the vines carefully, and dispose of them by burying them in a biodegradable container. *Do not burn* the plants; smoke and ash may cause a rash on exposed parts of the body, eyes, nasal passages, and lungs.

Poison ivy

All parts of poison ivy (*Rhus radicans*) and poison oak (*Rhus toxicodendron*) can cause serious dermatitis. These North American woody plants grow in almost any habitat and are quite similar in appearance. Each has alternating compound leaves of three leaflets each, berrylike fruits, and rusty brown stems. But poison ivy acts more like a vine than a shrub at times and can grow high into trees. Its grey fruit is not hairy, and its leaves are slightly lobed. On the other hand, poison oak is often shrubby, but it can climb. Its leaflets are lobed and resemble oak leaves; its fruit is hairy. Poison sumac (Rhus vernix) grows only in North American wet acid swamps. This shrub can grow as high as 12 feet (3.6 meters). The fruit it produces hangs in a cluster and is grayish-brown in color.

Poison sumac

Poison sumac has sharply-pointed, dark green compound, alternating leaves, and greenish-yellow inconspicuous flowers.

Who developed DDT?

Although DDT was synthesized as early as 1874 by Othmar Zeidler, it was the Swiss chemist Paul Müller (1899–1965) who recognized its insecticidal properties in 1939. He was awarded the 1948 Nobel Prize in Medicine for his development of *d*ichloro-*d*iphenyl-*t*richloro-ethene or DDT. This substance killed insects quickly but seemed to have little or no poisonous effect on plants and animals, unlike the arsenic-based compounds then in use. In the following twenty years it proved to be effective in controlling disease-carrying insects (malaria- and yellow fever-carrying mosquitoes, and typhus-carrying lice), and in killing many plant crop destroyers. Increasingly DDT-resistant insect species and the accumulative hazardous effects of DDT on plant and animal life cycles led to its disuse in many countries during the 1970s.

How can garden soil be used as potting soil?

The garden soil must be pasteurized and then mixed with coarse sand and peat moss. Soil may be pasteurized by putting the soil in a covered baking dish in the oven. When a meat thermometer stuck in the soil has registered 180°F (82°C) for 30 minutes, the soil is done.

Can a branch of the dogwood tree be forced into bloom?

Forcing dogwood (genus *Cornus*) into bloom is similar to forcing forsythia. Bring the dogwood branch indoors when the buds begin to swell. Put the branch in water and set it in a sunny window.

When was the first practical greenhouse built?

French botanist Jules Charles constructed a greenhouse in 1599 in Leiden, Holland, in which tropical plants for medicinal purposes were grown. The most popular plant there was the Indian date, also called the tamarind, whose fruit was made into a curative drink.

Did Johnny Appleseed really plant apple trees?

John Chapman (1774–1845), called Johnny Appleseed, established nurseries of apple seedlings and introduced trees from the Allegheny River to Indiana. Contrary to folklore, he did not prowl the countryside scattering seeds at random from a bag slung over his shoulders.

What were some of the accomplishments of **Dr. George Washington Carver?**

Because of the work of Dr. George Washington Carver (1864–1943) in plant diseases, soil analysis, and crop management, many southern farmers who adopted his methods increased their crop yields and profits. Carver developed recipes using cowpeas, sweet potatoes, and peanuts. He eventually made 118 products from sweet potatoes, 325 from peanuts, and 75 from pecans. He promoted soil diversification and the adoption of peanuts, soybeans, and other soil-enriching crops.

ANIMAL WORLD

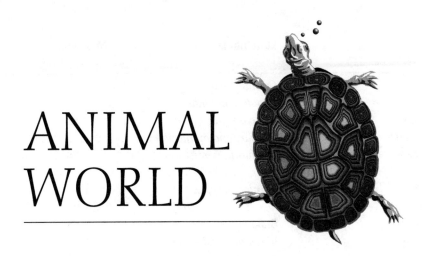

PHYSICAL CHARACTERISTICS, ETC.

See Also: Environment—Extinct and Endangered Plants and Animals

Which animal has the **longest gestation** period?

The animal with the longest gestation period is not a mammal; it is a viviparous amphibian, the Alpine black salamander, which can have a gestation period of up to 38 months at altitudes above 4,600 feet (1,402 meters) in the Swiss Alps; it bears two fully metamorphosed young.

How **long** do animals, in particular mammals, **live**?

Of the mammals, humans and fin whales live the longest. Below is the maximum lifespan for several animal species.

Animal	Latin name	Maximum life span in years
Marion's tortoise	*Testudo sumeirii*	152+
Quahog	*Venus mercenaria*	ca. 150
Common box tortoise	*Terrapene carolina*	138
European pond tortoise	*Emys orbicularis*	120+
Spur-thighed tortoise	*Testudo graeca*	116+
Fin whale	*Balaenoptera physalus*	116
Human	*Homo sapiens*	116
Deep-sea clam	*Tindaria callistiformis*	ca. 100

227

Animal	Latin name	Maximum life span in years
Killer whale	*Orcinus orca*	ca. 90
European eel	*Anguilla anguilla*	88
Lake sturgeon	*Acipenser fulvescens*	82
Freshwater mussel	*Margaritana margaritifera*	80 to 70
Asiatic elephant	*Elephas maximus*	78
Andean condor	*Vultur gryphus*	72+
Whale shark	*Rhiniodon typus*	ca. 70
African elephant	*Loxodonta africana*	ca. 70
Great eagle-owl	*Bubo bubo*	68+
American alligator	*Alligator mississipiensis*	66
Blue macaw	*Ara macao*	64
Ostrich	*Struthio camelus*	62.5
Horse	*Equus caballus*	62
Orangutan	*Pongo pygmaeus*	ca. 59
Bataleur eagle	*Terathopius ecaudatus*	55
Hippopotamus	*Hippopotamus amphibius*	54.5
Chimpanzee	*Pan troglodytes*	51
White pelican	*Pelecanus onocrotalus*	51
Gorilla	*Gorilla gorilla*	50+
Domestic goose	*Anser a. domesticus*	49.75
Grey parrot	*Psittacus erythacus*	49
Indian rhinoceros	*Rhinoceros unicornis*	49
European brown bear	*Ursus arctos arctos*	47
Grey seal	*Halichoerus gryphus*	46+
Blue whale	*Balaenoptera musculus*	ca. 45
Goldfish	*Carassius auratus*	41
Common toad	*Bufo bufo*	40
Roundworm	*Tylenchus polyhyprus*	39
Giraffe	*Giraffa camelopardalis*	36.25
Bactrian camel	*Camelus ferus*	35+
Brazilian tapir	*Tapirus terrestris*	35
Domestic cat	*Felis catus*	34
Canary	*Serinus caneria*	34
American bison	*Bison bison*	33
Bobcat	*Felis rufus*	32.3
Sperm whale	*Physeter macrocephalus*	32+
American manatee	*Trichechus manatus*	30
Red kangaroo	*Macropus rufus*	ca. 30
American buffalo	*Syncerus caffer*	29.5
Domestic dog	*Canis familiaris*	29.5
Lion	*Panthera leo*	ca. 29

Animal	Latin name	Maximum life span in years
African civet	*Viverra civetta*	28
Theraphosid spider	*Mygalomorphae*	ca. 28
Red deer	*Cervus elaphus*	26.75
Tiger	*Panthera tigris*	26.25
Giant panda	*Ailuropoda melanoleuca*	26
American badger	*Taxidea taxus*	26
Common wombat	*Vombatus ursinus*	26
Bottle-nosed dolphin	*Tursiops truncatus*	25
Domestic chicken	*Gallus g. domesticus*	25
Grey squirrel	*Sciurus carolinensis*	23.5
Aardvark	*Orycteropus afer*	23
Domestic duck	*Anas platyrhynchos domesticus*	23
Coyote	*Canis latrans*	21+
Canadian otter	*Lutra canadensis*	21
Domestic goat	*Capra hircus domesticus*	20.75
Queen ant	*Myrmecina graminicola*	18+
Common rabbit	*Oryctolagus cuniculus*	18+
White or beluga whale	*Delphinapterus leucuas*	17.25
Platypus	*Ornithorhynchus anatinus*	17
Walrus	*Odobenus rosmarus*	16.75
Domestic turkey	*Melagris gallapave domesticus*	16
American beaver	*Castor canadensis*	15+
Land snail	*Helix spiriplana*	15
Guinea pig	*Cavia porcellus*	14.8
Hedgehog	*Erinaceus europaeus*	14
Burmeister's armadillo	*Calyptophractus retusus*	12
Capybara	*Hydrochoerus hydrochaeris*	12
Chinchilla	*Chinchilla laniger*	11.3
Giant centipede	*Scolopendra gigantea*	10
Golden hamster	*Mesocricetus auratus*	10
Segmented worm	*Allolobophora longa*	10
Purse-web spider	*Atypus affinis*	9+
Greater Egyptian gerbil	*Gerbillus pyramidum*	8+
Spiny starfish	*Marthasterias glacialis*	7+
Millipede	*Cylindroiulus landinensis*	7
Coypu	*Myocastor coypus*	6+
House mouse	*Mus musculus*	6
Madagascanbrown-tailed mongoose	*Salanoia concolor*	4.75
Cane rat	*Thryonomys swinderianus*	4.3
Siberian flying squirrel	*Pteromys volans*	3.75
Common octopus	*Octopus vulgaris*	2 to 3

Animal	Latin name	Maximum life span in years
Pygmy white-toothed shrew	*Suncus etruscus*	2
Pocket gopher	*Thomomys talpoides*	1.6
Monarch butterfly	*Danaus plexippus*	1.13
Bedbug	*Cimex lectularius*	0.5 or 182 days
Black widow spider	*Latrodectus mactans*	0.27 or 100 days
Common housefly	*Musca domesticus*	0.04 or 17 days

What are the **largest** and **smallest** living animals?

Largest animals	Name	Length and weight
Sea mammal	Blue or sulphur-bottom whale (*Balaenoptera musculus*)	100–110 feet (30.5–33.5 meters) long; weighs 135–209 tons (122.4–189.6 tonnes)
Land mammal	African bush elephant (*Loxodonta africana*)	Bull is 10½ feet (3.2 meters) tall at shoulder; weighs 5.25–6.2 tons (4.8–5.6 tonnes)
Living bird	North African ostrich (*Struthio C. Camelus*)	8 to 9 feet (2.4–2.7 meters) tall; weighs 345 pounds (156.5 kilograms)
Fish	Whale shark (*Rhincodon typus*)	41 feet (12.5 meters) long; weighs 16½ tons (15 tonnes)
Reptile	Salt-water crocodile (*Crocodylus porosus*)	14–16 feet (4.3–4.9 meters) long; weighs 900–1,500 pounds (408–680 kilograms)
Rodent	Capybara (*Hydrochoerus hydrochaeris*)	3¼–4½ feet (1–1.4 meters) long; weighs 250 pounds (113.4 kilograms)

Smallest animals	Name	Length and weight
Sea mammal	Commerson's dolphin (*Cephalorhynchus commersonii*)	Weighs 50–70 pounds (236.7–31.8 kilograms)
Land mammal	Bumblebee or Kitti's hog-nosed bat (*Craseonycteris thong longyai*) or the pygmy shrew (*Suncus erruscus*)	1 inch (2.54 centimeters) long; weighs .062 to .07 ounces (1.6–2 grams) 1½–2 inches (3.8–5 centimeters) long; weighs 0.052–.09 ounces (1.5–2.6 grams)
Bird	Bee hummingbird (*Mellisuga helenea*)	2.25 inches (5.7 centimeters) long; weighs 0.056 ounces (1.6 grams)
Fish	Dwarf pygmy goby (*Trimmatam nanus*)	0.35 inches (8.9 millimeters) long
Reptile	Gecko (*Spaerodactylus parthenopion*)	0.67 inches (1.7 centimeters) long
Rodent	Pygmy mouse (*Baiomys taylori*)	4.3 inches (10.9 centimeters) long; weighs 0.24–0.28 ounces (6.8–7.9 grams)

Besides humans, which animals are then most intelligent?

According to Edward O. Wilson, a behavioral biologist, the ten most intelligent animals are the following:

1. Chimpanzee (two species)
2. Gorilla
3. Orangutan
4. Baboon (seven species, including drill and mandrill)
5. Gibbon (seven species)
6. Monkey (many species, especially macaques, the patas, and the Celebes black ape)
7. Smaller toothed whale (several species, especially killer whale)
8. Dolphin (many of the approximately 80 species)
9. Elephant (two species)
10. Pig

Do animals have color vision?

Most birds appear to have a well-developed color sense. Most mammals are color-blind. Apes and monkeys have the ability to tell colors apart. Dogs and cats seem to be color-blind and only see shades of black, white, and gray.

Which animals spend part of the winter in hibernation?

To avoid the harsh winter conditions, some animals migrate, while others build special nests, hide away in caves and burrows and enter a sleep-like state called hibernation. In hibernation heart rate and respiration slow and body temperature drops. Being so inactive, for several months these animals use little energy and can survive on the fat stores they built up earlier. Hibernators include certain birds, such as nighthawks and swifts; mammals, such as bats, chipmunks, ground squirrels, hamsters, hedgehogs, lemurs, groundhogs, marmots and picket mice; amphibians such as frogs and toads; and reptiles such as lizards, snakes, and turtles. The longest hibernator, with 9 months rest, is the Burrow ground squirrel (*Spermophilus parryl barrowensis*) of Point Barrow, Alaska.

Some animals, like dormice, become active at times and feed on nut and seed caches they stored away in the autumn. They are not true hibernators; nor are bears, skunks, and opossums. Although these animals lapse into prolonged sleep, they wake from time to time and do not undergo the extreme metabolic changes that true hibernators do.

Some animals go underground and become inactive to avoid the harsh summer conditions; this is called aestivation (estivation). For example, some lungfish will burrow into the dried-up mud of river or lake bottoms and wait for water to be replenished. **231**

Can animals regenerate parts of their bodies?

Regeneration does occur in animals; however, this capability progressively declines the more complex the animal species becomes. Among primitive invertebrates (lacking a backbone), regeneration frequently occurs. For example, a planarium (flatworm) can split symmetrically, each part resulting in a clone of the other. In higher invertebrates regeneration occurs in echinoderms (such as starfish) and arthropods (such as insects and crustaceans). Regeneration of appendages (limbs, wings, and antennae) occurs in insects (such as cockroaches, fruitflies, and locusts) and in crustaceans (such as lobsters, crabs, and crayfish). For example, regeneration of the crayfish's missing claw occurs after its next molt (shedding of its hard cuticle exterior shell/skin in order to grow and the subsequent hardening of a new cuticle exterior). However, sometimes the regenerated claw does not achieve the same size of the missing claw. But after every molt (occurring 2 to 3 times a year) the claw grows, and it eventually will be nearly as large as the original claw. On a very limited basis, some amphibians and reptiles can replace a lost leg or tail.

Do animals have **blood types**?

The number of recognized blood groups varies from species to species:

Animal	Number of blood groups
Pig	16
Cow	12
Chicken	11
Horse	9
Sheep	7
Dog	7
Rhesus monkey	6
Mink	5
Rabbit	5
Mouse	4
Rat	4
Cat	2

Do all animals have **red blood**?

The color of blood is related to the compounds that transport oxygen. Hemoglobin, containing iron, is red and is found in all vertebrates (animals having a backbone) and a few invertebrates (animals lacking a backbone). Annelids (segmented worms) have either a green pigment chlorocruorin or a red pigment, hemerythrin, in their blood. Some crustaceans (arthropods having divided bodies and generally having gills) have a blue pigment, hemocyanin, in their blood.

Which animals can **run faster than a human**?

The cheetah, the fastest mammal, can accelerate from zero to 45 miles (64 kilometers) per hour in two seconds; it has been timed at speeds of 70 miles (112 kilometers) per hour over short distances. In most chases, cheetahs average around 40 miles (64.3 kilometers) per hour. Humans can run at almost 28 miles (45 kilometers) per hour maximum. Most of the speeds given in the table below are for distances of ¼ mile (0.4 kilometer).

Animal	Maximum speed	
	Miles per hour	Kilometers per hour
Cheetah	70	112.6
Pronghorn antelope	61	98.1
Wildebeest	50	80.5
Lion	50	80.5
Thomson's gazelle	50	80.5
Quarter horse	47.5	76.4
Elk	45	72.4
Cape hunting dog	45	72.4
Coyote	43	69.2
Gray fox	42	67.6
Hyena	40	64.4
Zebra	40	64.4
Mongolian wild ass	40	64.4
Greyhound	39.4	63.3
Whippet	35.5	57.1
Rabbit (domestic)	35	56.3
Mule deer	35	56.3
Jackal	35	56.3
Reindeer	32	51.3
Giraffe	32	51.3
White-tailed deer	30	48.3
Wart hog	30	48.3
Grizzly bear	30	48.3
Cat (domestic)	30	48.3
Human	27.9	44.9

Do any animals **snore**?

Yes, some dogs definitely snore.

NAMES

What names are used for **male and female animals**?

Animal	Male name	Female name
Alligator	Bull	
Ant		Queen
Ass	Jack, jackass	Jenny
Bear	Boar or he-bear	Sow or she-bear
Bee	Drone	Queen or queen bee
Camel	Bull	Cow
Caribou	Bull, stag, or hart	Cow or doe
Cat	Tom, tomcat, gib, gibeat, boarcat, or ramcat	Tabby, grimalkin, malkin, pussy, or queen
Chicken	Rooster, cock, stag, or chanticleer	Hen, partlet, or biddy
Cougar	Tom or lion	Lioness, she-lion, or pantheress
Coyote	Dog	Bitch
Deer	Buck or stag	Doe
Dog	Dog	Bitch
Duck	Drake or stag	Duck
Fox	Fox, dog-fox, stag, reynard, or renard	Vixen, bitch, or she-fox
Giraffe	Bull	Cow
Goat	Buck, billy, billie, billie-goat, or he-goat	She-goat, nanny, nannie, or nannie-goat
Goose	Gander or stag	Goose or dame
Guinea pig	Boar	
Horse	Stallion, stag, horse, stud, slot, stable horse, sire, or rig	Mare or dam
Impala	Ram	Ewe
Kangaroo	Buck	Doe
Leopard	Leopard	Leopardess
Lion	Lion or tom	Lioness or she-lion
Lobster	Cock	Hen
Manatee	Bull	Cow

Animal	Male name	Female name
Mink	Boar	Sow
Moose	Bull	Cow
Mule	Stallion or jackass	She-ass or mare
Ostrich	Cock	Hen
Otter	Dog	Bitch
Owl		Jenny or howlet
Ox	Ox, beef, steer, or bullock	Cow or beef
Partridge	Cock	Hen
Peacock	Peacock	Peahen
Pigeon	Cock	Hen
Quail	Cock	Hen
Rabbit	Buck	Doe
Reindeer	Buck	Doe
Robin	Cock	
Seal	Bull	Cow
Sheep	Buck, ram, male-sheep, or mutton	Ewe or dam
Skunk	Boar	
Swan	Cob	Pen
Termite	King	Queen
Tiger	Tiger	Tigress
Turkey	Gobbler or tom	Hen
Walrus	Bull	Cow
Whale	Bull	Cow
Woodchuck	He-chuck	She-chuck
Wren		Jenny or jennywren
Zebra	Stallion	Mare

What names are used for **juvenile animals**?

Animal	Name for young
Ant	Antling
Antelope	Calf, fawn, kid, or yearling
Bear	Cub
Beaver	Kit or kitten
Bird	Nestling
Bobcat	Kitten or cub
Buffalo	Calf, yearling, or spike-bull
Camel	Calf or colt
Canary	Chick
Caribou	Calf or fawn
Cat	Kit, kitling, kitty, or pussy

Animal	Name for young
Cattle	Calf, stot, or yearling (m. bullcalf or f. heifer)
Chicken	Chick, chicken, poult, cockerel, or pullet
Chimpanzee	Infant
Cicada	Nymph
Clam	Littleneck
Cod	Codling, scrod, or sprag
Condor	Chick
Cougar	Kitten or cub
Cow	Calf (m. bullcalf; f. heifer)
Coyote	Cub, pup, or puppy
Deer	Fawn
Dog	Whelp
Dove	Pigeon or squab
Duck	Duckling or flapper
Eagle	Eaglet
Eel	Fry or elver
Elephant	Calf
Elk	Calf
Fish	Fry, fingerling, minnow, or spawn
Fly	Grub or maggot
Frog	Polliwog or tadpole
Giraffe	Calf
Goat	Kid
Goose	Gosling
Grouse	Chick, poult, squealer, or cheeper
Horse	Colt, foal, stot, stag, filly, hog-colt, youngster, yearling, or hogget
Kangaroo	Joey
Leopard	Cub
Lion	Shelp, cub, or lionet
Louse	Nit
Mink	Kit or cub
Monkey	Suckling, yearling, or infant
Mosquito	Larva, flapper, wriggler, or wiggler
Muskrat	Kit
Ostrich	Chick
Otte	Pup, kitten, whelp, or cub
Owl	Owlet or howlet
Oyster	Set seed, spat, or brood
Partridge	Cheeper
Pelican	Chick or nestling
Penguin	Fledgling or chick

Animal	Name for young
Pheasant	Chick or poult
Pigeon	Squab, nestling, or squealer
Quail	Cheeper, chick, or squealer
Rabbit	Kitten or bunny
Raccoon	Kit or cub
Reindeer	Fawn
Rhinoceros	Calf
Sea Lion	Pup
Seal	Whelp, pup, cub, or bachelor
Shark	Cub
Sheep	Lamb, lambkin, shearling, or yearling
Skunk	Kitten
Squirrel	Dray
Swan	Cygnet
Swine	Shoat, trotter, pig, or piglet
Termite	Nymph
Tiger	Whelp or cub
Toad	Tadpole
Turkey	Chick or poult
Turtle	Chicken
Walrus	Cub
Weasel	Kit
Whale	Calf
Wolf	Cub or pup
Woodchuck	Kit or cub
Zebra	Colt or foal

What names are used for **groups of animals**?

Animal	Group name
Ants	Nest, army, colony, state, or swarm
Bees	Swarm, cluster, nest, hive, or erst
Caterpillars	Army
Eels	Swarm or bed
Fish	School, shoal, haul, draught, run, or catch
Flies	Business, hatch, grist, swarm, or cloud
Frogs	Arm
Gnats	Swarm, cloud, or horde
Goldfish	Troubling
Grasshoppers	Cloud
Hornets	Nest

Animal	Group name
Jellyfish	Smuck or brood
Lice	Flock
Locusts	Swarm, cloud, or plague
Minnows	Shoal, steam, or swarm
Oysters	Bed
Sardines	Family
Sharks	School or shoal
Snakes	Bed, knot, den, or pit
Termites	Colony, nest, swarm, or brood
Toads	Nest, knot, or knab
Trout	Hover
Turtles	Bale or dole
Wasps	Nest, herd, or pladge

INSECTS, SPIDERS, ETC.

How many **species of insects** are there?

Estimates of the number of recognized insect species range from about 750,000 to upward of one million—but it is thought by some experts that this represents less than half of the number that exists in the world. About 7,000 new insect species are described each year, but unknown numbers are lost annually from the destruction of their habitats, mainly tropical forests.

What is the most **destructive insect** in the world?

The most destructive insect is the desert locust (*Schistocera gregaria*), the locust of the Bible, whose habitat ranges from the dry and semi-arid regions of Africa and the Middle East, through Pakistan and northern India. This short-horn grasshopper can eat its own weight in food a day, and during long migratory flights a large swarm can consume 20,000 tons (18,150,000 kilograms) of grain and vegetation a day.

What are the stages of **insect metamorphoses**?

Metamorphoses, marked structural changes in the growth processes, are either complete or incomplete. In *complete metamorphoses,* the insect (such as ants, moths, butterflies, termites, wasps, and beetles) goes through all the distinct stages of growth to reach adulthood. In *incomplete metamorphoses,* the insect (such as grasshoppers, crickets, and louse) does not go through all the stages of complete metamorphoses.

Complete metamorphoses

Egg—One egg is laid at a time or many (as much as 10,000).

Larva—What hatches from the eggs is called "larva." A larva can look like a worm.

Pupa—After reaching its full growth, the larva hibernates, developing a shell or "pupal case" for protection. A few insects (e.g., the moth) spin a hard covering called a "cocoon." The resting insect is called a pupa (except the butterfly is called a chrysalis), and remains in the hibernation state for several weeks or months.

Adult—During hibernation, the insect develops its adult body parts. When it has matured physically, the fully grown insect emerges from its case or cocoon.

Incomplete metamorphoses

Egg—One egg or many eggs are laid.

Early-stage nymph—Hatched insect resembles an adult, but smaller in size. However, those insects that would normally have wings have not yet developed them.

Late-stage nymph—At this time, the skin begins to molt (shed), and the wings begin to bud.

Adult—The insect is now fully grown.

Has the U.S. selected a **national insect**?

A group of advocates have petitioned the United States Congress to name the monarch butterfly as the national insect, but, to date, they have not been successful.

What is a **lepidopterist**?

A lepidopterist is a person who studies the order of insects called Lepidoptera, a large order including butterflies and moths.

How does a **butterfly** differ from a **moth**?

Characteristic	Butterflies	Moths
Antennae	Knobbed	Unknobbed
Active-time of day	Day	Night
Coloration	Bright	Dull
Resting position	Vertically above body	Horizontally beside body

Note: While these guidelines generally hold true, there are exceptions. In addition, moths have hairy bodies, and most have tiny hooks or bristles linking the fore-wing to the hind-wing; butterflies do not have either characteristic.

Which **butterfly gardens** can the public visit?

A number of gardens cultivate special plants that attract the various species of butterfly. The Cecil B. Day Butterfly Center at Callaway Gardens in Pine Mountain, Georgia, is perhaps the largest. Over 100 species visit the outdoor garden. Other gardens include Butterfly World at Tradewinds Park South in Coconut Creek, Florida; Butterfly World at Marine World Africa U.S.A. in Vallejo, California; the Cincinnati Zoo; and the Des Moines Botanical Gardens, with its tropical butterfly house. Butterfly World in Florida offers two screened aviaries housing two to three thousand butterflies, a collection of preserved specimens, and other displays; Butterfly World in California maintains an enclosed rain-forest environment in a 5,500 square-foot glass greenhouse.

Who discovered the **"dance of the bees"**?

In 1943, Karl von Frisch (1886–1982) published his study on the dance of the bees. It is a precise pattern of movements performed by returning forager (worker) honeybees in order to communicate the direction and distance of a food source to the other workers in the hive. The dance is performed on the vertical surface of the hive and two kinds of dances have been recognized: the round dance (performed when food is nearby) and the waggle dance (done when food is further away). In the waggle dance the bee moves in a figure 8. The angle between the figure 8 and the vertical indicates the direction. Since the publication on von Frisch's work, some doubt has been cast on the interpretation of the dance.

What are **"killer bees"**?

Africanized honeybees—the term entomologists prefer rather than killer bees—are a hybrid originating in Brazil, where African honeybees were imported in 1956. Breeders, hoping to produce a bee better suited to producing more honey in the tropics, found instead that African bees soon hybridized with and mostly displaced the familiar honeybees. Although they produce more honey, the Africanized honeybees also are more dangerous than European bees because they attack intruders in greater numbers. It is expected that the Africanized bees will spread into the United States, but the extent will be limited by cold weather. Some bee experts predict the bees will become less aggressive as they mingle and interbreed with North American bees.

How are **ants** distinguished from **termites**?

Both insect orders—ants (order Hymenoptera) and termites (order Isoptera)—have segmented bodies with multi-jointed legs. Listed below are some differences.

Characteristic	Ant	Termite
Wings	Two pairs with the front pair being much longer than the back pair	Two pairs of equal length
Antenna	Bends at right angle	Straight
Abdomen	Wasp-waist (pinched in)	No wasp-waist

Do male **mosquitoes** bite humans?

No; male mosquitoes live on plant juices, sugary saps, and liquids arising from decomposition. They do not have a biting mouth that can penetrate human skin as female mosquitoes do. In some species, the females need blood to lay their eggs. Blood lipids are converted into the needed iron and protein to produce great numbers (200) of eggs.

What is a "**daddy longlegs**"?

The name applies to two different kinds of invertebrates. The first is a harmless, non-biting long-legged arachnid that preys on small insects, dead insects, and fallen fruit. Also called a harvestman, it is often mistaken for a spider, but it lacks the segmented body shape that a spider has. Although it has the same number of legs (eight) as a spider, the harvestman's legs are far longer and thinner. These very long legs enable it to raise its body high enough to avoid ants or other small enemies. The term "daddy longlegs" also is used for a cranefly—a thin-bodied insect with long thin legs that has a snout-like proboscis with which it sucks water and nectar.

How many **eggs** does a **spider** lay?

The number of eggs varies according to the species. Some larger spiders lay over 2,000 eggs, but many tiny spiders lay one or two and perhaps no more than a dozen during their lifetime. Spiders of average size probably lay a hundred or so. Most spiders lay all their eggs at one time and enclose them in a single egg sac; others lay eggs over a period of time and enclose them in a number of egg sacs.

Which is stronger—steel or silk from a **spider's web**?

The only man-made material that comes close to matching the strength of a spider's silk is steel, but the strongest silk has a tensile strength five times greater than that of steel for the equivalent weight. Tensile strength is the longitudinal stress that a substance can bear without tearing apart.

How long does it take the average spider to weave a complete web?

The average orb-weaver spider takes 30 to 60 minutes to completely spin its web. The order Araneae (spiders) constitutes the largest division in the class Arachnida, containing about 32,000 species. Spiders use silk to capture their food in a variety of ways, ranging from the simple trip wires used by the large bird-eating spiders to the complicated and beautiful webs spun by the orb spiders. Some species produce funnel-shaped webs, and communal webs are built by communities of spiders.

A completed web features several spokes leading from the initial structure. The number and nature of the spokes depend on the species. A spider replaces any damaged threads by gathering up the thread in front of it and producing a new one behind it. The orb web must be replaced every few days because it loses its stickiness (and its ability to entrap food).

The largest aerial webs are spun by the tropical orb weavers of the genus *Nephila*, measuring up to 18 feet 9 inches (5.72 meters) in circumference. The smallest webs are spun by the *Glyphesis cottonae* and they cover about 0.75 square inches (4.84 square centimeters).

What causes the Mexican jumping bean to move?

The bean moth (*Carpocapa saltitans*) lays its eggs in the flowers or in the seed pods of the spurge, a bush known as *Euphorbia sebastiana*. The egg hatches inside the seed pod, producing a larva or caterpillar. The jumping of the bean is caused by the active shifting of weight inside the shell as the caterpillar moves. The jumps of the bean are stimulated by sunshine or by heat from the palm of the hand.

How long have cockroaches been on the Earth?

The earliest cockroach fossils are about 280 million years old. Some cockroaches measured 3 to 4 inches (7.62 to 10.16 centimeters) long. Cockroaches (order Dictyoptera) are nocturnal scavenging insects that eat not only human food but book-bindings, ink, and whitewash as well.

How do fleas jump so far?

The jumping power of fleas comes both from strong leg muscles and from pads of a rubber-like protein called resilin. The resilin is located above the flea's hind legs. To jump, the flea crouches, squeezing the resilin, and then it relaxes certain muscles. Stored energy from the resilin works like a spring, launching the flea. A flea can jump well both vertically and horizontally. Some species can jump 150 times their own length. To match that record, a human would have to spring over the length of 2¼ football fields—or the height of a 100-story building—in a single bound. The common flea (*Pulex irritans*) has been known to jump 13 inches (33 centimeters) in length and 7¼ inches (18.4 centimeters) in height.

How is the light in **fireflies** produced?

The light produced by fireflies (*Photinus pyroles*), or lightning bugs, is a kind of heatless light called bioluminescence caused by a chemical reaction in which the substance, luciferin, undergoes oxidation, when the enzyme luciferase is present. The flash is a photon of visible light that radiates when the oxidating chemicals produce a high-energy state and revert back to their normal state. The flashing is controlled by the insect's nervous system and takes place in special cells called photocytes; tracheal end organs are also involved in controlling the flashing rate. Air temperature also seems to be correlated with the flashing rate. The higher the temperature, the shorter the interval between flashes—8 seconds at 65°F (18°C) and 4 seconds at 82°F (27°C). The rhythmic flashes could be a means of attracting prey or enabling mating fireflies to signal in heliographic codes (that differ from one species to another) or perhaps even serve as a warning signal.

AQUATIC LIFE
See also: Biology—Fungi, Bacteria, Algae, etc.

How is the **age of fish** determined?

One way to determine the age of a fish is by its scales, which have growth rings just as trees do. Scales have concentric bony ridges, called "circuli," which reflect the growth patterns of the individual fish. The portion of the scale that is embedded in the skin contains clusters of these ridges (called "annuli"); each cluster marks one year's growth cycle.

At what **speeds** do fish **swim**?

The maximum swimming speed of a fish is somewhat determined by the shape of its body and its tail and by its internal temperature. The cosmopolitan sailfish (*Istiophorus platypterus*) is considered to be the fastest fish species, at least for short distances, swimming at greater than 60 miles (95 kilometers) per hour. However some American fishermen believe that the bluefin tuna (*Thunnus thynnus*) is the fastest, but the fastest speed recorded so far is 43.4 miles (69.8 kilometers) per hour. Data is extremely difficult to secure because of the practical difficulties in measuring the speeds. The yellowfin tuna (*Thunnus albacares*) and the wahoe (*Acanthocybium solandri*) are also fast, timed at 46.35 miles (74.5 kilometers) per hour and 47.88 miles (77 kilometers) per hour during 10- to 20-second sprints. Flying fish swim at 40+ miles (64+ kilometers) per hour, dolphins at 37 miles (60 kilometers) per hour, trout at 15 miles (24 kilometers) per hour, and blenny at 5 miles (8 kilometers) per hour. Humans can swim 5.19 miles (8.3 kilometers) per hour.

What is krill?

Krill (Euphausiids) is the name applied to planktonic crustaceans that constitute the diet of many whales, particularly baleen whales, and invertebrates such as fish, penguins, and seabirds. Blue whales, largest of the world's animals, live on euphausiids, and a whale's stomach can contain four tons of krill. Euphausiids are all marine. Most of the 85 species feed by filtering plankton from the water. Generally, they resemble primitive decapods such as shrimp and lobsters.

How are **coral reefs** formed?

Coral reefs grow only in warm, shallow water. The calcium carbonate skeletons of dead corals serve as a framework upon which layers of successively younger animals attach themselves. Such accumulations, combined with rising water levels, slowly lead to the formation of reefs that can be hundreds of meters deep and long. The coral animal, or polyp, has a columnar form; its lower end is attached to the hard floor of the reef; the upper end is free to extend into the water. A whole colony consists of thousands of individuals. The two kinds of corals, hard and soft, are distinguished by the type of skeleton secreted. The polyps of hard corals deposit around themselves a solid skeleton of calcium carbonate (chalk). The animal withdraws into a cup-like formation during the daytime; thus, most swimmers see only the skeleton of the coral.

What are **giant tube worms**?

These worms were found in 1977 when the submersible *Alvin* was exploring the ocean floor of the Galapagos Ridge (located 1.5 miles or 2.4 kilometers below the Pacific Ocean surface and 200 miles or 321.8 kilometers from the Galapagos Islands). *Riftia pachyptila Jones,* named after worm expert Meredith Jones of the Smithsonian Museum of Natural History, were discovered near the hydrothermal (hot water) ocean vents. Growing to lengths of five feet (1.5 meters), the worms lack both mouth and gut, and are topped with feathery plumes composed of over 200,000 tiny tentacles. The phenomenal growth of these worms is due to their internal food source—symbiotic bacteria, over 100 billion per ounce of tissue—that live within the worms' troposome tissues. To these troposome tissues, the tube worms transport absorbed oxygen from the water, together with carbon dioxide and hydrogen sulfide. Utilizing this supply, the bacteria living there and in turn produce carbohydrates and proteins that the worms need to thrive.

This was only one of *Alvin*'s discoveries during its historic voyage. Scientists expected to find a "desert" at these ocean depths where no light penetrated. Most of the world's organisms rely on photosynthesis (the use of light to make organic compounds) at the base of their food chains. But in these depths, giant tube worms, vent crabs, and mollusks thrive on chemoautotropic (chemically self-feeding) bacteria, which derive their life-sustaining energy either from the oxidation of substances spewing from the vents, or in symbiotic relationships, such as that with the giant tube worms.

What are the life stages of the **salmon**?

Alevin—A newly hatched translucent salmon with a yolk sac. Less than 1 inch (2.5 centimeters) long, it remains in the redd (gravel nest on the stream bottom) for several weeks until it exhausts the yolk sac.

Fry—One- to two-inch (2.5- to 5-centimeter) salmon that has depleted its yolk sac and has emerged from the redd.

Parr or Fingerling—Young two-inch (5-centimeter) salmon with visible markings known as parr marks. It feeds voraciously on insects and small freshwater crustaceans.

Smolt—Young salmon about two months old that has undergone physiological changes and is ready to migrate to sea. Only 30% of the smolts make it to the ocean.

Grilse—Atlantic salmon that spends only one year at sea before returning to spawn (reproduce).

Salmon—The mature stage of the fish when it returns to the river to spawn (reproduce), usually in the fall. It stores up oils and fats to sustain it on its upstream trip.

Kelt—Atlantic salmon that has spawned, and returned to the ocean. Pacific salmon die after spawning.

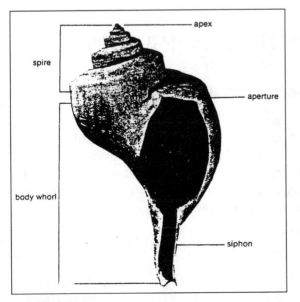

Whelk shell parts

What are the parts of a **whelk** shell?

Belonging to the family of Gastopods (meaning stomach foot), a whelk is one of the 70 thousand kinds of living mollusks (from the Latin *mollis*, meaning soft bodied) that produce shells of calcium carbonate. Its body has a prominent foot with a covering called a mantle. This mantle will eventually line the shell or exoskeleton the whelk builds.

The glands of the mantle secrete calcium carbonate (lime), which is formed into the protective shell. The calcium carbonate comes from the food or from the water that the whelk ingests. An active predator, feeding on other gastropods, worms, fish, and decaying vegetable matter, a whelk usually lives from two to five years, but some live for decades.

How much electricity does an electric eel generate?

An electric eel (*Electrophorus electricus*) has current-producing organs made up of electric plates on both sides of its vertebral column running almost its entire body length. The electrical charge—on the average of 350 volts, but as great as 550 volts—is released by the central nervous system. The shock consists of four to eight separate charges, which last only two- to three-thousandths of a second each. These shocks, used as a defense mechanism, can be repeated up to 150 times per hour without any visible fatigue to the eel. The most powerful electric eel, found in the rivers of Brazil, Colombia, Venezuela, and Peru, produces a shock of 400 to 650 volts.

REPTILES AND AMPHIBIANS

What is the difference between a **reptile** and an **amphibian**?

Reptiles are clad in scales, shields, or plates and their toes have claws. Young reptiles are miniature replicas of their parents in general appearance if not always in coloration and pattern. Reptiles include alligators, crocodiles, turtles, and snakes. Amphibians have moist, glandular skins, and their toes lack claws. Their young pass through a larval, usually aquatic stage before they metamorphose (change in form and structure) into the adult form. Amphibians include salamanders, toads, and frogs.

How fast can a **crocodile** move?

For short distances a crocodile can raise itself up like a lizard and charge like a galloping horse. A crocodile on land can move at speeds of 15 to 30 miles (24 to 40 kilometers) per hour, and in the water can move at the rate of 20 miles (32 kilometers) per hour.

How is the gender of baby **alligators** established?

The gender of an alligator is established by the temperature at which its egg is incubated. High temperatures of 90° to 93°F (32° to 34°C) result in males; low temperatures of 82° to 86°F (28° to 30°C) yield females. Gender is fixed during the second and third week of the egg's two-month incubation. Further temperature fluctuations before or after this time do not alter the gender of the young. The heat from decaying matter on top of the nest incubates the eggs.

What are a **turtle's** upper and lower **shell** called?

The turtle (order Testudines) uses its shell as a protective device. The upper shell is called the *dorsal carapace* and the lower shell is called the *ventral plastron*. The shell's sections are referred to as the *scutes*. The carapace and the plastron are joined at the sides.

Which **poisonous snakes** are native to the United States?

Copperheads, rattlesnakes, water moccasins (cottonmouths), and coral snakes are all poisonous. Copperheads (*Agkistrodon contortex*) tend to inhabit areas in the southeastern United States, Texas, and the lower Ohio Valley. Rattlesnakes (genus Crotalus) range throughout the United States, and cottonmouths (*Agkistrodon piscivorus*) range from Texas across to North Carolina. Coral snakes (*Microrus fulvius*), the most venomous snakes in the United States, tend to inhabit the southeastern states and Texas.

BIRDS

What names are used for **groups of birds**?

A group of birds in general is called a congregation, flight, flock, volery, or volley.

Bird	Group name
Bitterns	Siege or sedge
Budgerigars	Chatter
Chickens	Flock, run, brood, or clutch
Coots	Fleet or pod
Cormorants	Flight
Cranes	Herd or siege
Crows	Murder, clan, or hover
Curlews	Herd

247

Bird	Group name
Doves	Flight, flock, or dole
Ducks	Paddling, bed, brace, flock, flight, or raft
Eagles	Convocation
Geese	Gaggle or plump (on water), flock (on land), skein (in flight), or covert
Goldfinches	Charm, chattering, chirp, or drum
Grouses	Pack or brood
Gulls	Colony
Hawks	Cast
Hens	Brood or flock
Herons	Siege, sege, scattering, or sedge
Jays	Band
Larks	Exaltation, flight, or ascension
Magpies	Tiding or tittering
Mallards	Flush, sord, or sute
Nightingales	Watch
Partridges	Covey
Peacocks	Muster, ostentation, or pride
Penguins	Colony
Pheasants	Nye, brood, or nide
Pigeons	Flock or flight
Plovers	Stand, congregation, flock, or flight
Quails	Covey or bevy
Sparrows	Host
Starlings	Chattering or murmuration
Stork	Mustering
Swallows	Flight
Swans	Herd, team, bank, wedge, or bevy
Teals	Spring
Turkeys	Rafter
Turtle doves	Dule
Woodpeckers	Descent
Wrens	Herd

Which birds lay the **largest** and **smallest eggs?**

The elephant bird (*Aepyornis maximus*), an extinct flightless bird of Madagascar, also known as the giant bird or roc, laid the largest known bird eggs. Some of these eggs measured as much as 13.5 inches (34 centimeters) in length and 9.5 inches (24 centimeters) in diameter. The largest egg produced by any living bird is that of the North African ostrich (*Struthio camelus*). The average size is 6 to 8 inches (15 to 20 centimeters) in length and 4 to 6 inches (5 to 15 centimeters) in diameter.

The smallest mature egg, measuring less than 0.39 inches (1 centimeter) in length, is that of the vervain hummingbird (*Mellisuga minima*) of Jamaica.

Why do birds **migrate** annually?

Hormonal changes caused by natural breeding cycles trigger migration patterns. The major wintering areas for North American migrating birds are the southern United States and Central America. Migrating ducks follow four major flyways south: the Atlantic flyway, the Mississippi flyway, the central flyway, and the Pacific flyway. Some bird experts propose that the birds return north to breed for several reasons: (1) Birds return to nest, because there is a huge insect supply for their young. (2) The higher the Earth's latitude in the summer in the Northern Hemisphere, the longer the daylight available to the parents to find food for their young. (3) Less competition exists for food and nesting sites in the north. (4) In the north, there are fewer mammal predators for nesting birds (which are particularly vulnerable during the nesting stage). (5) Birds migrate south to escape the cold weather, so they return north when the weather improves.

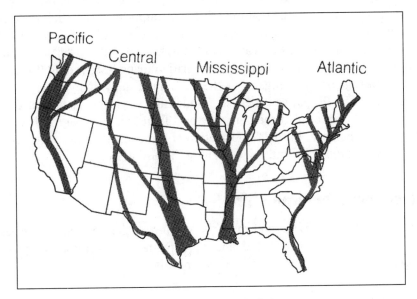

Flyways of North American ducks

When is **Buzzard Day** in Hinckley, Ohio?

Since 1957, Buzzard Day has been celebrated on the first Sunday after March 15. The buzzards are actually turkey vultures that end their yearly migration north in Hinckley, where they spend their summer. March 15 is usually the day the first "scouting" birds arrive.

When do the **swallows** come back to Capistrano in California?

The swallows arrive on March 19th and depart on October 23rd, when they migrate to the Southern Hemisphere.

Do all **birds fly**?

Among flightless birds, the penguins and the ratites are the best known. Ratites includes emus, kiwis, ostriches, rheas, and cassowaries. They are called ratite, because they lack a keel on the breastbone. All of these birds have wings but lost their power to fly millions of years ago. Many birds that live isolated on oceanic islands (for example, the great auk) apparently became flightless in the absence of predators and the consequent gradual disuse of their wings for escape.

Why do **geese** fly in formation?

Aerodynamicists have suspected that long-distance migratory birds, such as geese and swans, adapt the "V" flight formation in order to reduce the amount of energy needed for traveling. According to theoretical calculations, birds flying in a "V" formation can fly some 10% farther than a lone bird can. Formation flying lessens drag (the air pressure that pushes against the wings). The effect is similar to flying in a thermal upcurrent, where less total lift power is needed. In addition, when flying, each bird creates behind it a small area of disturbed air. Any bird flying directly behind it would be caught in this turbulence. In the "V" formation of the Canadian geese, each bird flies not directly behind the other, but aside or above the bird in front.

How **fast** does a **hummingbird** fly?

Hummingbirds fly at speeds up to 71 miles (80 kilometers) per hour. Small species beat their wings 50 to 80 times per second, and more frequently in courtship displays. For comparison, the following table lists some other birds:

Bird	Speed	
	Miles per hour	Kilometers per hour
Peregrine falcon	168–217	270.3–349.1
Swift	105.6	169.9
Merganser	65	104.6
Golden plover	50–70	80.5–112.6
Mallard	40.6	65.3
Wandering albatross	33.6	54.1
Carrion crow	31.3	50.4

Bird	Speed Miles per hour	Kilometers per hour
Herring gull	22.3–24.6	35.9–39.6
House sparrow	17.9–31.3	28.8–50.4
Woodcock	5	8

Do birds have **knees**?

The bird's knee is the joint between the thigh and the "drumstick" and is usually concealed by feathers in living birds. Many species have a kneecap or patella.

What are the natural **predators** of the **penguin**?

The leopard seal is the principal predator of both the adult and juvenile king penguin. The penguin may also be caught by a killer whale while swimming in open water. Eggs and chicks that are not properly guarded by adults are often devoured by skuas and sheathbills.

When was the **bald eagle** adopted as the national bird of the United States?

On June 20, 1782, the citizens of the newly independent United States of America adopted the bald or "American" eagle as their national emblem. At first the heraldic artists depicted a bird that could have been a member of any of the larger species, but by 1902, the bird portrayed on the seal of the United States of America had assumed its proper white plumage on head and tail. The choice of the bald eagle was not unanimous; Benjamin Franklin (1706–1790) preferred the wild turkey.

How does a **homing pigeon** find its way home?

Scientists suggest two hypotheses to explain the homing flight of pigeons. Neither has been proved to the satisfaction of all the experts. The first hypothesis involves an "odor map." The theory proposes that young pigeons learn this map by smelling different odors that reach their home in the winds from varying directions. They would, for example, learn that a certain odor is carried on winds blowing from the east. If a pigeon were transported eastward, the odor would tell it to fly westward to return home. The second hypothesis proposes that a bird may be able to extract its home's latitude and longitude from the Earth's magnetic field. It may prove in the future that neither theory explains the pigeon's navigational abilities or that some synthesis of the two theories is plausible.

What is the name of the bird that perches on the black rhinoceros' back?

The bird, a relative of the starling, is called an oxpecker (a member of the Sturnidae family). Found only in Africa, the yellow-billed oxpecker (*Buphagus africanus*) is widespread over much of western and central Africa, while the red-billed oxpecker (*Buphagus erythrorhynchus*) lives in eastern Africa from the Red Sea to Natal.

Seven to eight inches (17 to 20 centimeters) long with a coffee-brown body, the oxpecker feeds on more than 20 species of ticks that live in the hide of the black rhinoceros (*Diceros bicornis*), also called the hook-lipped rhino. The bird spends most of its time on the rhinoceros or on other animals, such as the antelope, zebra, giraffe, buffalo, etc. The bird has even been known to roost on the body of its host.

The relationship between the oxpecker and the rhinoceros is a type of symbiosis (a close association between two organisms in which at least one of them benefits) called mutualism. The rhinoceros' relief of its ticks and the bird's feeding clearly demonstrates mutualism (a condition in which both organisms benefit). In addition, the oxpecker, having much better eyesight than the nearsighted rhinoceros, alerts its host with its shrill cries and flight when danger approaches.

What can an **orphaned wild bird** eat?

An orphaned songbird needs to be fed every 20 minutes during daylight hours for several weeks. The food should be placed deep in its throat. A soft-billed bird (such as a warbler or catbird) may be given grated carrots, chopped hard-boiled eggs, cottage cheese, fresh fruit, or custard. A young hard-billed bird (such as a sparrow or finch) may be given the same food; but rape, millet, and sunflower seeds should be added to this diet when the bird becomes well-developed. A mixture of dry baby cereal and the yolk of a hard-boiled egg moistened with milk can also be given.

Is it best to stop **feeding hummingbirds** after Labor Day?

There is no evidence that feeding hummingbirds after Labor Day will prevent them from migrating to avoid the fatal effects of winter weather. In fact, it may help a weak-

ened straggler "refuel" for the long migration to warmer climates. Nectar feeders can be left out until the birds have migrated.

What is the proper size for the hole in a bird house for the **purple martin**?

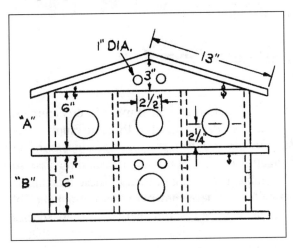

1" DIA.

13"

3"

2½"

2¼"

6"

"A"

"B"   6"

Purple martin birdhouse

The opening should be 2½ inches (6 centimeters) in diameter. Native Americans were the first to attract martins to artificial nesting sites by hanging dried gourds from trees. In the southeastern United States today, this method remains popular. As with all martin houses, the nesting box or gourd should be mounted on a pole at least 15 feet (4.5 meters) from any buildings or overhanging branches. Purple martins avoid nests that are vulnerable to climbing predators. They like to be near water. Purple martins are highly social; they readily accept multiple housing units. In the eastern United States these units are popular structures to attract colonies of birds. Martins feed on insects that humans consider pests.

How can **bluebirds** be encouraged to nest in a particular location?

Bluebirds may be attracted by providing nesting boxes and perches, cultivating an area of low or sparse vegetation, and planting nearby trees, vines, or shrubs such as blueberries, honeysuckle, and crabapples. Bluebirds prefer open countryside with low undergrowth. Parks, golf courses, and open lawns are their preferred habitats. In the last 40 years eastern bluebird populations have declined 90%, coinciding with the disappearance of farmland, widespread use of pesticides, and an increase in nest competitors (house sparrows and European starlings). Artificial nesting boxes sometimes provide more secure nesting places than do natural nest sites, because artificial structures can be built to resist predators. Bluebird boxes can be made with entrance holes (1½ inches or 3.8 centimeters in diameter) small enough to exclude starlings and can incorporate special raccoon guards on mounting poles. Mounted 3 to 6 feet (1 to 2 meters) above ground (to discourage predators), the nesting boxes should be no closer than 100 feet (30 meters). The box should be located within 50 feet (15 meters) of a tree so that the fledglings can perch. The box should have a 4-inch by 4-inch floor (10.6 by 10.6 centimeters), 8 to 12 inches (20.3 to 30.5 centimeters) in height, with the 1½ inch hole 6 to 10 inches (15.3 to 25.4 centimeters) above the floor.

MAMMALS

Which mammal has the shortest **gestation period**? Which one has the longest?

Gestation is the period of time between fertilization and birth in oviparous animals. The shortest gestation period known is 12 to 13 days, shared by three marsupials: the American or Virginian opossum (*Didelphis marsupialis*); the rare water opossum, or yapok (*Chironectes minimus*) of central and northern South America; and the eastern native cat (*Dasyurus viverrinus*) of Australia. The young of each of these marsupials are born while still immature, and complete their development in the ventral pouch of their mother. While 12 to 13 days is the average, the gestation period for these species is sometimes as short as 8 days. The longest gestation period for a mammal is that of the Asiatic elephant (*Elephas maximus*) with an average of 609 days, and a maximum of 760 days.

Mammal	Average gestation period	
	Days	**Months**
Armadillo	60 to 120	2 to 4
Ass	about 340	about 11½
Baboon	150 to 183	5 to 6
Bear, grizzly	210 to 265	7
Bear, polar	about 240	about 8
Bear, American black	210 to 215	7
Beaver	about 105	3½
Boar, wild	115	3¼
Bobcat	60 to 63	2
Capybara	150	5
Cat (domestic)	52	2
Cattle	about 238	9¼
Cheetah	91 to 95	3
Chimpanzee	230 to 240	7½
Dog	53 to 71	2
Dolphin	305 to 365	10 to 12
Dormouse	21 to 32	
Ermine	about 28	

| Mammal | Average gestation period | |
	Days	Months
Fox, red	60 to 63	2
Gerbil	21 to 28	
Gibbon	210 to 240	7 to 8
Giraffe	453 to 464	15¼
Goat	150	5
Gopher	12 to 20	
Guinea pig	63	2
Hamster	from 15 to 37	
Hippopotamus	about 240	8
Horse	340	about 11½
Kangaroo	183 to 325	6 to 11 (in pouch)
Koala	34 to 36	1¼
Leopard	90 to 105	3¼
Lion	100 to 119	3½
Llama	348 to 368	11½ to 12¼
Mouse	about 20 to 30	
Ocelot	70	2¼
Opossum, American	12 to 13	
Panda, giant	125 to 150	4½
Pig	101 to 129	3¼
Porpoise	183	6
Rabbit	30	
Raccoon	63	2
Reindeer	210 to 240	7 to 8
Rhinoceros, black		15
Sheep	135 to 160	4½ to 5
Shrew	13 to 24	
Skunk, striped	62 to 66	2
Squirrel	about 40	1⅓
Tiger	103	3½
Walrus		15 to 16 including 4 to 5 months of delayed implantation
Weasel	35 to 45	1½
Whale	305 to 365	10 to 12
Whale, beluga		14 to 15
Whale, sperm		14 to 15
Wolf	61 to 63	2
Zebra	340	about 11½

What names are used for **groups** or **companies** of **mammals**?

Mammal	Group name
Antelopes	Herd
Apes	Shrewdness
Asses	Pace, drove, or herd
Baboons	Troop
Bears	Sloth
Beavers	Family or colony
Boars	Sounder
Buffaloes	Troop, herd, or gang
Camels	Flock, train, or caravan
Caribou	Herd
Cattle	Drove or herd
Deer	Herd or leash
Elephants	Herd
Elks	Gang or herd
Foxes	Cloud, skulk, or troop
Giraffes	Herd, corps, or troop
Goats	Flock, trip, herd, or tribe
Gorillas	Band
Horses	Haras, stable, remuda, stud, herd, string, field, set, team, or stable
Jackrabbits	Husk
Kangaroos	Troop, mob, or herd
Leopards	Leap
Lions	Pride, troop, flock, sawt, or souse
Mice	Nest
Monkeys	Troop or cartload
Moose	Herd
Mules	Barren or span
Oxen	Team, yoke, drove, or herd
Porpoises	School, crowd, herd, shoal, or gam
Reindeer	Herd
Rhinoceri	Crash
Seals	Pod, herd, trip, rookery, or harem
Sheep	Flock, hirsel, drove, trip, or pack
Squirrels	Dray
Swine	Sounder, drift, herd, or trip
Walruses	Pod or herd
Weasels	Pack, colony, gam, herd, pod, or school
Whales	School, gam, mob, pod, or herd
Wolves	Rout, route, or pack
Zebras	Herd

How does the **breath-holding capability** of a human compare with other mammals?

Mammal	Average time in minutes
Human	1
Polar bear	1.5
Pearl diver (human)	2.5
Sea otter	5
Platypus	10
Muskrat	12
Hippopotamus	15
Sea cow	16
Beaver	20
Porpoise	15
Seal	15 to 28
Greenland whale	60
Sperm whale	90
Bottlenose whale	120

How does a human's **heartbeat's** rate compare with those of other mammals?

Mammal	Heart beats per minute
Human	70
Elephant	25
Mouse	600–700
Large dog	80
Small dog	120

Do any **mammals fly?**

Bats (order Chiroptera with 986 species) are the only truly flying mammals, although several gliding mammals are referred to as "flying" (such as flying squirrel and flying lemur). The "wings" of bats are double membranes of skin stretching from the sides of the body to the hind legs and tail, and are actually skin extensions of the back and belly. The wing membranes are supported by the elongated fingers of the forelimbs (or arms). Nocturnal (active at night), ranging in length from 1.5 inches to 1.33 feet (25 to 406 millimeters), living in caves or crevices, bats inhabit most of the temperate and tropical regions of both hemispheres. The majority of species feed on inserts and fruit, and some tropical species eat pollen and nectar of flowers and insects found inside them. Moderate-sized species usually prey on small mammals, birds, lizards, and frogs, and some eat fish. But true vampire bats (three species) eat the blood of animals by making an incision in the animal's skin; from these bats, animals can contract rabies. Most bats do not find their way around by sight, but have evolved a sonar system, called "echolocation," for locating solid objects. Bats emit vocal sounds through the nose or mouth while flying. These sounds, usually above the human hearing range, are reflected back as echoes. This process enables bats, when flying in darkness, to avoid solid objects and to locate the position of flying insects. Bats have the most acute sense of hearing of any land animal, detecting frequencies as high as 120 to 210 kilohertz. The highest frequency humans can hear is 20 kilohertz.

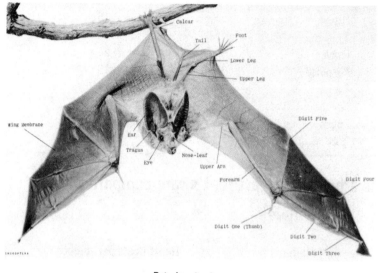

Bat wing structure

What are some animals that have **pouches**?

Marsupials (meaning "pouched" animals) differ from all other living mammals in their anatomical and physiological features of reproduction. Most female marsupials—kangaroos, bandicoots, wombats, banded anteaters, koalas, opossums, wallabies, tasmanian devils, etc.—possess an abdominal pouch (called a marsupium), in which their young are carried. However in some small terrestrial marsupials the marsupium is not a true pouch but merely a fold of skin around the mammae (milk nipples).

The short gestation period in marsupials (in comparison to other similarly sized mammals) allows their young to be born in an "undeveloped" state. Consequently, these animals have been viewed as "primitive" or second-class mammals. However, some now see that the reproductive process of marsupials poses an advantage over that of placental mammals. A female marsupial invests relatively few resources during the brief gestation period, more so during the lactation (nursing period) when the young are in the marsupium. If the female marsupial loses its young, it can sooner make a second attempt than a placental mammal in a comparable situation.

How long do **wombats** live and what do they eat?

Native to Australia and Tasmania, the common wombat or coarse-haired wombat (*Vombatus ursinus*) lives between 5 and 26 years (26 years in zoos). It dines mostly on grasses, roots, mushrooms, fresh shoots, and herbaceous plants. A wombat looks like a small bear in appearance, with a thick heavy body ranging from 2.3 to 4 feet (0.7 to 1.2 meters) and weighing 33 to 77 pounds (15 to 35 kilograms). Its rough fur ranges from yellowish buff, to gray, to dark brown or black. This marsupial resembles a rodent in its manner of feeding and in its tooth structure—all its teeth are rootless and ever growing to compensate for their wear. Shy, the wombat lives in a burrow and is an active digger.

Which **mammals lay eggs** and **suckle** their young?

The duck-billed platypus (*Ornithorhynchus anatinus*) and the echidna or spiny anteater (family Tachyglossidae), indigenous to Australia, Tasmania, and New Guinea, are the only two species of mammals that lay eggs (a non-mammalian feature) but suckle their young (a mammalian feature). These mammals (order Monotremata) resemble reptiles in that they lay rubbery shell-covered eggs that are incubated and hatched outside the mother's body. In addition they resemble reptiles in their digestive, reproductive, and excretory systems, and in a number of anatomical details (eye structure, presence of certain skull bones, pectoral [shoulder] girdle, and rib and vertebral structures). However they are classed as mammals because they have fur and a four-chambered heart, nurse their young from gland milk, are warm-blooded, and have some mammalian skeletal features.

How deep do **marine mammals** dive?

Below are listed the maximum depths and the longest durations of time underwater by various aquatic mammals:

Mammal	Maximum depth		Maximum time underwater
	Feet	Meters	
Weddell seal	1,968	600	70 minutes
Porpoise	984	300	6 minutes
Bottle-nosed whale	1,476	450	120 minutes
Fin whale	1,148	350	20 minutes
Sperm whale	>6,562	>2,000	75–90 minutes

How do the **great whales** compare in weight and length?

Whale	Average weight		Greatest length	
	Tons	**Kilograms**	**Feet**	**Meters**
Sperm	35	31,752	59	18
Blue	84	76,204	98.4	30
Finback	50	45,360	82	25
Humpback	33	29,937	49.2	15
Right	50 (est.)	45,360 (est.)	55.7	17
Sei	17	15,422	49.2	15
Gray	20	18,144	39.3	12
Bowhead	50	45,360	59	18
Bryde's	17	15,422	49.2	15
Minke	10	9,072	29.5	9

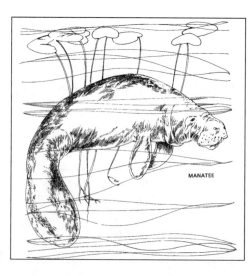

MANATEE

What is the name of the **seal-like animal in Florida**?

In winter the West Indian manatee (*Trichechus manatus*) migrates to temperate parts of Florida, such as the warm headwaters of the Crystal and Homosassa rivers in central Florida or the tropical waters of southern Florida. When the air temperature rises to 50°F (10°C), it will wander back along the Gulf coast and up the Atlantic coast as far as Virginia. Long-range offshore migrations to the coast of Guyana and South America have

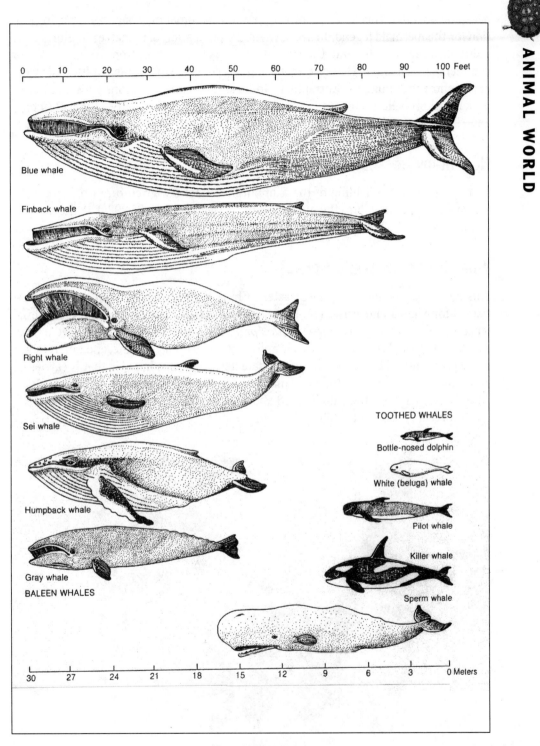

0   10   20   30   40   50   60   70   80   90   100 Feet

Blue whale

Finback whale

Right whale

Sei whale

TOOTHED WHALES

Bottle-nosed dolphin

White (beluga) whale

Humpback whale

Pilot whale

Killer whale

Gray whale

BALEEN WHALES

Sperm whale

30   27   24   21   18   15   12   9   6   3   0 Meters

Baleen and toothed whales

been documented. This large, plant-eating water mammal may have been the inspiration for the mermaid legend. In 1893, when the population of manatees in Florida was reduced to several thousand, the state gave it legal protection from being hunted or commercially exploited. However, many animals continue to be killed or injured by the encroachment of humans. Entrapment in locks and dams, collisions with barges and power boat propellers, etc., cause at least 30% of the manatee deaths (125 to 130 total annual number of deaths).

How many **horses** are there in the world?

According to Dr. D. Fielding of the Edinburgh School of Agriculture (in Edinburgh, Scotland, United Kingdom) the number of horses worldwide is 65,292,000, the number of donkeys is 41,599,000, and the number of mules is 15,462,000.

What is **Przewalski's horse**?

This native of Mongolia and northeastern China is the last truly wild horse species. Named for Nikolai Przewalski (1839–1888), the Russian colonel who reported its existence in 1870, Przewalski's horse (*Equus przwalskii*) is a stocky, short-legged animal with a dun-colored coat, a pale muzzle and belly, and dark legs, mane, and tail. Its short mane is bristly and erect. One unusual characteristic of this horse is that it carries 66 chromosomes rather then the customary 64 found in a domestic horse. Przewalski's horse was last seen in the wild in 1968 and is believed to be extinct; 1,000 or so specimens survive in zoos and wildlife parks.

Przewalski's horse

How many letters are permissible in the name of a **thoroughbred horse**?

Before the name of a thoroughbred can become "official" it must be submitted to the Jockey Club for approval. One of their requirements limits the name to not more than three pronounceable words with a maximum of eighteen letters.

What is the difference between an **African elephant** and an **Indian elephant**?

The African species (*Loxodonta africana*) of this family Elephantidae has a larger body, ears, and tusks, a concave rather than convex back, usually three toenails instead of four on its hind feet, and two finger-like lips rather than one at the tip of its trunk. The ear tops of the Asian elephant (*Elephas maximus*) turn forward whereas those of the African elephant turn backward.

Is there a **cat** that lives in the **desert**?

The sand cat (*Felis margarita*) is the only member of the cat family tied directly to desert regions. Found in North Africa, the Arabian peninsula, and the deserts of Turkmenistan, Uzbekistan, and western Pakistan, the sand cat has adapted to extremely arid desert areas. The padding on the soles of its feet is well-suited to the loose sandy soil, and it can live without drinking "free" water. Having sandy colored or grayish-ochre dense fur, grows to a body length of 17.6 to 22.2 inches (45 to 57 centimeters). Mainly nocturnal (active at night) the cat feeds on rodents, hares, birds, and reptiles.

The Chinese desert cat (*Felis bieti*) does not live in the desert, as its name implies, but inhabits the steppe country and mountains. Likewise the Asiatic desert cat *(Felis silvestris ornata)* inhabits the open plains of India, Pakistan, Iran, and Asiatic Russia.

Which **bear** lives in a **tropical rain forest**?

The Malayan sun bear (*Ursus malayanus*) is one of the rarest animals in the tropical forests of Sumatra, Malay Peninsula, Borneo, Burma, Thailand, and southern China. The smallest bear, with a length of 3.3 to 4.6 feet (1 to 1.4 meters) and weighing 60 to 143 pounds (27 to 65 kilograms), it has a strong, stocky body. With powerful paws and long, curved claws, it is an expert tree climber. The sun bear tears at tree bark to expose insects, larvae, and the nests of bees and termites. Fruit, coconut palm, and small rodents, too, are part of its diet. Sleeping and sunbathing during the day, it is active at night. Unusually shy and retiring, cautious and intelligent, it is declining in population as the forests are being destroyed.

Do **camels** store water in their humps?

A camel's hump or humps do not store water, since they are fat reservoirs. The camel's ability to go long periods without drinking water, up to ten months if there is plenty of green vegetation and dew to feed on, results from a number of physiological adaptations. One major factor is that camels can lose up to 40% of their body weight with no ill effects. A camel can also withstand a variation of its body temperature by as much as 14°F (-10°C). A camel can drink 30 gallons of water in ten minutes and up to 50 gallons over several hours. A one-humped camel is called a dromedary or Arabian camel; a Bactrian camel has two humps and lives in the wild on the Gobi desert. Today, the Bactrian is confined to Asia, but most of the Arabian camels are on African soil.

Why do **nine-banded armadillos** always have four offspring of the same gender?

The one feature that distinguishes the nine-banded armadillo (*Dasypus novemcinctus*) is that the female almost always gives birth to four young of the same sex. This consistency results from the division of the one fertilized egg into four parts to produce quadruplets.

How many quills does a **porcupine** have?

For its defensive weapon, the average porcupine relies on its 30,000 quills, or specialized hairs, comparable in hardness and flexibility to slivers of celluloid and so sharply pointed that they can penetrate any hide. The quills that do the most damage are the short ones that stud the porcupine's muscular tail. With a few lashes, the porcupine can send a rain of quills that pierce tiny scale-like barbs into the skin of its adversary. The quills work their way inward because of their barbs and the involuntary muscular action of the victim. Sometimes the quills can work themselves out, but other times the quills penetrate vital organs, and the victim dies.

Slow-footed and stocky, porcupines spend much of their time in trees, using their formidable incisors to strip off bark and foliage for their food; they supplement their diet with fruits, grasses, etc. Porcupines have a ravenous appetite for salt; as herbivores (plant-eating animals) they consume a diet that provides insufficient salt. So natural salt licks, animal bones left by carnivores (meat-eating animals), yellow pond lilies, and other items having a high salt content (including paints, plywood adhesives, and sweated-on clothing, etc., of humans) carry a strong appeal for porcupines.

What is a **capybara**?

The capybara (*Hydrochaerus hydrochaeris*) is the largest of all living rodents. Also called the water hog, water pig, water cary, or carpincho, it looks like a huge guinea pig. Its body length can reach 3¼ to 4½ feet (1 to 1.3 meters), and it usually weighs between 120 to 130 pounds (54 to 59 kilograms) or more. A native of northern South America, this rodent leads a semi-aquatic life, feeding on aquatic plants and grasses. A subspecies, native to Panama, is smaller and weighs between 60 to 75 pounds (27 to 34 kilograms).

What is **chamois**?

The chamois (of the family Bovidae) is a goat-like animal living in the mountainous areas of Spain, central Europe (the Alps and Apennines), south central Europe, the Balkans, Asia Minor, and the Caucasus. Agile and surefooted, with acute senses, it can jump 6.5 feet (2 meters) in height and 20 feet (6 meters) in distance, and run at speeds of 31 miles (50 kilometers) per hour. Its skin has been made into "shammy" leather for cleaning glass and polishing automobiles; although more commonly today the shammy or chamois skins sold are simply specially treated sheepskin.

How long does a **gray wolf** live?

The gray wolf (*Canis lupus*) also known as the timber wolf, is the largest and most widespread species of the family Canidae. It can live in the wild for less than 10 years, but under human care, up to 20 years. In many areas it has been hunted and killed, because it is believed to pose a threat to humans and domesticated animals (cattle, sheep, and reindeer). It may soon suffer the fate of the red wolf (*Canis rufus*) that once flourished in the southeastern and south central United States. Two of the red wolf's three subspecies are extinct, and the third is found only in a small area of the coastal plains and marshes of eastern Texas and Western Louisiana. The endangered gray wolf is declining faster in the New World than in the Old. In the United States, it is limited to Alaska (10,000), Northern Minnesota (1,200), and Isle Royale, Michigan (20), with perhaps a few packs in Wisconsin, northern Michigan, and the Rocky Mountain area; in Canada, they number 15,000. Strongly social, living in packs, and weighing between 75–175 pounds (43–80 kilograms), in physical appearance the gray wolf resembles a large domestic dog, such as the Alaskan malamute.

What is the chemical composition of a **skunk's spray**?

The chief odorous components of the spray have been identified as crotyl mercaptan, isopentyl mercaptan, and methyl crotyl disulfide in the ratio of 4:4:3. The liquid is an oily, pale-yellow, foul-smelling spray that can cause severe eye inflammation. This defensive weapon is discharged—either as a fine spray or a short stream of rain-sized drops—from two tiny nipples located just inside the skunk's anus. Although the liquid's range is 1.83 to 2.75 yards (2 to 3 meters), its smell can be detected 1.55 miles (2.5 kilometers) downwind.

HUMAN BODY

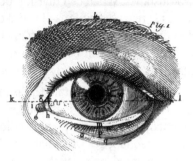

FUNCTIONS, PROCESSES, AND CHARACTERISTICS

Which **chemicals** constitute the human body?

About 24 elements are used by the body in its functions and processes.

Major elements		
Element	**Percentage**	**Function**
Oxygen	65.0	Part of all major nutrients of tissues; vital to energy production
Carbon	18.5	Essential life element of proteins, carbohydrates, and fats; building blocks of cells
Hydrogen	9.5	Part of major nutrients; building blocks of cells
Nitrogen	3.3	Essential part of proteins, DNA, RNA; essential to most body functions
Calcium	1.5	Form nonliving bone parts; a messenger between cells
Phosphorous	1.0	Important to bone building; essential to cell energy

Potassium, sulfur, sodium, chlorine, and magnesium each occur at 0.35% or less. The human body also contains traces of iron, cobalt, copper, manganese, iodine, zinc, fluorine, boron, aluminum, molybdenum, silicon, chromium, and selenium.

How many **chromosomes** are in a human body cell?

A human being normally has 46 chromosomes (23 pairs) in all but the sex cells. One of each pair is inherited from the mother's egg; the other, from the father's sperm. When the sperm and egg unite in fertilization, they create a single cell, or zygote, with 46 chromosomes. When cell division occurs, the 46 chromosomes are duplicated and the process is repeated billions of times over, with each of the cells containing the identical set of chromosomes. Only the gametes, or sex cells, are different. In their cell division, the members of each pair of chromosomes are separated and distributed to different cells. Each gamete has only 23 chromosomes.

Chromosomes contain thousands of genes, which carry information for a specific trait. That information is in the form of a chemical code, and the chemical compound that encodes this genetic information is deoxyribonucleic acid or DNA. A gene can be seen as a sequence of DNA that is coded for a specific protein. These proteins determine specific physical traits (such as height, body shape, color of hair, eyes, skin, etc.), body chemistry (blood type, metabolic functions, etc.), and some aspects of behavior and intelligence. More than 150 human disorders are inherited, and genes are thought to determine susceptibility to many diseases.

How many cells are in the human body?

Sources give figures that vary from 50 trillion to 75 trillion cells.

What is the average lifespan of cells in the human body?

Cell type	Length of time
Blood cells	
red blood cells	120 days
lymphocytes	Over 1 year
other white cells	10 hours
platelets	10 days
Bone cells	25–30 years
Brain cells	Lifetime
Colon cells	3–4 days
Skin cells	19–34 days
Spermatozoa	2–3 days
Stomach cells	2 days

What is the antigen-antibody reaction?

The immune system operates with two main components: white blood cells and antibodies, both circulating in the blood. The antigen-antibody reaction forms the basis for immunity. When an antigen (*antibody generator*), such as a harmful bacterium, virus, fungus, parasite, or other foreign substance invades the body, a specific antibody is generated to attack the antigen. The antibody is produced by beta-lymphocytes (B-cells) in the spleen or lymph nodes. An antibody may either destroy the antigen directly, or it may "label" it so that a white blood cell (called a microphage, or scavenger cell) can engulf the foreign intruder. After a human has been exposed to an antigen, a later exposure to the same antigen will produce a faster immune system reaction. The necessary antibodies will be produced more rapidly and in larger amounts. Artificial immunization uses this antigen-antibody reaction to protect the human body from certain diseases, by exposing the body to a safe dose of antigen to produce effective antibodies as well as a "readiness" for any future attacks of the harmful antigen.

How do **T cells** differ from **B lymphocytes**?

T cells, responsible for dealing with most viruses, some bacteria and fungi, and for cancer surveillance, are one of the two main classes of lymphocytes. As one variety of white blood cells, lymphocytes are part of the body's immune system; the immune system fights invading organisms that have penetrated the body's general defenses. T lymphocytes, or T cells, compose about 80% of the lymphocytes circulating in the blood. They have been "educated" in the thymus to perform particular functions. Killer T cells are sensitized to multiply when they come into contact with antigens (foreign proteins) on abnormal body cells (cells that have been invaded by viruses, cells in transplanted tissue, or tumor cells). These killer T cells attach themselves to the abnormal cells and release chemicals (lymphokines) to destroy them. Helper T cells assist killer cells in their activities and control other aspects of the immune response. When B lymphocytes, which compose about 10% of total lymphocytes, contact the antigens on abnormal cells, the lymphocytes enlarge and divide to become plasma cells. Then the plasma cells secrete vast numbers of immunoglobulins or antibodies into the blood, which attach themselves to the surfaces of the abnormal cells, to begin a process that will lead to the destruction of the invaders.

What are **endorphins**?

They are chemical substances produced in the brain. They act as opiates, in that they interfere with the sensation of pain by binding to opiate receptor sites involved in pain perception. This action increases the threshold for pain.

Is it true that people **need less sleep** as they get older?

As a person ages, the time spent in sleeping changes. The following table shows how long a night's sleep generally lasts.

Age	Sleep time (in hours)
1–15 days	16–22
6–23 months	13
3–9 years	11

269

Age	Sleep time (in hours)
10–13 years	10
14–18 years	9
19–30 years	8
31–45 years	7.5
45–50 years	6
50+ years	5.5

What is **REM sleep**?

REM sleep is *rapid eye movement* sleep. It is characterized by faster breathing and heart rates than NREM (*nonrapid eye movement*) sleep. The eyes move rapidly, and dreaming, often with elaborate story lines, occurs. The only people who do not experience REM sleep are those who have been blind from birth. REM sleep occurs in about 4 to 5 periods during a typical night, varying from 5 minutes to about an hour, and growing progressively longer as sleep continues.

Scientists do not understand why dreaming is important, but they think the brain is either cataloging the information it picked up during the day and throwing out the data it does not want, or is creating scenarios to work through situations causing emotional distress. Regardless of the reasons, most people who are deprived of sleep or dreams become disoriented, unable to concentrate, and may even have hallucinations.

Why do people **snore**?

Snoring is produced by vibrations of the soft palate, usually caused by any condition that hinders breathing through the nose. It is more common when a person is sleeping on the back.

How **loud** can a **snore** sound?

Research has indicated that a snore can reach 69 decibels, as compared with 70 to 90 decibels for a pneumatic drill.

How many **calories** does a person burn while **sleeping**?

A 150-pound (68-kilogram) person burns 1.0 calories per minute during bed rest. Approximate caloric values of other activities for a person weighing 150 pounds are given below. Values may vary, depending on the vigor of the exercise, air temperature, clothing, etc.

Activity	Calories used per hour
Basketball	500
Bicycling (5½ mph)	210
(13 mph)	660
Bowling	220–270
Calisthenics	300
Digging	360–420
Gardening	200
Golfing (using power cart)	150–220
(pulling cart)	240–300
(carrying clubs)	300–360
Football	500
Handball (social)	600–660
(competitive)	>660
Hoeing	300–360
Housework	180
Jogging (5–10 mph)	500–800
Lawn mowing (power)	250
(hand)	420–480
Raking leaves	300–360
Sitting	100
Skiing (cross–country)	600–660
(downhill)	570
Snow shoveling	420–480
Square dancing	350
Standing	140
Swimming moderately	500–700
Tennis (doubles)	300–360
(singles)	420–480
Vacuuming	240–300
Volleyball	350
Walking (2 mph)	150–240
(3.5 mph)	240–300
(4 mph)	300–400
(5 mph)	420–480

How many **calories** does the average college student expend in a day?

The average college student weighing 115 pounds (52 kilograms) expends 2,093 calories a day. The total is derived from the total number of calories used per pound (listed below) multiplied by the weight of the student.

Activity	Hours spent in activity	Calories per pound per hour	Total calories used per pound
Asleep	8	0.4	3.2
Lying still, awake	1	0.5	0.5
Dressing and undressing	1	0.9	0.9
Studying in class, eating, studying, talking	8	0.7	5.6
Walking	1	1.5	1.5
Standing	1	0.8	0.8
Driving a car	1	1.0	1.0
Running	0.05	4.0	2.0
Playing ping–pong	0.05	2.7	1.3
Writing	2	0.7	1.4
TOTAL	24	———	18.2

Who were the doctor and patient involved in the first studies on digestion performed by direct observation of the patient's stomach?

Alexis St. Martin, a French Canadian, was accidentally wounded by a shotgun blast in 1822. Fortunately, Dr. William Beaumont (1785–1853), an Army surgeon, was nearby and began treatment of the wound immediately. St. Martin's recuperation lasted nearly three years, and the enormous wound healed except for a small opening leading into his stomach. A fold of flesh covered this opening; when this was pushed aside the interior of the stomach was exposed to view. Dr. Beaumont began a series of experiments and observations that formed the basis of our modern knowledge of digestion. Today, the use of x-rays and other medical instruments provides the same diagnostic function.

When a person **swallows** solid or liquid food, what prevents it from going down the windpipe?

Once food is chewed, voluntary muscles move it to the throat. In the pharynx (throat), automatic involuntary reflexes take over. The epiglottis closes over the larynx (voice box), which leads to the windpipe. A sphincter at the top of the esophagus relaxes, allowing the food to enter the digestive tract.

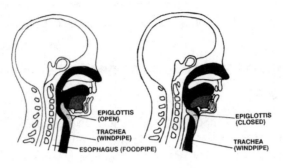

EPIGLOTTIS
(OPEN)

TRACHEA
(WINDPIPE)

ESOPHAGUS (FOODPIPE)

EPIGLOTTIS
(CLOSED)

TRACHEA
(WINDPIPE)

Epiglottis opened and closed

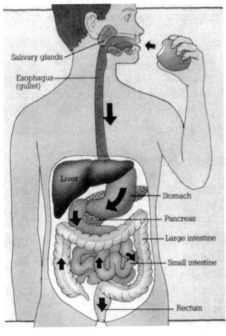

Salivary glands

Esophagus
(gullet)

Liver

Stomach

Pancreas

Large intestine

Small intestine

Rectum

Digestion process

How long does it take food to digest?

The stomach holds a little under two quarts (1.9 liters) of semi-digested food that stays in the stomach three to five hours. The stomach releases food slowly to the rest of the digestive tract. Fifteen hours or more after the first bite started down the alimentary canal, the final residue of the food is passed along to the rectum and is excreted through the anus as feces.

Who is called the father of physiology?

As an experimenter, Claude Bernard (1813–1878) enriched physiology by his introduction of numerous new concepts into the field. The most famous of these concepts is that of the *milieu intérieur* or internal environment. The complex functions of the various organs are closely interrelated and are all directed to maintaining the constancy of internal conditions despite external changes. All cells exist in this aqueous (blood and lymph) internal environment, which bathes the cells and provides a medium for the elementary exchange of nutrients and waste material.

273

Who coined the term **homeostasis**?

Walter Bradford Cannon (1871–1945), who elaborated on Claude Bernard's (1813–1878) concept of the *milieu intérieur* (interior environment), used the term *homeostasis* to describe the body's ability to maintain a relative constancy in its internal environment.

How much **heat** is **lost through the head** when a person is not wearing a hat?

Between 7% and 55% of total body heat can be lost through the head. The amount of blood going to the head is controlled by cardiac output and the harder the body works, the more blood is circulated to the head, where heat from the blood is quickly radiated away.

In the United States, what is the **average height and weight** for a man and a woman?

The average female is 5 feet 3¾ inches (1.62 meters) tall and weighs 135 pounds (61.24 kilograms). The average male is 5 feet 9 inches (1.75 meters) tall and weighs 162 pounds (73.48 kilograms). Between 1960 and 1990 the average American male became two inches (5.08 centimeters) taller and 27 pounds (12.25 kilograms) heavier, while the average American woman grew two inches (5.08 centimeters) taller and gained one pound (0.45 kilograms).

What are **desirable weights** for men and women?

Below is listed the desirable weight ranges for adult men (21 years and older) and women (25 years and older). Height is measured with shoes on: women with two–inch heels and men with one–inch heels. Weight is for clothed figures. For nude weight, subtract two to four pounds for women and five to seven pounds for men. Women 18 to 25 years old can use the table by subtracting one pound for each year under the age of 25.

Men			
Height Feet Inches	Small frame	Medium frame	Large frame
5 2	112–120	118–129	126–141
5 3	115–123	121–133	129–144
5 4	118–126	124–136	132–148
5 5	121–129	127–139	135–152
5 6	124–133	130–143	138–156
5 7	128–137	134–147	142–161
5 8	132–141	138–152	147–166

Height Feet Inches		Small frame	Medium frame	Large frame
5	9	136–145	142–156	151–170
5	10	140–150	146–160	155–174
5	11	144–154	150–165	159–179
6	0	148–158	154–170	164–184
6	1	152–162	158–175	168–189
6	2	156–167	162–180	173–194
6	3	160–171	167–185	178–199
6	4	164–175	172–190	182–204

Women

Height Feet Inches		Small frame	Medium frame	Large frame
4	10	92–98	96–107	104–119
4	11	94–101	98–110	106–122
5	0	96–104	101–113	109–125
5	1	99–107	104–116	112–128
5	2	102–110	107–119	115–131
5	3	105–113	110–122	118–134
5	4	108–116	113–126	121–138
5	5	111–119	116–130	125–142
5	6	114–123	120–135	129–146
5	7	118–127	124–139	133–150
5	8	122–131	128–143	137–154
5	9	126–135	132–147	141–158
5	10	130–140	136–151	145–163
5	11	134–144	140–155	149–168
6	0	138–148	144–159	153–173

Considering that 64% of all Americans are overweight, who is the **heaviest person** that ever lived?

John Brower Minnoch (1941–1983) of Bainbridge Island, Washington, weighed 976 pounds (443 kilograms) in 1976 and was estimated to have weighed more than 1,387 pounds (630 kilograms) when he was rushed to the hospital in 1978 with heart and respiratory failure. Much of his weight was due to fluid retention. After two years on a hospital diet, he was discharged at 476 pounds (216 kilograms). He had to be readmitted, however, after reportedly gaining 197 pounds (87 kilograms) in seven days. In 1983, when he died, he weighed 798 pounds (362 kilograms).

The heaviest woman ever recorded was Rosie Carnemolla (b. 1944) of Poughkeepsie, New York, who weighed 850 pounds (386 kilograms). When Mrs. Percy Pearl Washington, who suffered from polydipsia (excessive thirst), died in a Milwaukee hospital in 1972, the scales only registered 800 pounds (363 kilograms) maximum, but she was credited with weighing 880 pounds (400 kilograms).

What are the types of human **body shapes**?

The best known example of body typing (classifying body shape in terms of physiological functioning, behavior, and disease resistance), was devised by American psychologist William Herbert Sheldon (1898–1977). Sheldon's system, known as somatotyping, distinguishes three types of body shapes, ignoring overall size: endomorph, mesomorph, and ectomorph. The extreme endomorph tends to be spherical: a round head, a large fat abdomen, weak penguin-like arms and legs, with heavy upper arms and thighs but slender wrists and ankles. The extreme mesomorph is characterized by a massive cubical head, broad shoulders and chest, and heavy muscular arms and legs. The extreme ectomorph has a thin face, receding chin, high forehead, a thin narrow chest and abdomen, and spindly arms and legs. Sheldon's system includes mixed body types, determined by component ratings. Sheldon assumed a close relationship between body build and behavior and temperament. This system of body typing has many critics.

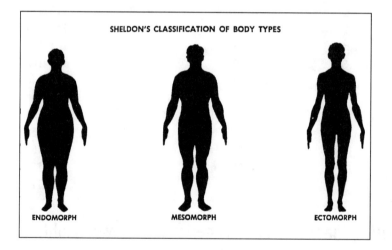

SHELDON'S CLASSIFICATION OF BODY TYPES

ENDOMORPH   MESOMORPH   ECTOMORPH

Who were the original **Siamese twins**?

Chang and Eng (1811–1874) were the original Siamese twins. Although born in Siam (now Thailand), they were actually three-quarters Chinese.

What causes **protruding ears**?

If a fold paralleling a part of the outer roll of the ear is sparse, but not absent, the ear will protrude. This trait usually runs in certain families.

In addition to left- or right-handedness, what other left–right preferences do people have?

Most people have a preferred eye, ear, and foot. In one study, for example, 46% were strongly right-footed, while 3.9% were strongly left-footed; furthermore 72% were strongly right-handed and 5.3% strongly left–handed. However estimates vary about the proportion of left-handers to right–handers, but it may be as high as one in ten. Some 90% of healthy adults use the right hand for writing; two-thirds favor the right hand for most activities requiring coordination and skill. There are no male–female differences in these proportions.

Who is the world's oldest person?

The greatest authenticated age to which any human has lived is 120 years 237 days in the case of Shigechiyo Izumi (1871–1986) of Japan.

How much of the body remains after cremation?

The solids left after cremation are about 3% of the former body weight, so a corpse weighing 150 pounds (68 kilograms) produces about five pounds (2 kilograms), or about 200 cubic inches, of remains.

What is cryonic suspension?

Cryonic suspension, the controversial process of freezing and storing bodies for later revival, has been practiced since the late 1960s. Ordinarily, people are placed in cryonic suspension only after they are pronounced legally dead. Suspended animation, as the long-term storage of humans is generally known, has been a topic of scientific speculation since the beginning of the modern era.

BONES, MUSCLES, AND NERVES

How many bones are in the human body?

Babies are born with about 300 to 350 bones, but many of these fuse together between birth and maturity to produce an average adult total of 206. Bone counts vary according to the method used to count them, because some systems treat as multiple bones a structure that other systems treat as a single bone with multiple parts.

Location	Number
Skull	22
Ears (pair)	6
Vertebrae	26
Sternum	3
Throat	1
Pectoral girdle	4
Arms (pair)	60
Hip bones	2
Legs (pair)	58
Total	206

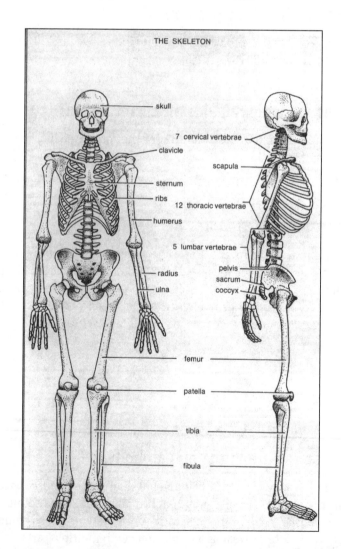

THE SKELETON

skull

7 cervical vertebrae

clavicle

scapula

sternum

ribs

12 thoracic vertebrae

humerus

5 lumbar vertebrae

pelvis
sacrum
coccyx

radius

ulna

femur

patella

tibia

fibula

What is the most commonly **broken bone**?

The clavicle (collar bone) is one of the most frequently fractured bones in the body. Fractured clavicles are caused by either a direct blow or a transmitted force resulting from a fall on the outstretched arm.

Why do humans have **wisdom teeth**?

Although they are useless to humans now, they must have had some purpose in the past. Nature rarely produces extraneous features. Perhaps when primitive man ate tough meats, the extra molars in the mouth, now known as wisdom teeth, helped to chew up the meat fibers. As humans evolved, the brain became larger and the face position moved further downward and inward. Also the protruding jawbones of early humans gradually moved backward, leaving no room for these third molars or wisdom teeth.

How many muscles are in the human body?

There are about 656 muscles in the body, although some authorities make this figure even as high as 850 muscles. No exact figure is available because authorities disagree about which are separate muscles and which ones slip off larger ones. Also, there is a wide variability from one person to another, though the general plan remains the same.

Muscles are used in three body systems: the skeletal muscles move various parts of the body, are striped or striated fibers and are called voluntary muscles because the person controls their use; smooth muscles, found in the stomach and intestinal walls, vein and artery walls, and in various internal organs, are called involuntary muscles, and are not generally controlled by the person; cardiac muscles, or the heart muscles, contain striped and involuntary muscles.

Why does excessive exercise cause **muscles** to become **stiff and sore**?

If muscles are worked too hard, the cells run out of oxygen. This starts a fermentation process that produces lactic acid. The build-up of lactic acid in the muscles causes soreness and stiffness.

What are the **hamstring muscles**?

There are three hamstring muscles, located at the back of the thigh. They flex the leg on the thigh during activities like kneeling.

279

What is an ecorche?

An ecorche is a flayed figure, a three-dimensional representation of the human body, usually made of plaster, with the envelope of skin and fat removed. Its intent is to depict the surface muscles with precise anatomical correctness.

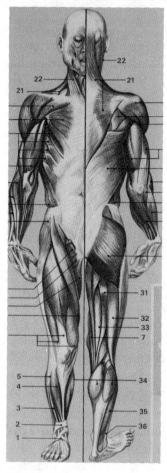

Muscles of the body (ecorche)

How many muscles does it take to produce a smile and a frown?

There are 17 smiling muscles. The average frown needs 43.

How much force does a human bite generate?

All the jaw muscles working together can close the teeth with a force as great as 55 pounds (25 kilograms) on the incisors or 200 pounds (90.7 kilograms) on the molars. A force as great as 268 pounds (122 kilograms) for molars has been reported.

What is the largest nerve in the body?

The sciatic nerve is the largest in the human body—about as thick as a lead pencil—0.78 inches (1.98 centimeters). It is a broad, flat nerve composed of nerve fibers, and it runs from the spinal cord down the back of the each leg.

ORGANS AND GLANDS

What is the largest organ in the human body?

The largest and heaviest human organ is the skin, with a total surface area of about 20 square feet (1.9 square meters) for an average person or 25 square feet (2.3 square meters) for a large person; the skin weighs 5.6 pounds (2.7 kilograms) on the average. Although generally it is not thought of as an organ, skin fits the medical definition of

an organ. An organ is a collection of various tissues integrated into a distinct structural unit and performing specific functions. After the skin, the second largest and heaviest organ is the liver, weighing 2.5 to 3.3 pounds (1.1 to 1.5 kilograms). The liver is the body's chemical factory and regulates the levels of most of the body's chemicals.

What is the basic unit of the **brain**?

Neurons are the nerve cells that are the major constituent of the brain. At birth the brain has the maximum number of neurons—20 billion to 200 billion neurons. Thousands are lost daily, never to be replaced and apparently not missed, until the cumulative loss builds up in very old age.

How hard does the **heart** work?

The heart squeezes out about 2½ ounces (70.8 grams) of blood at every beat. It daily pumps at least 2,500 gallons (9,450 liters) of blood, which weighs 20 tons (18,144 kilograms). On the average the adult heart beats 70 to 75 times a minute. The rate of the heartbeat is determined in part by the size of the organism. Generally the smaller the size, the faster the heartbeat. Thus women's hearts beat 6 to 8 times per minute more than men's hearts do. At birth the heart of a baby can beat as fast as 130 times per minute.

Are the **lungs** identical?

No, the right lung is shorter than the left by 1 inch (2.5 centimeters); however, its total capacity is greater. The right lung has three lobes, the left lung has two. The maximum capacity of the lung averages about 6,500 cubic centimeters or about 1.7 gallons (6.4 liters).

What was the likely **purpose** of the human **appendix**?

Experts can only theorize on its use. It may have had the same purpose it does in present-day herbivores, where it harbors colonies of bacteria that help in the digestion of cellulose in plant material. Another theory suggests that tonsils and the appendix might manufacture the antibody-producing white blood cells called B-lymphocytes. The third theory is that the appendix may "attract" body infections to localize the infection in one spot that is not critical to body functioning. The earliest surgical removal of the appendix was by Claudries Amyand (1680–1740) in England in 1736.

Which **gland** is the **largest**?

The liver is the body's largest gland and its second-largest organ, after the skin. At 2.5 to 3.3 pounds (1.1 to 1.5 kilograms) the liver is seven times larger than it needs to be to **281** perform its estimated 500 functions. It is the main chemical factory of the body. A ducted gland that produces bile to break down fats and reduce acidity in the digestive process, the liver is also a part of the circulatory system. It cleans poisons from the blood and regulates blood composition.

Whose brain is larger: that of Neanderthal man or modern man?

The capacity of the skull of "classic" Neanderthal man was often larger than that of modern man. The capacity was between 1,350 to 1,700 cubic centimeters with the average being 1,400 to 1,450 cubic centimeters. The mean cranial capacity of modern man is 1,370 cubic centimeters, with a range of 950 to 2,200. However, brain size alone is not an index of intelligence.

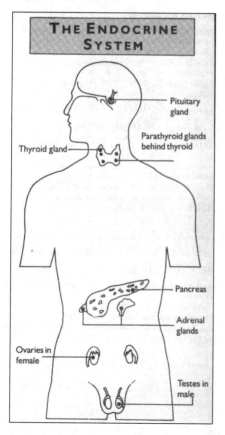

THE ENDOCRINE SYSTEM

Pituitary gland

Parathyroid glands behind thyroid

Thyroid gland

Pancreas

Adrenal glands

Ovaries in female

Testes in male

What are the seven endocrine glands?

The major endocrine glands include the pituitary, thyroid, parathyroids, adrenals, pancreas, testes, and ovaries. These glands secrete hormones into the blood system, that generally stimulate some change in metabolic activity:

Pituitary—secretes ACTH to stimulate the adrenal cortex; produces aldosterone to control sodium and potassium reabsorption by the kidneys; FSH to stimulate gonad function and prolactin to stimulate milk secretion of breasts; TSH to stimulate the thyroid gland to produce thyroxin; LH to stimulate ovulation in females and testerone production in males; GH to stimulate general growth. Stores oxytocin for uterine contraction.

Thyroid gland—secretes thyroxine (T_3) and triiodothyronine (T_4) to stimulate metabolic rate, especially in growth and development, and secretes calcitonin to lower blood-calcium levels.

Parathyroids—secrete hormone PTH to increase blood-calcium levels; activates vitamin D and stimulates calcium reabsorption in kidneys.

Adrenals—secrete epinephrine and norepinephrine to help the body cope with stress, raise blood pressure, heart rate, metabolic rate, raise blood sugar levels, etc. Aldosterone secreted by the adrenal cortex maintains sodium-phosphate balance in kidneys; cortisol helps the body adapt to stress, mobilizes fat, and raises blood sugar level.

Pancreas—secretes insulin to control blood sugar levels, stimulates glycogen production, fat storage, and protein synthesis. Glucagon secretion raises blood sugar level and mobilizes fat.

Ovaries and testes—secrete estrogens, progesterone, or testosterone to stimulate growth and reproductive processes.

What regulates **body temperature** in humans?

The hypothalamus gland controls internal body temperature. The gland responds to sensory impulses from temperature receptors in the skin and in the deep body regions. The hypothalamus establishes a "set point" for the internal body temperature, then constantly compares this with its own actual temperature. If the two do not match, the hypothalamus activates either temperature-decreasing or temperature-increasing procedures to bring actual temperature into alignment with the set point.

BODY FLUIDS

What are the **four humors** of the body?

The four constituent humors of the body were identified as blood, phlegm, yellow bile, and black bile, originating in the heart, brain, liver, and spleen, respectively. Empedocles of Agrigentum (504–433 B.C.E.) probably originated the theory in which he equated the body fluids to the four elements of nature: earth, fire, air, and water. These humors could determine the health of the body and the personality of the person as well. To be in good health the humors should be in harmony within the body. Ill health could be remedied by treatments to realign the humors and reestablish the harmony.

What is the **normal pH** of blood, urine, and saliva?

Normal pH of arterial blood is 7.4; pH of venous blood is about 7.35. Normal urine pH averages about 6.0. Saliva has a pH between 6.0 and 7.4.

How similar are **seawater** and **blood**?

Component	Seawater (grams/liter)	Blood (grams/liter)
Na (sodium)	10.7	3.2–3.4
K (potassium)	0.39	0.15–0.21
Ca (calcium)	0.42	0.09–0.11
Mg (magnesium)	1.34	0.012–0.036
Cl (chloride)	19.3	3.5–3.8
SO_4 (sulfate)	2.69	0.16–0.34
CO_3 (carbonate)	0.073	1.5–1.9
Protein		70.0

How much blood is in the average human body?

A man weighing 154 pounds (69.8 kilograms) would have about 5.5 quarts of blood. A woman weighing 110 pounds (49.8 kilograms) would have about 3.5 quarts.

How many miles of **blood vessels** are contained in the body?

If they could be laid end to end, human blood vessels would extend about 60,000 miles (97,000 kilometers).

How does the body introduce **oxygen** to the **blood** and where does this happen?

Blood entering the right side of the heart (right auricle or atrium) contains carbon dioxide, a waste product of the body. The blood travels to the right ventricle, which pushes it through the pulmonary artery to the lungs. In the lungs, the carbon dioxide is removed and oxygen is added to the blood. Then the blood travels through the pulmonary vein carrying the fresh oxygen to the left side of the heart, first to the left auricle where it goes through a one-way valve into the left ventricle, which must push the oxygenated blood to all portions of the body (except the lungs) through a network of arteries and capillaries. The left ventricle must contract with six times the force of the right ventricle, so its muscle wall is twice as thick as the right.

What is the amount of **carbon dioxide** found in normal **blood**?

Carbon dioxide normally ranges from 19 to 25 mm per liter in arterial blood and 22 to 30 mm per liter in venous blood.

Which of the major **blood types** are the most common in the United States?

Blood type	Frequency in U.S.
O+	37.4%
O-	6.6%
A+	35.7%
A-	6.3%
B+	8.5%
B-	1.5%
AB+	3.4%
AB-	0.6%

In the world, the preponderance of one blood group varies greatly by locality. Group O is generally the most common (46%), but in some areas Group A predominates.

Which **blood type** is the rarest?

The rarest blood type is Bombay blood (subtype h-h), found only in a Czechoslovakian nurse in 1961 and in a brother and sister named Jalbert living in Massachusetts in 1968.

What are the **blood group combinations** that can normally be used to help determine paternity?

If the mother is	and the child is	the father can be	but not
O	O	O, A, or B	AB
O	A	A or AB	O or B
O	B	B or AB	O or A
A	O	O, A, or B	AB
A	A	any group	
A	B	B or AB	O or A
A	AB	B or AB	O or A
B	O	O, A, or B	AB
B	B	any group	
B	A	A or AB	O or B
B	AB	A or AB	O or B
AB	AB	A, B, or AB	O

SKIN, HAIR, AND NAILS

How much **skin** does an average person have?

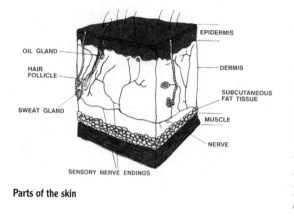

EPIDERMIS

OIL GLAND

HAIR
FOLLICLE

DERMIS

SUBCUTANEOUS
FAT TISSUE

SWEAT GLAND

MUSCLE

NERVE

SENSORY NERVE ENDINGS

Parts of the skin

The average human body is covered with about 20 square feet or 2 square meters of skin. Weighing almost 6 pounds (2.7 kilograms), the skin is composed of two main layers: the epidermis (outer layer) and the dermis (inner layer). The epidermis layer is replaced continually as new cells, produced in the stratum basale, mature and are pushed to the surface by the newer cells beneath; the entire epidermis is replaced in about 27 days. The dermis, the lower layer, contains nerve endings, sweat glands, hair follicles, and blood vessels. The upper portion of the dermis has small fingerlike projections, called "papillae," which extend into the upper layer. The capillaries in these papillae deliver oxygen and nutrients to the epidermis cells and also function in temperature regulation. The patterns of ridges and grooves visible on the skin of the soles, palms, and fingertips are formed from the tops of the dermal papillae.

Can a person get a **suntan** through a window?

No. The ultraviolet radiation that causes suntan does not pass through glass. A tan is the skin's natural defense against these harmful sun rays. Pigment cells in the lower skin level make more pigment and melanin to absorb the ultraviolet rays. As the pigment spreads into the top layer of the skin, it becomes darker.

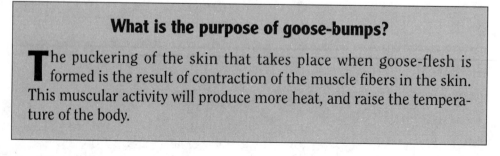

What is the purpose of goose-bumps?

The puckering of the skin that takes place when goose-flesh is formed is the result of contraction of the muscle fibers in the skin. This muscular activity will produce more heat, and raise the temperature of the body.

What is **dermatoglyphics**?

It is the study of the skin ridge patterns on fingers, toes, palms of hands, and soles of feet. The patterns are used as a basis of identification and also have diagnostic value because of association between certain patterns and chromosomal abnormalities.

Who first classified **fingerprints** as a means of identification?

It is generally acknowledged that Francis Galton (1822–1911) was the first to classify fingerprints. However, his basic ideas were further developed by Sir Edward Henry (1850–1931), who devised a system based on the pattern of the thumb print. In 1901 in England, Henry established the first fingerprint bureau with Scotland Yard, called the Fingerprint Branch.

How can **tattoos** be removed?

Tattoos can be removed by skin grafting, by infrared coagulation treatment, or by sal-abrasion in which table salt is scrubbed into the anaesthetized skin. The hypertonic irritant solution disperses the tattoo pigment particles. This procedure will produce scars.

How many hairs does the average person have on his or her head?

The amount of hair covering varies from one individual to another. An average person has about 100,000 hairs on their scalp. Most redheads have about 90,000 hairs, blonds have about 140,000, and brunettes fall in between these two figures. Most people shed between 50 to 100 hairs daily.

How much does human **hair grow in a year**?

Each hair grows about 5 inches (12.7 centimeters) every year.

What **information** can a forensic scientist determine from a human hair?

A single strand of human hair can identify the age and gender of the owner, drugs and narcotics the individual has taken, and, through DNA evaluation and sample comparisons, from whose head the hair came.

287

Do the **nails** and **hair** of a **dead person** continue to grow?

Between 12 and 18 hours after death, the body begins to dry out. That causes the tips of the fingers and the skin of the face to shrink, creating the illusion that the nails and hair have grown.

How fast do **finger nails** grow?

Healthy nails grow about 0.8 inches (2 centimeters) each year. The middle fingernail grows the fastest, because the longer the finger, the faster its nail growth. Fingernails grow four times as fast as toenails.

SENSES AND SENSE ORGANS

What are the **floaters** that move around on the eye?

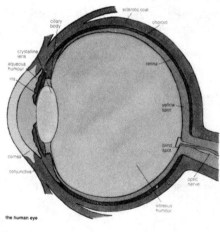

the human eye

Parts of the eye

Floaters are semi–transparent specks perceived to be floating in the field of vision. Some originate with red blood cells that have leaked out of the retina. The blood cells swell into spheres, some forming strings, and float around the areas of the retina. Others are shadows cast by the microscopic structures in the *vitreous humor,* a jellylike substructure located behind the retina. A sudden appearance of a cloud of dark floaters, if accompanied by bright light flashes, could indicate retinal detachment.

What is the difference in the functions of the **rods and cones** found in the eyes?

Rods are concerned with light and dark distinction and cones with color perception.

How often does the human **eye blink**?

The rate of blinking varies, but on the average the eye blinks once every five seconds or 17,000 times each day or 6¼ million times a year.

What are **phosphenes**?

If the eyes are shut tightly, the lights apparently seen are phosphenes. Technically, the luminous impressions are due to the excitation of the retina caused by pressure on the eyeball.

What does it mean to have **20/20 vision**?

Many people think that with 20/20 vision the eyesight is perfect, but it actually means that the eye can see clearly at 20 feet what a normal eye can see clearly at that distance. Some people can see even better—20/15, for example. With their eagle eyes, they can view objects from 20 feet away with the same sharpness that a normal-sighted person would have to move in to 15 feet to achieve.

Are more people **nearsighted** or **farsighted**?

About 30% of Americans are nearsighted to some degree. About 60% are farsighted. If the light rays entering the pupils of the eye converge exactly on the retina, then a sharply focused picture is relayed to the brain. But if the eyeball is shaped differently, the focal point of the light rays is too short or too long, and vision is blurred. Convex lenses for farsightedness correct a too-long focal point. Concave lenses correct near-sightedness when the focal point of light rays, being too short, converge in front of the retina.

Who invented **bifocal lenses**?

The original bifocal lens was invented in 1784 by Benjamin Franklin (1706–1790). At that time, the two lenses were joined in a metallic frame. In 1899, J.L. Borsch welded the two lenses together. One-part bifocal lenses were developed by Bentron and Emerson in 1910 for the Carl Zeiss Company.

Why do all **newborn babies** have **blue eyes**?

The color of the iris gives the human eye its color. The amount of dark pigment, melanin, in the iris is what determines its color. In newborns the pigment is concentrated in the folds of the iris. When a baby is a few months old, the melanin moves to the surface of the iris and gives the baby his or her permanent eye color.

How do **colors** affect one's **moods**?

According to the American Institute for Biosocial Research, "colors are electromagnetic wave bands of energy." Each color has its own wavelength. The wave bands stimulate chemicals in your eye, sending messages to the pituitary and pineal glands. These master endocrine glands regulate hormones and other physiological systems in the body. Stimulated by response to colors, glandular activities can alter moods, speed up heart rates, and increase brain activity.

What are the three **bones in the ear** called?

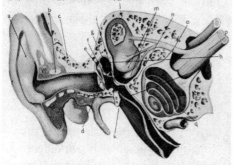

Diagram of ear (a) auricle (b) cartilaginous part of auditory meatus (c) bony part of auditory meatus (d) tympanic membrane or eardrum, (e), (f), (g) chain of ossicles (hammer, anvil and Sylvius bones) in (h) tympanic cavity, (i) stirrup bone, (j) semicircular duct, (k) oval window, (l) Eustachian tube, (m) utricle and saccule (n) perilymph, (o) endolymphatic duct, (p) auditory nerve (q) cohlear duct

The ear's three bones are the malleus, which means hammer; incus, which means anvil; and stapes, which means stirrup. The bones look somewhat like the objects for which they are named. The stapes is the smallest bone in the body, measuring 1.02 to 1.34 inches (2.6 to 3.4 centimeters) and weighing 0.00071 to 0.0015 ounces (0.002 to 0.004 grams). These three tiny bones in the middle ear conduct sound vibrations from the outer to inner ear.

How broad is the **range of sound frequency** that most people can hear?

Sounds with frequencies from about 20 to 20,000 hertz can be heard by most people. A hertz is a measure of sound frequencies. Environmental sound is measured in decibels (db) to calculate its loudness. Hearing starts with 0 db; every increase of 10 units is equivalent to a tenfold increase. In comparison, leaves rustling = 10 db, a typical office is 50 db, pneumatic drills = 80 db, riveting machines = 110 db, and a jet takeoff at 200 feet (61 meters) measures 120 db. Noise above 70 db is harmful to hearing; noise at 140 db is physically painful.

What are the primary sensations of **taste**?

The four primary taste categories are sweet, sour, salty and bitter. The intensity and location of these areas on the tongue varies from person to person. Some of the nine thousand taste buds are located in the other areas of the mouth as well. The lips (usually very salt-sensitive), the inner cheeks, the underside of the tongue, the back of the throat, and the roof of the mouth are some examples. The sense of taste is intimately associated with the sense of smell, so that foods taste bland to someone suffering from a cold. Also important are the appearance, texture, and temperature of food.

HEALTH AND MEDICINE

HEALTH HAZARDS, RISKS, ETC.

Which **"risk factors"** affects one's **health**?

Such characteristics as age, gender, work, family history, behavior, and body chemistry are some of the risk factors to consider when calculating whether one is at risk for various conditions. Some risk factors are statistical, describing trends among large groups of people but not giving information about what will happen to individuals. Other risk factors might be described as causative—exposure to them has a direct effect on whether or not the person will become sick.

What are the leading **causes of stress**?

In 1967, when they conducted a study of the correlation between significant life events and the onset of illness, Dr. Thomas H. Holmes and Dr. Richard H. Rahe from the University of Washington compiled a chart of the major causes of stress, with assigned point values. They published their findings on stress effects as "The Social Readjustment Scale," printed in *The Journal of Psychosomatic Research.* The researchers calculated that a score of 150 points indicated a 50–50 chance of the respondent developing an illness or a "health change." A score of 300 would increase the risk to 90%. Of course, many factors enter into an individual's response to a particular event, so this scale, partially represented below, can only be used as a guide.

Event	Point value
Death of spouse	100
Divorce	73

Event	Point value
Marital separation	65
Jail term or death of close family member	63
Personal injury or illness	53
Marriage	50
Fired at work	47
Marital reconciliation or retirement	45
Pregnancy	40
Change in financial state	38
Death of close friend	37
Mortgage over $10,000	31
Foreclosure of mortgage or loan	30
Outstanding personal achievement	28
Trouble with boss	23
Change in work hours or conditions or change in residence or schools	20
Vacation	13
Christmas	12
Minor violations of the law	11

What are the **odds against** being **struck by lightning**?

606,944 to 1 against.

What are the **odds** on being **killed on a motorcycle**?

1,250 to 1 against.

Is there more violence on the streets and in mental institutions when there is a **full moon**?

A review of 37 studies that attempted to correlate the phases of the moon with violent crime, suicide, crisis center hotline calls, psychiatric disorders, and mental hospital admissions found that there is absolutely no relation between the moon and the mind.

Can **owning a pet** be beneficial to your health?

As a result of several studies, researchers now believe that regular contact with pets can reduce heart rate, blood pressure, and levels of stress. In a study of 93 heart attack patients, only 1 of 18 pet owners died compared to 1 of 3 patients who did not have pets. Pets offer constancy, stability, comfort, security, affection, and intimacy.

What are the leading causes of death in the United States?

Of the 2,162 thousand deaths in 1990, the leading cause of death was heart disease. Below are listed the four major causes of death in the United States:

Cause	Number	Percent of total deaths
Heart disease	725,010	33.5
Cancer	506,000	23.4
Cerebrovascular diseases	145,340	6.7
Accidents	93,550	4.3

What are the leading causes of death for young persons?

For persons under the age of 21, accidents are the leading cause of death. Motor vehicle accidents account for almost half of all accidents involving this age group. However, the home appears to be the most dangerous place for infants (0 to 1 years). From birth to age one, the most prominent cause of accidental death is choking on food or objects placed in the mouth. Motor vehicle accidents rank second, and suffocation (from smothering by bedclothes, plastic bags, etc.) is the third leading cause. Fourth are fires, burns, etc.; the fifth is drowning. For children aged 1 to 4 years, the ranking is (1) motor vehicle accidents; (2) drowning; (3) fires and burns; (4) choking; and (5) falls. For ages 5 through 14, the order is (1) motor vehicle accidents; (2) drowning; (3) firearms; (4) poisoning; and (5) fires, burns, etc. From ages 15 to 24, motor vehicle accidents ranked first, drownings and poisoning tied for second and third place, and falls and fires, burns, etc,. tied for fourth and fifth positions.

Did raising the speed limit on rural interstate highways from 55 to 65 miles per hour have an effect on the accident and death rate?

There was an estimated 20% to 30% increase in deaths on those roads and a 40% increase in serious injuries when the speed limit was raised from 55 miles (88.5 kilometers) per hour to 65 miles (104.5 kilometers) per hour on rural interstate highways. **293**

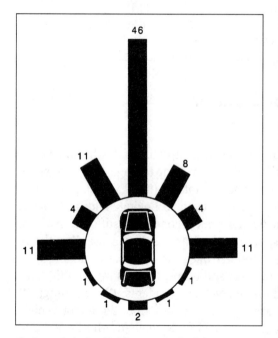

Car impact in fatal accidents: percentage by direction

Which direction of impact results in the greatest number of fatalities in **automobile crashes**?

Frontal crashes are responsible for the largest percentage of fatalities in passenger cars.

Which **sport** has the highest rate of **injuries** and what kind of injury is most common?

Football players suffer more injuries than other athletes, collectively. They have 12 times as many injuries as do basketball players, who have the next highest rate of injury. Knee problems are the most common type of injury. Two–thirds of basketball players' injuries and one–third of football players' injuries are knee-related.

Do **electric and magnetic fields** produced by power transmission lines present a health hazard?

No study has yet produced evidence to allow a firm conclusion on this question. Scientists cannot agree on the significance of the inconsistent and puzzling findings.

On electromagnetic fields, in general, a 1990 study by the United States *Environmental Protection Agency* (EPA) noted a *possible* statistically significant link between cancer and exposure to *extremely low frequency* (ELF) electromagnetic fields. ELF waves are nonionizing electromagnetic radiation that closely resemble the body's micropulsations. Biological studies have yet to prove this cancer connection. So far the ELF risks are controversial. Effects (from those studies supporting the existence of ELF hazards) run the gamut from headaches, to miscarriages, to cancer. In addition, *electromagnetic radiation* (EMR)—from devices such as electric blankets, video display terminals, microwaves, toasters, and hair dryers, powered by *alternating current* (AC) power lines—has been the target of study to better understand the health hazards that might be posed.

Can **ozone** be harmful to humans?

Ozone (O_3) in the lower atmosphere contributes to air pollution. It is formed by chemical reactions between sunlight and oxygen in the air in the presence of impurities. Such ozone can damage rubber, plastic, and plant and animal tissue. Exposure to certain concentrations can cause headaches, burning eyes, and irritation of the respiratory tract in many individuals. Asthmatics and others with impaired respiratory systems are particularly susceptible. Exposure to low concentrations for only a few hours can significantly affect normal individuals while exercising. Symptoms include chest pain, coughing, sneezing, and pulmonary congestion.

Does using **aluminum cooking utensils** for food preparation have any effect on health?

This question has not yet been resolved. Studies indicate that patients with Alzheimer's disease show an increased level of aluminum in their brains. However, research scientists do not consider the evidence currently available to be strong enough to definitely state that aluminum plays a causal role in this disease.

Aluminum can be ingested from a variety of other sources—antacids, some baking powders, food additives such as potassium alum (the flour whitener) and aluminum calcium silicate (a table salt additive to keep salt free-running), and aluminum foil and anti-perspirants. Most of the aluminum taken into the body is excreted and the remainder is stored in the lungs, brain, liver, and thyroid gland. There are some aluminum-induced diseases. The aluminum content in the water in dialysis machines has induced dementia (a form of mental disorder) in some patients undergoing this treatment. Aluminum hydroxide in antacids can cause tiredness, weakness, and loss of appetite; aluminum chloride in anti-perspirants can cause allergic dermatitis (itching, inflamed skin).

295

Why is exposure to **asbestos** a health hazard?

Exposure to asbestos has long been known to cause asbestosis. This is a chronic, restrictive lung disease caused by the inhalation of tiny mineral asbestos fibers, which scar lung tissues. Asbestos has also been linked with cancers of the larynx, pharynx, oral cavity, pancreas, kidneys, ovaries, and gastrointestinal tract. The American Lung Association reports that prolonged exposure doubles the likelihood that a smoker will develop lung cancer. It takes cancer 15 to 30 years to develop from asbestos. These fibers were used in building materials between 1900 and the early 1970s as insulation for walls and pipes, as fireproofing for walls and fireplaces, in soundproofing and acoustic ceiling tiles, as a strengthener for vinyl flooring and joint compounds, and as a paint texturizer. Asbestos poses a health hazard only if the tiny fibers are released into the air, but this can happen with any normal fraying or cracking. Asbestos removal aggravates this normal process and multiplies the danger level—it should only be handled by a contractor trained in handling asbestos. Once released, the particles can hang suspended in the air for more than 20 hours.

Why is **radon** a health hazard?

Radon is a colorless, odorless, tasteless, radioactive gaseous element produced by the radioactive decay of radium. Its three naturally occurring isotopes are found in many natural materials, such as soil, rocks, well water, and building materials. Because the gas is continually released into the air, it makes up the largest source of radiation that humans encounter. Some believe that radon may be a significant cause of cancer, especially lung cancer. It has been estimated that radon may cause as much as 10%, or 5,000 to 20,000 cases, of lung cancer deaths annually. Smokers seem to be at a higher risk than non-smokers. The *E*nvironmental *P*rotection *A*gency (EPA) recommends that the level should not be more than 4 picocuries per liter. The estimated national average is 1.5 picocuries per liter. Because EPA's "safe level" is equivalent to 200 chest x–rays per year, others believe that lower levels should be established. The *A*merican *S*ociety of *H*eating, *R*efrigeration, and *A*ir–Conditioning *E*ngineers (ASHRAE) recommends 2 picocuries/liter. The EPA estimates that nationally 8% to 12% of all houses are above the 4 picocuries/liter; whereas in another survey in 1987, it was estimated that 21% of homes were above this level.

How is human exposure to **radiation** measured?

The *r*adiation *a*bsorbed *d*ose (rad) and the *r*oentgen *e*quivalent *m*an (rem) were used for many years to measure the amount and effect of ionizing radiation absorbed by humans. While officially replaced by the gray and the sievert, both rad and rem are still used in many reference sources. The rad equals the energy absorption of 100 ergs per gram of irradiated material (an erg is a unit or work or energy). The rem is the absorbed dose of ionizing radiation that produces the same biological effect as 1 rad of x-rays or gamma rays (which are equal). The rem of x-rays and gamma rays is therefore equal to the rad; for each type of radiation, the number of rads is multiplied by a specific factor to find the number of rems. The millirem, 0.001 rems, is also frequently used; the average radiation dose received by a person in the United States is about 180 millirems per year.

In the SI system (*Système International d'Unités,* or International System of Units), the gray and the sievert are used to measure radiation absorbed; these units have largely superseded the older rad and rem. The gray (Gy), equal to 100 rads, is now the base unit. It is also expressed as the energy absorption of 1 joule per kilogram of irradiated material. The sievert (Sv) is the absorbed dose of radiation that produces the same biological effect as 1 gray of x-rays or gamma rays. The sievert is equal to 100 rems, and has superseded the rem. The becquerel (Bq) measures the radioactive strength of a source, but does not consider effects on tissue. One becquerel is defined as one disintegration (or other nuclear transformation) per second.

What is the estimated **radiation** dose to which the **average person** in the United States is exposed each year?

Of the approximately 160 millirems of radiation exposure per year, 50% comes from natural sources and 50% from human activities. Medical diagnosis and therapy account for more than 90% of the man-made dose. Cosmic rays, solar radiation, upper atmospheric radiation, and natural radioactivity are sources of ionizing radiation (which creates ions through ejection of electrons from atoms of the body's molecules, and may cause biological damage). Medical x–rays and radioactive isotopes are man-made sources of ionizing radiation.

How much **radiation** does the average **dental x–ray** emit?

Dental x-ray examinations are estimated to contribute 0.15 millirems per year to the average genetically significant dose of radiation, a small amount when compared to other medical x–rays.

How does the **United States Environmental Protection Agency (EPA)** classify **carcinogens**?

A carcinogen is an agent that can produce cancer (a malignant growth or tumor that spreads throughout the body, destroying tissue). The EPA classifies chemical and physical substances by their toxicity to humans.

Group A. Human carcinogen

This classification indicates that there is sufficient evidence from epidemiological studies to support a cause–effect relationship between the substance and cancer.

Group B. Probable human carcinogen

B_1: Substances are classified as B_1 carcinogens on the basis of sufficient evidence from animal studies, and limited evidence from epidemiological studies.

B_2: Substances are classified as B_2 carcinogens on the basis of sufficient evidence from animal studies, with inadequate or nonexistent epidemiological data.

Group C. Possible human carcinogen

For this classification, there is limited evidence of carcinogenicity from animal studies and no epidemiological data.

Group D. Not classifiable as to human carcinogenicity

The data from human epidemiological and animal studies are inadequate or completely lacking, so no assessment as to the substance's cancer-causing hazard is possible.

Group E. Evidence of noncarcinogenicity for humans

Substances in this category have tested negative in at least two adequate (as defined by the EPA) animal cancer tests in different species and in adequate epidemiological and animal studies. Classification in group E is based on available evidence; substances may prove to be carcinogenic under certain conditions.

What is "good" and "bad" **cholesterol?**

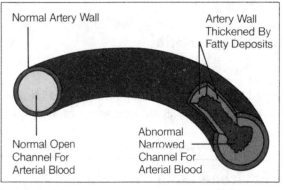

Atheroma buildup

Chemically a lipid, cholesterol is an important constituent of body cells. This fatty substance, produced mostly in the liver, is involved in bile salt and hormone formation, and in the transport of fats in the bloodstream to the tissues throughout the body. Both cholesterol and fats are transported around as lipoproteins (units having a core of cholesterol and fats in varying proportions, with an outer wrapping of carrier protein [phospholoids and apoproteins]). An overabundance of cholesterol in the bloodstream can be an inherited trait, can be triggered dietary intake, or can be the result of a metabolic disease, such as diabetes mellitus. Dietary fats (from meat, oil, and dairy products) strongly affect the cholesterol level. High cholesterol levels in the blood may lead to a narrowing of the inner lining of the coronary arteries from the build-up of a fatty tissue called atheroma. This increases the risk of coronary heart disease or stroke. However, if most cholesterol in the blood is in the form of high density lipoproteins (HDL), then it seems to protect against arterial disease. HDL picks up cholesterol in the arteries and brings it back to the liver for excretion or reprocessing. HDL is referred to as "good cholesterol." Conversely, if most cholesterol is in the form of low density lipoproteins (LDL), or very low density lipoproteins (VLDL), then arteries can become clogged. "Bad cholesterol" is a term used to refer to LDL and VLDL.

How does **blood alcohol level** affect the body and behavior?

The effects of drinking alcoholic beverages depend on the amount of actual ethyl alcohol consumed and body weight. The level of alcohol in the blood is calculated in milligrams (1 milligram equals 1/30,000 of an ounce) of pure (ethyl) alcohol per deciliter (3.5 fluid ounces), commonly expressed in percentages.

Number of drinks	Blood alcohol level	Effect
1	0.02–.03%	Changes in behavior, coordination, and ability to think clearly
2	0.05%	Sedation or tranquilized feeling
3	0.08–0.10%	Legal intoxication in many states
5	0.15–0.20%	Person is obviously intoxicated and may show signs of delirium
12	0.30–0.40%	Loss of consciousness
24	0.50%	Heart and respiration become so depressed that they cease to function and death follows

What percentage of the American public are **smokers**?

Smokers constitute a minority, and a shrinking one—apparently around 30% of the adult American population (1985). Smoking has only recently been categorized as harmful to health. Studies have linked cigarette smoking to lung cancer, oral cavity cancers, chronic bronchitis, emphysema, and coronary heart disease.

What is the **composition of cigarette smoke**?

Cigarette smoke contains about four thousand chemicals. Carbon dioxide, carbon monoxide, methane, and nicotine are some of the major components, with lesser amounts of acetone, acetylene, formaldehyde, propane, hydrogen cyanide, toluene, and many others.

Why does the risk of **cancer** diminish rapidly after one quits the **cigarette smoking** habit?

Exposure of a premalignant cell to a promoter converts the cell to an irreversibly malignant state. Promotion is a slow process, and exposure to the promoter must be sustained for a certain period of time. This requirement explains why the risk of cancer diminishes rapidly after one quits smoking; both cancer initiators and promoters appear to be contained in tobacco smoke.

Does cigarette **smoking** impair **bone healing**?

In addition to effects that include heart disease, a variety of cancers, and the harm it does the body's ability to heal skin wounds, cigarette smoking also slows down the healing time for bone fractures. A 5-centimeter bone gap, which would ordinarily heal in 10 months in a non-smoker, will take 15 months in a smoker.

FIRST AID, POISONS, ETC.

How long does it take to **bleed to death**?

Serious bleeding requires immediate attention and care. If a large blood vessel is severed or lacerated, a person can bleed to death in one minute or less. Rapid loss of one quart or more of the total blood volume often leads to irreversible shock and death.

What is the **Heimlich maneuver**?

This effective first-aid technique to resuscitate choking and drowning victims was introduced by Dr. Henry J. Heimlich (b. 1920) of Xavier University, in Cincinnati, Ohio. It is a technique for removing a foreign body from the trachea or pharynx, where it is preventing flow of air to the lungs. When the victim is in the vertical position, the maneuver consists of applying subdiaphragmatic pressure by wrapping one's arms around the victim's waist from behind, making a fist with one hand and placing it against the victim's abdomen between the navel and the rib cage, clasping one's fist with the other hand and pressing in and up with a quick, forceful thrust. Repeat several times if necessary. When the victim is in the horizontal position (which some experts recommend), the rescuer straddles the victim's thighs.

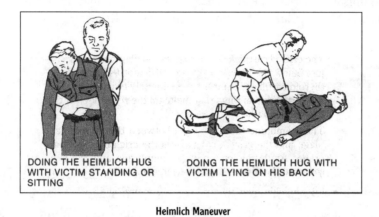

DOING THE HEIMLICH HUG WITH VICTIM STANDING OR SITTING

DOING THE HEIMLICH HUG WITH VICTIM LYING ON HIS BACK

Heimlich Maneuver

How is **activated charcoal** used medically?

Activated charcoal is an organic substance, such as burned wood or coal, that has been heated to approximately 1,000°F (537°C) in a controlled atmosphere. The result is a fine powder containing thousands of pores that can rapidly absorb toxins and poisons. Activated charcoal is used medically in the treatment of drug overdoses and poisonings.

What is the **deadliest natural toxin**?

Botulinal toxin, produced by the bacterium *Clostridium botulinum,* is the most potent poison for humans. It has an estimated lethal dose in the bloodstream of 10-9 milligrams per kilogram. It causes botulism, a severe neuroparalytic disease that travels to the junctions of skeletal muscles and nerves, where it blocks the release of the neurotransmitter acetylcholine, causing muscle weakness and paralysis, and impairing vision, speech, and swallowing. Death occurs when the respiratory muscles are paralyzed; this usually occurs during the first week of illness. Mortality from botulism is about 25%.

Because the bacterium can form the toxin only in the absence of oxygen, canned goods and meat products wrapped in airtight casings are potential sources of botulism. The toxin is more likely to grow in low-acid foods, such as mushrooms, peas, corn, or beans, rather than high-acid foods like tomatoes. However, some new tomato hybrids are not acidic enough to prevent the bacteria from forming the toxin. Foods being canned must be heated to a temperature high enough and for a long enough time to kill the bacteria present. Suspect food includes any canned or jarred food product with a swollen lid or can. Ironically, this dreaded toxin in tiny doses is being used to treat **301** disorders that bring on involuntary muscle contractions, twisting, etc. The United States *F*ood and *D*rug *A*dministration (FDA) has approved the toxin for: treatment of strabismus (misalignment of the eyes), blepharospasm (forcible closure of eyelids), and hemifacial spasm (muscular contraction on one side of the face).

Have childhood deaths from poisoning decreased since the introduction of childproof containers?

After childproof packaging was required on all drugs and medications beginning in 1973, the childhood poisoning death rate declined dramatically. A 50% decrease was noted in 1973–1976 and the decline has continued. Other factors in this decline have been the development of poison control centers, changes in products to reduce poisonous agents, and the introduction of single-dose packages.

How deadly is **strychnine**?

The fatal dose of strychnine or deadly nightshade (the plant from which it is obtained) is 15 to 30 milligrams. It causes severe convulsions and respiratory failure. If the patient lives for 24 hours, recovery is probable.

Which part of **mistletoe** is poisonous?

Mistletoe's white berries contain toxic amines, which cause acute stomach and intestinal irritation with diarrhea and a slow pulse. Mistletoe should be considered a potentially dangerous Christmas decoration, especially in the presence of children.

What is the **poison on arrows** used by South American tribes to kill prey and enemies?

The botanical poison used by the Aucas and similar tribes in the South American jungles is curare. It is a sticky, black mixture with the appearance of liquorice and is processed from either of two different vines. One is a liana (*Chondodendron tomentosum*); the other is a massive tree-like vine (*Strychonos quianensis*).

How deadly is **Amanita phalloides**?

The poisonous mushroom *Amanita phalloides* has a fatality rate of about 50%. Ingestion of part of one mushroom may be sufficient to cause death. Over 100 fatalities occur each year from eating poisonous mushrooms, with more than 90% caused by the *Amanita phalloides* group.

Which first aid measures can be used for a **bite** by a **black widow spider**?

Black widow spider

The black widow spider (*Latrodectus mactans*) is common throughout the United States. Its bite is severely poisonous, but no first aid measures are of value. A victim's age, body size, and degree of sensitivity determine the severity of symptoms, which include an initial pinprick with a dull numbing pain, followed by swelling. An ice cube may be placed over the bite to relieve pain. Between 10 and 40 minutes after the bite, severe abdominal pain and rigiditiy of stomach muscles develops. Muscle spasms in the extremities, ascending paralysis, and difficulty in swallowing and breathing follow. The mortality rate is less than 1%; but anyone who has been bitten should see a doctor; the elderly, infants, and those with allergies are most at risk, and should be hospitalized.

How can a **tick** be removed?

A tick, common in woods and forests througout the United States, fastens itself to a host with its teeth, then secretes a cementlike material to reinforce its hold. This flat brown speckled 0.25-inch insect can transmit diseases such as Rocky Mountain spotted fever and Lyme disease. To remove a tick, cover it with mineral, salad, or machine oil to block its breathing pores. If it does not disengage after about 30 minutes, use tweezers and grasp the tick as close to the skin as possible. Pull it away with a steady pressure, or lift the tick slightly upward and pull it parallel to the skin until the tick detaches. Do not twist or jerk the tick. Wash the bite site and hands well with soap and water and apply alcohol. If necessary apply a cold pack to reduce pain.

How can the amount of **lead in tap water** be reduced in an older house having lead-containing pipes?

The easiest way is to let the tap run until the water becomes very cold before using it for human consumption. By letting the tap run, water that has been in the lead-containing pipes for a while is flushed out. Also, cold water, being less corrosive than warm, contains less lead from the pipes. Lead (Pb) accumulates in the blood, bones, and soft tissues of the body as well as the kidneys, nervous system, and blood-forming organs. Excessive exposure to lead can cause seizures, mental retardation, and behavior disorders. Infants and children are particularly susceptible to low doses of lead and suffer from nervous system damage.

Another source of lead poisoning is old flaking lead paint. Lead oxide and other lead compounds were added to paints before 1950 to make the paint shinier and more durable. 14% of the lead ingested by humans comes from the seam soldering of food cans, according to the United States *Food* and *Drug* *A*dministration (FDA). The FDA has proposed a reduction in this lead to 50% over the next five years. Improperly glazed pottery can be a source of lead poisoning. Acidic liquids such as tea, coffee, wine, and juice can break down the glazes so that the lead can leak out of the pottery. The lead is ingested little by little over a period of time. People can also be exposed to lead in the air. Lead gasoline additives, nonferrous smelters, and battery plants are the most significant contributors of atmospheric lead emissions.

How did **lead** contribute to the **fall of the Roman Empire?**

Some believe Romans from the period around 150 B.C.E. may have been victims of lead poisoning. Symptoms of lead poisoning include sterility, general weakness, apathy, mental retardation, and early death. The lead could have been ingested in water taken from lead-lined water pipes or from food cooked in their lead-lined cooking pots. It is thought that this inadvertent toxic food additive may have contributed to the decline and fall of the Roman Empire. However, this is only speculation, since only a few Roman emperors displayed derangement. If lead pipes contributed to lead poisoning, then many Romans should have exhibited the symptoms, and all the emperors who lived in the same dwelling should have been equally susceptible.

DISEASES, DISORDERS, AND OTHER HEALTH PROBLEMS

Which **disease** is the **most common?**

The most common noncontagious disease is periodontal disease, such as gingivitis or inflammation of the gums. Few people in their lifetime can avoid these effects of tooth decay. The most common contagious disease in the world is coryza, or the common cold.

Which **disease** is the **deadliest?**

The most deadly infectious disease was the pneumonic form of the plague, the so-called Black Death of 1347–1351, with a mortality rate of 100%. Today, the disease with the highest mortality (almost 100%) is rabies in humans, when it prevents the victim from swallowing water. This disease is not to be confused with the symptoms resulting from being bitten by a rabid animal. With immediate attention after an animal bite, the rabies virus can be prevented from invading the nervous system and the survival rate in this circumstance is 95%. The virus AIDS (*acquired immunodeficiency syndrome*), first reported in 1981, is caused by the HIV (*human immunodeficiency virus*). The *World Health Organization* (WHO) reported 345,553 HIV-positive cases worldwide as of April 1, 1991, and estimated that there are 8 to 10 million people worldwide infected with HIV, with 1 million of these being in the United States. The United States harbors 171,876 cases of HIV-positive, with the number of deaths totalling 108,731. There has not yet been a case of recovery from AIDS.

What are the symptoms and signs of **AIDS**?

The early symptoms (called AIDS-*related* complex, or ARC, symptoms) of AIDS infection include night sweats, prolonged fevers, severe weight loss, persistent diarrhea, skin rash, persistent cough, and shortness of breath. The diagnosis changes to AIDS (*acquired immunodeficiency syndrome*) when the immune system is affected and the patient becomes susceptible to opportunistic infections and unusual cancers, such as herpes viruses (herpes simplex, herpes zoster, cytomegalovirus infection), *candida albicans* (fungus) infection, cryptosporidium enterocolitis (protozoan intestinal infection), *pneumocystis carinii pneumonia* (PCP, a common AIDS lung infection), toxoplasmosis (protozoan brain infection), *progressive multifocal leukoencephalopathy* (PML, a central nervous system disease causing gradual brain degeneration), *mycobacterium avium intracellulare* infection (MAI, a common generalized bacterial infection), and Kaposi's sarcoma (a malignant skin cancer characterized by blue-red nodules on limbs and body, and internally in the gastrointestinal and respiratory tracts, where the tumors cause severe internal bleeding). More than 75% of AIDS victims die within two years of diagnosis.

The signs of AIDS include generalized swollen glands, emaciation, blue or purple-brown spots on the body, especially on the legs and arms, prolonged pneumonia, and oral thrush.

How is the term **zoonosis** defined?

A zoonosis is any infectious or parasitic disease of animals that can be transmitted to humans. Examples include cat-scratch fever, psittacosis from parrots, and trichinosis from pigs.

What is meant by **vectors** in medicine?

A vector is an animal that transmits a particular infectious disease. A vector picks up disease organisms from a source of infection, carries them within or on its body, and later deposits them where they infect a new host. Mosquitoes, fleas, lice, ticks, and flies are the most important vectors of disease to humans.

Which species of mosquito causes **malaria** and **yellow fever** in humans?

The bite of the female mosquito of the genus *Anopheles* can contain a parasite of the genus *Plasmodium,* which causes malaria, a serious tropical infectious disease affecting 200 to 300 million people worldwide. More than one million African babies and children die from the disease annually. The *Aedes aegypti* transmits yellow fever, a serious infectious disease characterized by jaundice, giving the patient yellowish skin; 10% of yellow fever patients die.

What was the contribution of **Dr. Gorgas** to the building of the **Panama Canal**?

Dr. William C. Gorgas (1854–1920) brought the endemic diseases of Panama under control by destroying mosquito breeding grounds, virtually eliminating yellow fever and malaria. His work was probably more essential to the completion of the canal than any engineering technique.

How is **Lyme disease** carried?

The cause of the disease is the spirochete *Borrelia burgdorferi* which is transmitted to humans by the small tick *Ixodes dammini* or other ticks in the Ixodidae family. The tick injects spirochete–laden saliva into the victim's bloodstream or deposits fecal matter on the skin. This multisystemic disease usually begins in the summer with a skin lesion called *erythema chronicum migrans* (ECM), followed by more lesions, a molar rash, conjuctivitis, and urticaria. The lesions are eventually replaced by small red blotches. Other common symptoms in the first stage include fatigue, intermittent headache, fever, chills, and achiness.

In stage two, which can occur weeks or months later, cardiac or neurologic abnormalities sometimes develop. In the last stage (weeks or years later) arthritis develops with marked swelling, especially in the large joints. If tetracycline, penicillin, or erythromycin is given in the early stages, the later complications can be minimized. High dosage of intravenously given penicillin can also be effective on the late stages.

Why is **Legionnaire's disease** known by that name?

Legionnaire's disease was first identified in 1976 when a sudden, virulent outbreak of pneumonia took place at a hotel in Philadelphia, Pennsylvania, where delegates to an American Legion Convention were staying. The cause was eventually identified as a previously unknown bacterium that was given the name *Legionnella pneumophilia*. The bacterium probably was transmitted by an airborne route. It can spread through the cooling tower or evaporation condensers in air-conditioning systems, and has been known to flourish in soil and excavation sites. Usually the disease occurs in late summer or early fall and its severity ranges from mild to life-threatening, with a mortality rate as high as 15%. Symptoms include diarrhea, anorexia, malaise, headache, generalized weakness, recurrent chills and fever, accompanied by cough, nausea, and chest pain. Antibiotics such as Erythroycin® are administered along with other therapies (fluid replacement, oxygen, etc.) that treat the symptoms.

Which name is now used as a synonym for **leprosy**?

Hansen's disease is the name of this chronic, systemic infection characterized by progressive lesions. Caused by a bacterium, *Mycabacterium leprae,* that is transmitted through airborne respiratory droplets, the disease is not highly contagious. Continuous close contact is needed for transmittal. Antimicrobial agents, such as sulfones (dapsone in particular), are used to treat the disease.

Who was **Typhoid Mary**?

Mary Mallon (1855–1938), a cook who lived in New York City around 1900, was identified as a chronic carrier of the typhoid bacilli. She was the cause of at least 53 outbreaks of typhoid fever. The name "Typhoid Mary" became known around the world as a symbol for a chronic carrier of disease.

How many types of **herpes virus** are there?

There are five human herpes viruses all characterized by an eruption of small, usually painful, skin blisters:

Herpes simplex type 1—causes recurrent cold sores and infections of the lips, mouth, and face. The virus is contagious and spreads by direct contact with the lesions or fluid from the lesions. Cold sores are usually recurrent at the same sites and reoccur where there is an elevated temperature at the affected site; such as a fever or prolonged sun exposure. Occasionally this virus may occur on the fingers, with a rash of blisters. If the virus gets into the eye, it could cause conjunctivitis, or even a corneal ulcer. On rare occasions, it can spread to the brain to cause encephalitis.

Herpes simplex type 2—causes genital herpes and infections acquired by babies at birth. The virus is contagious and can be transmitted by sexual intercourse. The virus produces small blisters in the genital area that burst to leave small painful ulcers, which heal within 10 days to three weeks. Headache, fever, enlarged lymph nodes, and painful urination are other symptoms.

Varicella-zoster (herpes zoster)—causes chicken pox and shingles. Shingles can be caused by the dormant virus in certain sensory nerves that re-emerge with the decline of the immune system (because of age, certain diseases, and the use of immunosuppressants), excessive stress, or use of corticosteroid drugs. The painful rash of small blisters dry and crust over, eventually leaving small pitted scars. The rash tends to occur over the rib area or a strip on one side of the neck or lower body. Sometimes it involves the lower half of the face, and can affect the eyes. Pain that can be both severe and long-lasting affects about half of the sufferers, and is caused by nerve damage.

Epstein-Barr—causes infectious mononucleosis (acute infection—bringing high fever, sore throat, and swollen lymph glands, especially in the neck—that occurs mainly during adolescence) and is associated with Burkitt's lymphoma (malignant tumors of the jaw or abdomen that occur mainly in African children and in tropical areas).

Cytomegalovirus—usually no symptoms but enlarges the cells it infects; it can cause birth defects when a pregnant mother infects her unborn child.

Herpes gestationis is a rare skin-blister disorder occurring only in pregnancy and is not related to the herpes simplex virus.

Who were the "Blue People" in Appalachia?

The "Blue People" were descendants of Martin Fugate, a French immigrant to Kentucky. He carried a recessive gene that limited or stopped the body's production of the enzyme diaphorase. Diaphorase breaks down methemoglobin into hemoglobin in red blood cells. When the enzyme is not present, a disproportionate amount of methemoglobin remains in the blood, giving the cells a bluish tint, rather than the normal pink associated with Caucasians. The condition is strictly one of pigment and does not deprive the person of oxygen.

Despite bluish color, there are no known health risks associated with the deficiency. Fugate's family suffered from the condition because of excessive inbreeding. When both spouses had the recessive gene, their children would be blue. As the family became mobile following World War II and moved out of their Kentucky valley, the inbreeding ceased. As of 1982 there were only two to three members of the family with the condition.

How are warts caused?

A wart is a lump on the skin produced when one of the 30 types of papillomavirus invades skin cells and causes them to multiply rapidly. There are several different types

of warts: common warts, usually on injury sites; flat warts on hands, accompanied by itching; digitate warts having fingerlike projections; filiform warts on eyelids, armpits, and necks; plantar warts on the soles of the feet; and genital warts, pink cauliflower-like areas that, if occurring in a woman's cervix, could predispose her to cervical cancer. Each is produced by a specific virus and most usually symptomless. Wart viruses are spread by touch or by contact with the skin shed from a wart.

What is **lactose intolerance?**

Lactose, the principal sugar in cow's milk, and found only in milk and dairy products, requires the enzyme lactase for human digestion. Lactose intolerance occurs when the lining of the walls of a person's small intestine does not produce normal amounts of this enzyme. Lactose intolerance causes abdominal cramps, bloating, diarrhea, and excessive gas when more than a certain amount of milk is ingested. Most people are less able to tolerate lactose as they grow older.

A person with lactose intolerance need not eliminate dairy products totally from the diet. The problem can often be managed ny decreasing the consumption of milk products and drinking milk only during meals. Another alternative is to buy a commercial lactose preparation that can be mixed into milk. These preparations convert lactose into simple sugars that can be easily digested.

Which **medical condition** is associated with **Abraham Lincoln**'s awkward appearance?

Abraham Lincoln (1809–1865) probably had Marfan's syndrome (*Arachnodactyly*), which abnormally lengthens the bones. It is a rare inherited degenerative disease of the connective tissue. Besides the excessively long bones, the syndrome manifests itself in chest deformities, scoliosis (spine curvature), an arm span that can exceed height, eye problems (especially myopia or nearsightedness), abnormal heart sounds, and sparse subcutaneous fat. In 1991, researchers identified the gene behind the disease.

What is **carpal tunnel syndrome?**

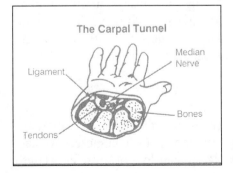

The Carpal Tunnel

Median Nerve

Ligament

Bones

Tendons

Carpal tunnel syndrome occurs when a branch of the median nerve in the forearm is compressed at the wrist as it passes through the tunnel formed by the wrist bones (or carpals), and a ligament that lies just under the skin. The syndrome occurs most often in middle age and more so in women than in men. The symptoms are intermittent at first, then become constant. Numbness and tingling begin in the thumb and first two fingers; **309** then the hand and sometimes the whole arm becomes painful. Treatment involves wrist splinting, weight loss, and control of edema, or swelling; treatments for arthritis may also help. A surgical procedure, in which the ligament at the wrist is cut, can relieve pressure on the nerve.

What is **monkey-paw**?

A contracture of the hand resulting from median nerve palsy; the thumb cannot be opposed to the tips of the fingers.

Why is **neurofibromatosis** called **"Elephant Man's disease"**?

Neurofibromatosis, or Von Recklinghausen's disease (for the German doctor who first described the disease in detail), severely deformed the Englishman Joseph Merrick (1862–1889), who was referred to as "the Elephant Man." Neurofibromatosis is an uncommon inherited developmental disorder of the nervous system, bones, muscles, and skin. Occurring in one in every 3,000 births, its symptoms generally appear during childhood or adolescence. The disorder is characterized by numerous soft, fibrous swellings that grow on nerve trunks of the extremities, head, neck, and body; *cafe au lait* spots (pale, coffee-colored patches) on the skin of the trunk and pelvis; scoliosis; and short stature. In most cases, neurofibromatosis affects only the skin. But if the soft fibrous swellings occur in the central nervous system, they can cause epilepsy, affect vision and hearing, and can also shorten life. Surgical removal of neurofibromas is necessary only if they cause complications, or are likely to do so.

What is **Lou Gehrig's disease**?

Sometimes called Lou Gehrig's disease, *a*myotrophic *l*ateral *s*clerosis (ALS) is a motor neuron disease of middle or late life. It results from a progressive degeneration of nerve cells controlling voluntary motor functions that ends in death 3 to 10 years after onset. There is no cure for it. At the beginning of the disease, the patient notices weakness in the hands and arms, with involuntary muscle quivering and possible muscle cramping or stiffness. Eventually all four extremities become involved. As nerve degeneration progresses, disability occurs and physical independence declines until the patient, while mentally and intellectually aware, can no longer swallow or move.

What is **narcolepsy**?

Although most people think of a narcoleptic as a person who falls asleep at inappropriate times, victims of narcolepsy also share other symptoms, including excessive daytime sleepiness, hallucinations, and cataplexy (a sudden loss of muscle strength following an emotional event). Persons with narcolepsy experience an uncontrollable desire to sleep, sometimes many times in one day. Episodes may last from a few minutes to several hours.

How does **jet lag** affect one's body?

The physiological and mental stress encountered by airplane travellers when crossing four or more time zones is commonly called *jet lag*. Patterns of hunger, sleep, and elimination, along with alertness, memory, and normal judgement, may all be affected. More than 100 biological functions that fluctuate during the 24-hour cycle (circadian rhythm) can become desynchronized. Most people's bodies adjust at a rate of about one hour per day. Thus after four time zone changes, the body will require about four days to return to its usual rhythms. Flying eastward is often more difficult than flying westward, which adds hours to the day.

What is **Factor VIII**?

Factor VIII is one of the enzymes involved in the clotting of blood. Hemophiliacs lack this enzyme and are at high risk of bleeding to death unless they receive supplemental doses of Factor VIII. The lack of Factor VIII is due to a defective gene, which shows a sex-linked inherited pattern, affecting males (one in 10,000). Females can carry the gene. Hemorrhage into joints and muscles usually make up the majority of bleeding episodes.

What is **"Christmas Factor"**?

In the clotting of blood, Factor IX, or the Christmas Factor, is a coagulation factor present in normal plasma, but deficient in the blood of persons with hemophilia B, or Christmas disease. The Factor and disease were labelled after a man named Christmas who, in 1952, was the first patient in whom this genetic disease was shown to be distinct from hemophilia (another genetic blood-clotting disease, in which the blood does not have Factor VIII).

What is the medical term for a **heart attack**?

Myocardial infarction is the term used for a heart attack, in which part of the heart muscle's cells die as a result of reduced blood flow through one of the main arteries (many times due to atherosclerosis); the outlook for the patient is dependant on the size and location of the blockage and extent of damage, but 33% of patients die within 20 days after the attack; it is a leading cause of death in the United States.

How does a **stroke** occur?

A stroke, or *cerebrovascular accident* (CVA), is a medical emergency produced by a blood clot that lodges in an artery and blocks the flow of blood to a portion of the brain. It produces symptoms ranging from paralysis of limbs and loss of speech to unconsciousness and death. Less commonly, a stroke may be the result of bleeding into the substance of the brain. Strokes are fatal in about 33% to 50% of the cases, and 500,000 cases a year make it the third most common cause of death. Certain factors increase the risk of stroke, such as high blood pressure (hypertension), which weakens artery walls, and atherosclerosis (thickening of the lining of artery walls), which narrows arteries.

How are the forms of **cancer** classified?

The many forms of cancer (malignant tumors) are classified into four groups: carcinomas, which involve the skin and skin-like membranes of the internal organs; sarcomas, which involve the bones, muscles, cartilage, and fat; leukemias, which involve the white blood cells; and lymphomas, which involve the lymphatic system.

What are **HeLa cells**?

HeLa cells, used in many biomedical experiments, were obtained from a cervical carcinoma developing in a woman named *Henriette La*cks. Her epithelial tissue, obtained by biopsy became the first continuously cultured human malignant cells.

What are **dust mites**?

Dust mites are microscopic arachnids (members of the spider family), commonly found in house dust. Dust mite allergen is probably one of the most important causes of asthma (breathlessness and wheezing caused by the narrowing of small airways of the lungs) in North America, as well as the major cause of common allergies (exaggerated reactions of the immune system to exposure of offending agents).

If the sap of the **poison ivy** plant touches the skin, will a rash develop?

Studies show that 85% of the population will develop an allergic reaction if exposed to poison ivy, but this sensitivity varies with each individual according to circumstance, age, genetics, and previous exposure. The poison comes mainly from the leaves whose allergens touch the skin. A red rash with itching and burning will develop; skin blisters will also develop, usually within 6 hours to several days after exposure. Thorough washing of the affected area with mild soap within five minutes of exposure can be effective; sponging with alcohol and applying a soothing and drying lotion, such as calamine lotion, is the prescribed treatment for light cases. If the affected area is large, fever, headache, and generalized body weakness may develop. For severe reactions, a physician should be consulted, who may prescribe a corticosteroid drug. Clothing that touched the plants should also be washed.

What is "tango foot"?

Officially known as *tibialis anticus,* tango foot is a strain of the leg muscles resulting from dancing the tango and maxixe.

What is dyslexia and what causes it?

Dyslexia covers a wide range of language difficulties. In general, a person with dyslexia cannot group the meaning of sequences of letters, words, or symbols or the concept of direction. The condition can affect people of otherwise normal intelligence. Dyslexic children may reverse letter and word order, make bizarre spelling errors, and may not be able to name colors or write from dictation. It may be caused by minor visual defects, emotional disturbance, or failure to train the brain. New evidence shows that a neurological disorder may be the underlying cause. Approximately 90% of dyslexics are male.

The term *dyslexia* (of Greek origin) was first suggested by Professor Rudolph of Stuttgart, Germany in 1887. The earliest references to the condition date as far back as 30 C.E. when Valerius Maximus and Pliny described a man who lost his ability to read after being struck on the head by a stone.

What is anorexia?

Anorexia simply means a loss of appetite. Anorexia nervosa is a psychological disturbance that is characterized by an intense fear of being fat. It usually affects teenage or young adult females. This persistent "fat image," however untrue in reality, leads the patient to self–imposed starvation and emaciation (extreme thinness) to the point where one–third of the body weight is lost. There are many theories on the causes of this disease, which is difficult to treat and can be fatal. Between 5% and 10% of patients hospitalized for anorexia nervosa later die from starvation or suicide. Symptoms include a 25% or greater weight loss (for no organic reason) coupled with a morbid dread of being fat, an obsession with food, an avoidance of eating, compulsive exercising and restlessness, binge eating followed by induced vomiting, and/or use of laxatives or diuretics.

Why do deep-sea divers get the bends?

Bends is a painful condition in the limbs and abdomen. It is caused by the formation and enlargement of bubbles of nitrogen in blood and tissues as a result of rapid reduction of air pressure. This condition can develop when a diver ascends from a dive too rapidly after being exposed to increased pressure underwater. Severe pain will develop in the muscles and joints of the arms and legs. More severe symptoms include vertigo, nausea, vomiting, choking and shock, and sometimes death. Bends is also known as caisson disease, tunnel disease, and diver's paralysis.

What is **progeria**?

Progeria is premature old age. There are two distinct forms of the condition, both of which are extremely rare. In Hutchinson-Gilford syndrome, aging starts around age 4, and by 10 or 12, the affected child displays all the external features of old age, including gray hair, baldness, and loss of fat, resulting in thin limbs and sagging skin on the trunk and face. There are also internal degenerative changes, such as atherosclerosis (fatty deposits lining the artery walls). Death usually occurs at puberty. Werner's syndrome, or adult progeria, starts in early adult life and follows the same rapid progression as the juvenile form. The cause of progeria is unknown.

How many people in America are estimated to have **Alzheimer's disease**?

Researchers have concluded that 10% of people over 65 have Alzheimer's disease, including nearly 50% of people over 85. The National Institute on Aging has estimated that four million Americans have Alzheimer's disease. Some experts, however, do not agree with these figures.

Alzheimer's disease is a progressive condition in which nerve cells in the brain degenerate and the brain substance shrinks. Although the cause is unknown, some theorize that it is toxic poisoning by a metal such as aluminum. Others believe it to be of genetic origin. There are three stages; in the first, the person becomes forgetful; in the second, the patient experiences severe memory loss and disorientation, lack of concentration, loss of ability to calculate and express thoughts (dysphasia), anxiety, and sudden personality changes. In the third stage the patient is severely disoriented and confused, suffers from hallucinations and delusions, suffers severe memory loss, and the nervous system declines, with regression into infantile behavior, violence, etc., requiring hospital care.

How is an **iatrogenic illness** defined?

It is an adverse mental or physical condition caused by the effects of treatment by a physician or surgeon. The term implies that the condition could have been avoided by
judicious care on the part of the physician.

What is **pelvic inflammatory disease?**

Pelvic inflammatory disease (PID), is a term used for a group of infections in the female organs, including inflammations of the Fallopian tubes, cervix, uterus, and ovaries. It is the most common cause of female infertility today. PID is most often found in sexually active women under the age of 25 and almost always results from gonorrhea or chlamydia, but women who use IUDs are also at risk. Aerobic or anaerobic organisms, such as *Neisseria gonorrhoeae,* cause acute cervicitis, (a foul-smelling vaginal discharge accompanied by itching, burning, and pelvic discomfort, often resulting in infertility and spontaneous abortion) or common bacteria such as staphylocci, chlamydiae, and coliforms, produce acute endometritis (oozing vaginal discharge with ulteration, lower abdominal pain and spasms, fever, etc.). Chronic endometritis is becoming increasingly common because of the widespread use of IUDs. Untreated PID could be fatal.

What is the chemical composition of **kidney stones?**

About 80% are calcium, mainly calcium oxalate and/or calcium phosphate; 5% are uric acid; 2% are amino acid cystine; the remainder are magnesium ammonium phosphate. About 20% of these stones are infective stones, linked to chronic urinary infections, and contain a combination of calcium, magnesium, and ammonium phosphate produced from the alkalinity of the urine and bacteria action on urea (a substance in urine).

How are **burns** classified?

Type	Causes and Effects
First-degree	Sunburn; steam. Reddening and peeling. Affects epidermis (top layer of skin). Heals within a week.
Second-degree	Scalding; holding hot metal. Deeper burns causing blisters. Affects dermis (deep skin layer). Heals in two to three weeks.
Third-degree	Fire. A full layer of skin is destroyed. Requires a doctor's care and grafting.
Circumferential	Any burns (often electrical) that completely encircle a limb or body region (such as the chest), which can impair circulation or respiration; requires a doctor's care; fasciotomy (repair of connective tissues) is sometimes required.
Chemical	Acid, alkali. Can be neutralized with water (for up to half an hour). Doctor's evaluation recommended.
Electrical	Destruction of muscles, nerves, circulatory system, etc., below the skin. Doctor's evaluation and ECG monitoring required.

If more than 10% of body surface is affected in second- and third-degree burns, shock can develop when large quantities of fluid (and its protein) are lost. When skin is burned, it cannot protect the body from airborne bacteria.

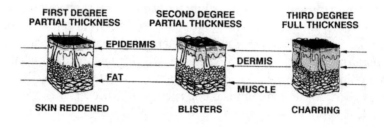

Was the disease **pellagra** ever common in the United States?

Pellagra, characterized by diarrhea, skin problems, and dementia, occurred frequently years ago in rural cotton mill towns, mining and sawmill camps, and in orphanages, prisons, and poorhouses. In these places, poverty and tradition reduced the diet to corn meal, fat salt pork, and carbohydrates, with very few green vegetables, especially during the winter months. Pellagra is brought on by an inadequate diet lacking in vitamins, especially vitamin B.

What is **pica**?

Pica refers to the craving for unnatural or non-nutritious substances. It is named after the magpie (*Pica pica*), which has a reputation for sticking its beak into all kinds of things to satisfy its hunger or curiosity. Pica can manifest itself in both genders, all races, and in all parts of the world, but is especially noted in pregnant women.

What is the medical name for a **sinking feeling** in the pit of the stomach?

Epigastric sensation.

What is **Chinese restaurant syndrome**?

*M*onosodium *g*lutamate (MSG), a commonly used flavor-enhancer, is thought to cause flushing, headache, and numbness about the mouth in susceptible people. Because many Chinese restaurants use MSG in food preparation, these symptoms may appear after a susceptible person eats Chinese food.

What is the **phobia** of number 13 called?

Triskaidekaphobia is a superstitious fear of the number 13. A phobia (a tense and irrational fear of an object, situation, or organism) can develop for a wide range of triggers. The list below demonstrates the variety:

Phobia subject	Phobia term
Animals	Zoophobia
Beards	Pogonophobia
Books	Bibliophobia
Churches	Ecclesiaphobia
Dreams	Oneirophobia
Flowers	Anthophobia
Food	Sitophobia
Graves	Taphophobia
Infection	Nosemaphobia
Lakes	Limnophobia
Leaves	Phyllophobia
Lightning	Astraphobia
Men	Androphobia
Money	Chrometophobia
Music	Musicophobia
Sex	Genophobia
Shadows	Sciophobia
Spiders	Arachnophobia
Sun	Heliophobia
Touch	Haptophobia
Trees	Dendrophobia
Walking	Basiphobia
Water	Hydrophobia
Women	Gynophobia
Work	Ergophobia
Writing	Graphophobia

Was **Napoleon** poisoned?

The most common opinion today is that Napoleon Bonaparte (1769–1821), Emperor of France from 1804–1815, died of a cancerous, perforated stomach. A significant minority of doctors and historians have made other claims ranging from various diseases to malign neglect to outright homicide. A Swedish toxicologist, Sten Forshufvud, advanced the theory that Napoleon died of arsenic poisoning, administered by an agent of the French Royalists who was planted in Napoleon's household during his final exile on the island of St. Helena.

HEALTH CARE

Which **symbol** is used to represent **medicine**?

Symbol of Asklepius

Caduceus

The staff of Aesculapius has represented medicine since 800 B.C.E. It is a single serpent wound around a staff. The caduceus, the twin-serpent magic wand of the god Hermes or Mercury, came into use after 1800 and is commonly used today. The serpent has traditionally been a symbol of healing, and it is an old belief that eating part of a serpent would bring the power of healing to the ingester. Early Greeks saw in the serpent regenerative powers expressed by the serpent's periodic sloughing of its skin, and thus venerated the serpent. Later the Greek god of medicine, Asklepius, called Aesculapius by the Romans, was believed to perform his functions in the form of a serpent. Sometimes this god is represented in art as an old man with a staff, around which is coiled a serpent.

What is the **Hippocratic Oath**?

The Hippocratic Oath is an oath demanded of physicians who are entering practice, and can be traced back to the Greek physician and teacher Hippocrates (ca.460–ca.377 B.C.E.). The oath reads as follows:

> I swear by Apollo the physician, by Aesculapius, Hygeia, and Panacea, and I take to witness all the gods, all the goddesses, to keep according to my ability and my judgement the following Oath:
>
> To consider dear to me as my parents him who taught me this art; to live in common with him and if necessary to share my goods with him; to look upon his children as my own brothers, to teach them this art if they so desire without fee or written promise; to impart to my sons and the sons of the master who taught me and the disciples who have enrolled themselves and have agreed to rules of the profession, but to these alone, the precepts and the instruction. I will prescribe regiment for the good of my patients according to my ability and my judgement and never to harm anyone. To please no one will I prescribe a deadly drug, nor give advice which may cause his death. Nor will I give a

woman a pessary to procure abortion. But I will preserve the purity of my life and my art. I will not cut for stone, even for patients in whom the disease is manifest; I will leave this operation to be performed by practitioners (specialists in this art). In every house where I come I will enter only for the good of my patients, keeping myself far from all intentional ill-doing and all seduction, and especially from the pleasures of love with women or with men, be they free or slaves. All that may come to my knowledge in the exercise of my profession or outside of my profession or in daily commerce with men, which ought not to be spread abroad, I will keep secret and will never reveal. If I keep this oath faithfully, may I enjoy life and practice my art, respected by all men and in all times; but if I swerve from it or violate it, may the reverse be my lot.

The oath varies slightly in wording among different sources.

Who was the **father of modern medicine**?

Thomas Sydenham (1624–1689), who was also called the English Hippocrates, reintroduced the Hippocratic method of accurate observation at the bedside, and recording of observations, to build up a general clinical description of individual diseases. He is also considered one of the founders of epidemiology.

What was the **first medical college** in the United States?

The College of Philadelphia Department of Medicine, now the University of Pennsylvania School of Medicine, was established on May 3, 1765. The first commencement was held June 21, 1768, when the first medical diplomas were presented to the ten members of the graduating class.

Who was the **first woman physician** in the United States?

Elizabeth Blackwell (1821–1910), received her degree in 1849 from Geneva Medical College in New York. After overcoming many obstacles, she set up a small practice that expanded into the New York Infirmary for Women and Children, which featured an all-female staff.

When was the **first blood bank** opened?

Several sites claim the distinction. Some sources list the first blood bank as opening in 1940 in New York City under the supervision of Dr. Richard C. Drew (1904–1950). Others list an earlier date of 1938 in Moscow at the Sklifosovsky Institute (Moscow's **319** central emergency service hospital), founded by Professor Sergei Yudin. The term blood bank was coined by Bernard Fantus, who set up a centralized storage depot for blood in 1937 at the Cook County Hospital in Chicago, Illinois.

Who is generally regarded as the father of medicine?

Hippocrates (ca.460–ca.377 B.C.E.), a Greek physician, holds this honor. Greek medicine, previous to Hippocrates, was a mixture of religion, mysticism, and necromancy. Hippocrates established the rational system of medicine as a science, separating it from religion and philosophy. Diseases developed from natural causes and natural laws; they were not the "wrath of the gods." Hippocrates believed that the four elements (earth, air, fire, and water) were represented in the body by four body fluids (blood, phlegm, black bile, and yellow bile), or "humors." When they existed in harmony within the body, the body was in good health. The duty of the physician was to help nature to restore the body's harmony. Diet, exercise, and moderation in all things kept the body well, and psychological healing (good attitude toward recovery), bed rest, and quiet were part of his therapy. Hippocrates was the first to recognize that different diseases produced different symptoms; and he described them in such detail that the descriptions generally would hold today. His descriptions not only included diagnosis but prognosis.

What is the difference between **homeopathy, osteopathic medicine, naturopathy,** and **chiropractic**?

Developed by a German physician, Christian F.S. Hahneman (1755–1843), the therapy of homeopathy treats patients with small doses of any of two thousand substances. Based on the principle that "like cures like," the medicine selected is one that produces the same symptoms in a healthy person that the disease is producing in the sick person.

Chiropractic is based on the belief that disease results from the lack of normal nerve function. Relying on physical manipulation and adjustment of the spine for therapy, rather than on drugs or surgery, this therapy, used by the ancient Egyptians, Chinese, and Hindus, was rediscovered in 1895 by Daniel David Palmer (1845–1913).

Developed in the United States by Dr. Andrew Skill, osteopathic medicine recognizes the role of the musculoskeletal system in healthy function of the human body.

The physician is fully licensed and uses manipulation techniques as well as traditional diagnostic and therapeutic procedures. Osteopathy is practiced as part of standard Western medicine.

Naturopathy is based on the principle that disease is due to the accumulation of waste products and toxins in the body. Practitioners believe that health is maintained by avoiding anything artificial or unnatural in the diet or in the environment.

What is the number of active **physicians per 100,000 population** in the United States?

There were 223.4 active physicians per 100,000 population in the United States in 1990. This figure is projected to rise to 247.8 per 100,000 population by the year 2000.

How many **nursing homes** are in the United States?

In 1990, there were 15,607 nursing homes in the United States; 11,486 of them were profit-making facilities and 4,121 were non-profit facilities.

How many **health maintenance organizations (HMOs)** are in the United States, and has this number changed over the past decade?

In 1990 the total number of health maintenance organizations (HMOs) of all types was 556. This represented a 135% increase from 236 in 1980.

What is the difference between the degrees **doctor of dental surgery (DDS)** and **doctor of medical dentistry (DMD)**?

The title depends entirely on the school's preference in terminology. The degrees are equivalent.

What is the difference between an **ophthalmologist, optometrist**, and **optician**?

An ophthalmologist is a physician who specializes in care of the eyes. Ophthalmologists conduct examinations to determine the quality of vision and the need for **321** corrective glasses or contact lenses. They also check for the presence of any disorders, such as glaucoma or cataracts. Ophthalmologists may prescribe glasses, contact lenses, and medication, or perform surgery, as necessary.

An optometrist is a specialist trained to examine the eyes and to prescribe, supply, and adjust glasses or contact lenses. Because they are not physicians, optometrists may not prescribe drugs or perform surgery. An optometrist refers patients requiring these types of treatment to an ophthalmologist.

An optician is a person who fits, supplies, and adjusts glasses or contact lenses. Because their training is limited, opticians may not examine or test eyes or prescribe glasses or drugs.

How many people **visit a dentist regularly**?

Only about half the population visits a dentist as often as once a year.

DIAGNOSTIC EQUIPMENT, TESTS, ETC.

What is the meaning of the **medical abbreviation NYD**?

Not yet diagnosed.

How is **blood pressure** measured?

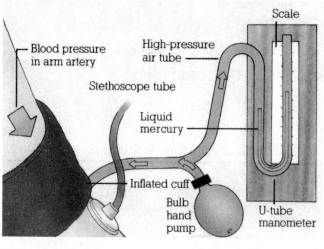

A sphygmomanometer is the device used to measure blood pressure. It was invented in 1881 by an Austrian named Von Bash. It consists of a cuff with an inflatable bladder that is wrapped around the upper arm, a rubber bulb to inflate the bladder, and a device that indicates the pressure of blood. Measuring arterial tension (blood pressure) of a person's circulation is achieved when the cuff is applied to the arm over the artery and pumped to a pressure that occludes or blocks it. This gives the systolic measure, or the maximum pressure of the blood, which occurs during contraction of the ventricles of the heart. Air is then released from the cuff until the blood is first heard passing through the opening artery (called Korotkoff sounds). This gives diastolic pressure, or the minimum value of blood pressure that occurs during the relaxation of the arterial-filling phase of the heart's activity.

What is the meaning of the numbers in a **blood pressure reading?**

When blood is forced into the aorta, it exerts a pressure against the walls; this is referred to as blood pressure. In the reading the upper number, the systolic, measures the pressure during the period of the heart's ventricular contraction. The lower number, the diastolic, measures the pressure when blood is entering the relaxed chambers of the heart. While these numbers can vary due to age, sex, weight, and other factors, the normal blood pressure is around 110/60 to 140/90 millimeters of mercury.

What are the normal **test ranges** for total **cholesterol, triglycerides,** low density lipoproteins **(LDL),** and high density lipoproteins **(HDL)?**

Normal blood cholesterol levels can range from 150 to 289 milligrams per deciliter. In order to lower the risk of coronary heart disease, the recommended total cholesterol level is less than 200 milligrams per deciliter. A range of 200 to 239 is considered to be "borderline" and anything above 240 milligrams per deciliter is considered to be high. LDL–cholesterol in the blood seems to encourage the formation of artery-clogging fatty deposits, so its level should be below 130 milligrams per deciliter (130 to 159 is considered borderline and greater than 160 is high risk). However, the average LDL–cholesterol level is around 150 milligrams per deciliter, with the normal range occurring between 100 and 200. The average HDL–cholesterol level is 55 milligrams per deciliter with the norm range being between 30 and 80. Anything less than 35 is considered a risk factor for a heart attack. The ratio of total cholesterol to HDL–cholesterol (which washes out excess cholesterol from the blood) should be 3.5 to 4.5. Triglycerides (fatty substances consisting of fatty acids and glycerol that are usually stored as fat as an energy reserve) range from 30 to 145 milligrams per deciliter.

What is **nuclear magnetic resonance imaging?**

*M*agnetic *r*esonance *i*maging (MRI), sometimes called *n*uclear *m*agnetic *r*esonance *i*maging (NMRI), is a non-invasive, non-ionizing diagnostic technique. It is useful in detecting small tumors, blocked blood vessels, or damaged vertebral disks. Because it does not involve the use of radiation, it can often be used where x-rays are dangerous. **323**

Large magnets beam energy through the body, causing hydrogen atoms in the body to resonate. This produces energy in the form of tiny electrical signals. A computer detects these signals, which vary in different parts of the body and according to whether an organ is healthy or not. The variation enables a picture to be produced on a screen and interpreted by a medical specialist.

What distinguishes MRI from computerized x-ray scanners is that most x-ray studies cannot distinguish between a living body and a cadaver, while MRI "sees" the difference between life and death in great detail. More specifically, it can discriminate between healthy and diseased tissues with more sensitivity than conventional radiographic instruments like x-rays or CAT scans. CAT (computerized *axial tomography*) scanners have been in use since 1973 and are actually glorified x-ray machines. They offer three-dimensional viewing but are limited because the object imaged must remain still.

The concept of using MRI to detect tumors in patients was proposed by Raymond Damadian in a 1927 patent application. The fundamental MRI imaging concept used in all present-day MRI instruments was proposed by Paul Lauterbar in an article in *Nature* in 1973. The main advantages of MRI are that it not only gives superior images of soft tissues (like organs), but can measure dynamic physiological changes in a noninvasive manner (without penetrating the body in any way).

Ultrasound is another type of 3-D computerized imaging. Using brief pulses of ultrahigh frequency acoustic waves (lasting 0.01 second), this diagnostic tool can produce a sonar map of the imaged object.

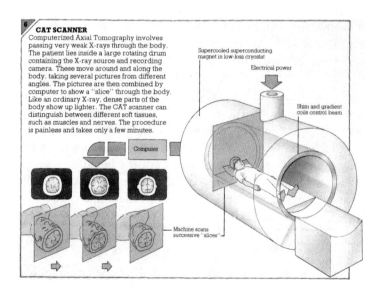

6
CAT SCANNER
Computerized Axial Tomography involves passing very weak X-rays through the body. The patient lies inside a large rotating drum containing the X-ray source and recording camera. These move around and along the body, taking several pictures from different angles. The pictures are then combined by computer to show a "slice" through the body. Like an ordinary X-ray, dense parts of the body show up lighter. The CAT scanner can distinguish between different soft tissues, such as muscles and nerves. The procedure is painless and takes only a few minutes.

Supercooled superconducting magnet in low-loss cryostat

Electrical power

Shim and gradient coils control beam

Computer

Machine scans successive "slices"

What is the instrument a doctor uses to check **reflexes**?

A plessor or plexor or percussor is a small hammer, usually with a soft rubber head, used to tap the body directly. Also called a reflex hammer or a percussion hammer, it is used by a doctor to elicit reflexes by tapping on tendons. In the most common test, the patient sits on a surface high enough to allow his or her legs to dangle freely, and the physician lightly taps the patellar tendon, just below the kneecap. This stimulus briefly stretches the quadriceps muscle on top of the thigh. The stretch causes the muscle to contract, which makes the leg kick forward. The time interval between the tendon tap and the start of the leg extension is about 50 microseconds. That interval is too short for the involvement of the brain, so the muscle movement is totally reflexive. This test indicates the status of an individual's reflex control of movement.

DRUGS, MEDICINES, ETC.

What is **pharmacognosy**?

It is the science of natural drugs and their physical, botanical, and chemical properties.

What is the meaning of the **abbreviations** often used by a doctor when writing a **prescription**?

Latin phrase	Shortened form	Meaning
quaque hora	qh	every hour
quaque die	qd	every day
bis in die	bid	twice a day
ter in die	tid	three times a day
quarter in die	qid	four times a day
pro re nata	prn	as needed
ante cibum	a.c.	before meals
post cibum	p.c.	after meals
per os	p.o.	by mouth
nihil per os	n.p.o.	nothing by mouth
signetur	sig	let it be labeled
statim	stat	immediately
ad libitum	ad lib	at pleasure
hora somni	h.s.	at bedtime
cum	c	with
sine	s	without

Latin phrase	Shortened form	Meaning
guttae	gtt	drops
semis	ss	a half
et	et	and

What are the common **medication measures**?

Approximate equivalents of apothecary measures are given below.

Apothecary volume

Volume	Equivalent
1 minim	0.06 milliliters or 0.02 fluid drams or 0.002 fluid ounces
1½ minims	0.1 milliliters
15 minims	1 milliliter
480 minims	1 fluid ounce
1 dram	3.7 milliliters or 60 minims
1 t (teaspoon)	60 drops
3 t (teaspoons)	½ ounce
1 T (tablespoon)	½ ounce
2 T (tablespoons)	1 ounce
1 C (cup)	8 ounces or 30 milliliters

Apothecary weights

Weight	Equivalent
1 grain	60 milligrams or ½ dram
60 grains	1 dram or 3.75 grams
8 drams	1 ounce or 30 grams

How long can a **prescription drug** be kept?

Generally a prescription drug should not be kept more than one year. Some over-the-counter medications display an expiration date on their box or container. A cream should not be used if it has separated into its components. Although some general guidelines are given below, "When in doubt, throw it out."

Remedy	Maximum shelf life (years)
Cold tablets	1 to 2
Laxatives	2 to 3
Minerals	6 or more
Nonprescription painkiller tablets	1 to 4
Prescription antibiotics	2 to 3

Remedy	Maximum shelf life (years)
Prescription antihypertension tablets	2 to 4
Travel sickness tablets	2
Vitamins (protected from heat, light, and moisture)	6 or more

Which **drugs** are **prescribed the most**?

Brand name	Therapy
Zantac®	Treats ulcers
Cardizem®	Treats cardiovascular problems
Tagamet®	Treats ulcers
Vasotec®	Treats cardiovascular problems
Ceclor®	Antibiotic
Mevacor®	Lowers serum cholesterol
Naprosyn®	Anti-arthritic drug
Capoten®	Treats cardiovascular problems
Tenormin®	Treats cardiovascular problems
Procardia®	Treats cardiovascular problems

What is meant by the term **"orphan drugs"**?

Orphan drugs treat diseases that affect fewer than 200,000 Americans. With little chance of making money, a drug company is not likely to undertake the necessary research and expense of finding drugs that might treat these diseases. The Orphan Drug Act of 1983 offers a number of incentives to drug companies to encourage development of these drugs. The act has provided hope for millions of people with rare and otherwise untreatable conditions.

From what plant is **taxol** extracted?

Taxol is produced from the bark of the western or Pacific yew (*Taxus brevifolia*). It has been shown to inhibit the growth of HeLa cells (human tumor cells) and is a promising new treatment for several kinds of cancer. Taxol, however, is in short supply and may not be able to be synthesized on a commercial scale due to its molecular configuration.

Who discovered **penicillin**?

British bacteriologist Sir Alexander Fleming (1881–1955) discovered the bacteria-killing property of penicillin in 1928. Fleming noticed that no bacteria grew around bits of the *penicillium notatum* fungus that accidentally fell into a bacterial culture in his laboratory. However, although penicillin was clinically used, it was not until 1941 that **327**

Dr. Howard Florey (1898–1968) purified and tested it. The first large-scale plant to produce penicillin was constructed under the direction of Dr. Ernest Chain (1906–1979). By 1945, penicillin was commercially available. In that year, Chain, Florey, and Fleming received a joint Nobel Prize for their work on penicillin.

How many of the medications used today are derived from plants?

Of the more than 250,000 known plant species, less than 1% have been thoroughly tested for medical applications. Yet out of this tiny portion have come 25% of our prescription medicines. The United States National Cancer Institute has identified three thousand plants from which anti-cancer drugs are or can be made. This includes ginseng (*Panax quinquefolius*), Asian mayapple (*Podophyllum hexandrum*), western yew (*Taxus brevifolia*), and rosy periwinkle. 70% of these three thousand come from rain forests, which are also a source of countless other drugs treating diseases and infections. Rain forest plants are rich in so-called secondary metabolites, particularly alkaloids, which biochemists believe the plants produce to protect them from disease and insect attack.

Who discovered the antibiotic **streptomycin**?

Russian-born microbiologist Selman A. Waksman (1888–1973) coined the term antibiotic and subsequently discovered streptomycin in 1943. In 1944 Merck and Company agreed to produce it to be used against tuberculosis and tuberculosis meningitis. Because the drug can adversely affect hearing and balance, careful control of dosage is needed.

Who developed the **poliomyelitis vaccine** in America?

Immunologist Jonas E. Salk (b. 1914) developed the first vaccine (made from a killed-virus) against poliomyelitis. In 1952 he prepared and tested the vaccine and in 1954 massive field tests were successfully undertaken. Two years later immunologist Dr. Albert Sabin (b. 1906) developed an oral vaccine made from inactivated live viruses of three polio strains.

Who was the first to use **chemotherapy**?

Chemotherapy is the use of chemical substances to treat diseases, specifically malignant diseases. The drug must interfere with the growth of bacterial, parasitic, or tumor cells, without significantly affecting host cells. Especially effective in types of cancer as leukemia and lymphoma, chemotherapy was introduced in medicine by the German physician Paul Ehrlich (1854–1915).

What are **monoclonal antibodies**?

Monoclonal antibodies are artificially produced antibodies, designed to neutralize a specific foreign protein (antigen). Cloned cells (genetically identical) are stimulated to produce antibodies to the target antigen. Most monoclonal antibody work so far has involved cloned cells from mice infected with cancer. In some cases monoclonal antibodies they are used to destroy cancer cells directly; in others they carry other drugs to combat cancer cells.

Why are **anabolic steroids** harmful?

Anabolic (protein-building) steroids are drugs that mimic the effects of testosterone and other male sex hormones. They can build muscle tissue, strengthen bone, and speed muscle recovery following exercise or injury. They are sometimes prescribed to treat osteoporosis in postmenopausal women and some types of anemia. Some athletes use anabolic steroids to build muscle strength and bulk, and to allow a more rigorous training schedule. Weight lifters, field event athletes, and body builders are most likely to use anabolic steroids. The drugs are banned from most organized competitions because of the dangers they pose to health and to prevent an unfair advantage.

Adverse effects include hypertension, acne, edema, and damage to liver, heart, and adrenal glands. Psychiatric symptoms can include hallucinations, paranoid delusions, and manic episodes. In men, anabolic steroids can cause infertility, impotence, and premature balding. Women can develop masculine characteristics, such as excessive body and facial hair growth, male-pattern balding, disruption of menstruation, and deepening of the voice. Children and adolescents can develop problems in growing bones, leading to short stature.

How is **patient-controlled analgesia (PCA)** administered?

This drug delivery system dispenses a preset intravenous (IV) dose of a narcotic analgesic for reduction of pain when the patient pushes a switch on an electric cord. The device consists of a computerized pump with a chamber containing a syringe holding up to 60 milliliters of a drug. The patient administers a dose of narcotic when the need for pain relief arises. A lockout interval device automatically inactivates the system if the patient tries to increase the amount of narcotic within a preset time period.

How is the drug **Clomid®** used to treat infertility?

In female infertility, Clomid® acts by stimulating gonadotropin release, which then stimulates ovulation. In men, the drug is used to stimulate sperm production. Multiple births may occur as a result of the treatment.

How does **RU-486, the abortion pill,** cause a woman to abort a fetus?

A pill containing RU-486 (mifepristone) causes the abortion of a fetus by depriving the fertilized embryo of a compatible uterine environment, to terminate the pregnancy within 45 days of fertilization. It is currently used in China and France.

What is **China white**?

China white is a powerful synthetic heroin, and even small amounts can be fatal. The chemical name is 3-methylfentanyl.

Which tests detect **cocaine** use?

Urine tests detect cocaine used by a person within the past 24 to 36 hours. Depending on hair length, hair analysis can detect cocaine used more than a year ago, because hair grows only about 0.5 inch (1.27 centimeters) per month.

Where is **marijuana grown legally** in the United States?

The federal marijuana farm at the University of Mississippi was established in 1968. This farm supplies most of the marijuana (*Cannabis sativa*) used for official medical research in America.

How long do **chemicals from marijuana** stay in the body?

When marijuana is smoked, *tetra*hydrocannabinol (THC), its active ingredient, is absorbed primarily in the fat tissues. The body transforms the THC into metabolites that can be detected by urine tests for up to a week. Tests involving radioactively labeled THC have traced the metabolites for up to a month. The retention of labeled THC in humans is about 40% at three days and 30% at one week.

SURGERY AND OTHER NON-DRUG TREATMENTS

What are the most **common surgical procedures** performed in the United States?

In 1990, these were the most-performed operations:

1. *Caesarean section* (with 909,000 performed)
2. *Gall bladder removal* (with 498,000 performed)
3. *Heart catheterization* (with 481,000 performed)
4. *Hysterectomy* (with 437,000 performed)

What are the most frequent surgical procedures for which **second opinions** are most commonly advised?

Back surgery
Breast surgery
Bunion removal
Cataract removal
Dilatation and curettage
Gall bladder removal
Heart surgery
Hemorrhoid removal
Hernia repair
Hip reconstruction
Hysterectomy
Knee surgery
Nasal septum repair
Prostate removal
Tonsillectomy
Varicose vein removal

What is the most popular type of **plastic surgery**?

The 1988 statistics from the American Academy of Facial Plastic and Reconstructive Surgery indicate that the most popular plastic surgical procedure is rhinoplasty or a "nose job." Sometimes in conjunction with this surgery mentoplasty (chin implantation) is done as well.

Procedure	Number performed
Nose surgery (rhinoplasty)	99,358
Head/neck tumors	70,247
Head/neck trauma	43,012
Injectable fillers (zyderm)	42,529
Head/neck reconstructive	41,929
Eyelid surgery (blepharoplasty)	38,732
Scar revision	30,187

Procedure	Number performed
Facial fractures	26,711
Facelift	21,025
Face sanding (dermabrasion)	15,393

What is the **longest operation** on record?

The longest surgical operation was performed on Mrs. Gertrude Levandowski of Burnips, Michigan, for removal of an ovarian cyst. She lost 308 pounds during the operation; her weight was 616 pounds before it started. The operation lasted 96 hours and took place from Feb. 4 to 8, 1951 in Chicago, Illinois.

Who was the **first black surgeon** to do **heart surgery**?

Dr. Daniel Hale Williams (1858–1931) was a pioneer in open heart surgery. In 1893, he was able to save a knifing victim by opening up the patient's chest with the help of a surgical team, exposing the beating heart and sewing the knife wound a fraction of an inch from the heart. This feat was performed without the aid of x-rays, blood transfusions, or anesthetics.

Does **catgut** really come from cats?

Catgut, an absorbable sterile strand, is obtained from collagen derived from healthy mammals. It was originally prepared from the submucous layer of the intestines of sheep. It is used as a surgical ligature.

How much do **organ transplant** operations generally cost in the United States and how successful are they?

Two American doctors, Joseph E. Murray and E. Donnall Thomas, shared the 1990 Nobel Prize for Medicine for their breakthroughs in organ transplantation. Murray performed the first successful organ transplant in 1954 when he transferred a healthy kidney from one identical twin to the other, and Thomas pioneered bone marrow transplants as a cure for leukemia. Murray recognized that close genetic makeup helped to overcome the body's natural rejection process of transplanted organs. In 1967 cardiac patients with failing hearts were encouraged when Dr. Christian Barnard (b. 1922) performed the first successful human heart transplant. By 1970, however, the mortality rates of heart transplant patients were so high that leading medical centers were considering dropping such operations.

The primary difficulty was the refusal of the body's immune system to accept the

organ and not treat it as a "foreign invader." The promise of transplants could not be realized until an effective immunosuppressant could be discovered. In 1979 Dr. Jean Borel of the Sandoz, Ltd., pharmaceutical company won a United States patent and clinical testing rights for his immunosuppressant cyclosporine (derived from soil fungal compound). Although there was some concern that the drug could induce T-lymphocytes in the immune system to produce a form of cancer called lymphoma, the drug was successful in increasing transplant survival rates. With the drug and improved techniques, kidney transplant survival rates went from 53%–83% to 73%–90% and one-year heart transplant survival rates went from 63% to 81%. One-year liver transplant survival percentages went from 32% to 54%–71%. By 1990 many once-difficult operations became relatively routine. For example, 300,000 cornea transplants have been done with a success rate of 90%.

Dr. Thomas Starzl developed FK-506, a new immunosuppressant, that is 50 to 100 times more powerful than cyclosporine. In addition, organs now can be preserved in ViaSpan®, a DuPont organ preservative that keeps organs awaiting transplant viable for a day or longer; now time and distance are less serious obstacles to matching organs with potential recipients. On April 15, 1991, according to the United Network for Organ Sharing, 22,909 Americans were waiting for donor organs: 18,366 for kidneys; 2,007 for hearts; 1,386 for livers; 570 for pancreases; 411 for lungs; and 169 for heart–lung. Below are listed some types of transplants, their costs, and success rates (as of 1988).

Organ	Number of transplants	Cost range	Mean cost	Success rate (%)
Heart	1,689	$50,000–287,000	$148,000	82.8
Heart–lung	67	135,000–250,000	210,000	56.9
Kidney	8,942	25,000–130,000	51,000	92–97
Liver	2,191	66,000–367,000	235,000	76.0
Lung	91	N.A.	225,000	48.4
Pancreas	418	51,000–135,000	70,000	89.1

Who received the **first heart transplant**?

On December 3, 1967, in Capetown, South Africa, Dr. Christiaan Barnard (b. 1922) and a team of 30 associates performed the first heart transplant. In a five-hour operation the heart of Denise Ann Darval, age 25, an auto accident victim, was transplanted into the body of Louis Washansky, a 55-year-old wholesale grocer. Washansky lived for 18 days before dying of pneumonia.

The first heart transplant performed in the United States was on a 2½-week-old baby boy at Maimonides Hospital, Brooklyn, New York, on December 6, 1967, by Dr. Adrian Kantrowitz (b. 1918). The baby boy lived 6½ hours. The first adult to receive a heart transplant in the United States was Mike Kasperak, age 54, at the Stanford Medical Center in Palo Alto, California, on January 6, 1968. Dr. Norman Shumway performed the operation. Mr. Kasperak lived 14 days. From 1968 to February, 1990, there were 7,831 heart transplants performed. In 1989 alone, 1,621 transplants were done.

When was the **first artificial heart** used?

On December 2, 1982, Dr. Barney B. Clark (1921–1983), a 61–year–old retired dentist, became the first human to receive a permanently implanted artificial heart. It was known as the Jarvik-7 after its inventor, Dr. Robert Jarvik (b. 1946). The 71/2-hour operation was performed by Dr. William DeVries (b. 1943), a surgeon at the University of Utah Medical Center. Dr. Clark died on March 23, 1983, 112 days later. In Louisville, Kentucky, William Schroeder (1923–1986) survived 620 days with an artificial heart (November 25, 1984, to August 7, 1986). On January 11, 1990, the United States Food and Drug Administration (FDA) recalled the Jarvik-7, which had been the only artificial heart approved by the FDA for use.

How much does the **cardiac pacemaker** weigh?

The modern pacemaker generator is a hermetically sealed titanium metal can weighing from 1 to 4.5 ounces (30 to 130 grams) and powered by a lithium battery that can last 2 to 15 years. Used to correct an insufficient or irregular heartbeat, the cardiac pacemaker corrects the low cardiac rhythm through electrical stimulation, which increases the contractions of the heart muscle. The contraction and expansion of the heart muscle produces a heartbeat, one of three billion in an average lifetime, which pumps blood throughout the body. A pacemaker has gold or platinum electrodes, conducting wires, and a pacing box (miniature generator). In both kinds of pacemakers—those implanted in the chest or external—the electrode is attached to the heart's right ventricle (chamber), either directly through the chest or threaded through a vein.

Why are **eye transplants** not available?

The eye's retina is part of the brain, and the retina's cells are derived from brain tissue. Retinal cells and the cells that connect them to the brain are the least amenable to being manipulated outside the body.

Who was the **first test-tube baby**?

Louise Brown, born on July 25, 1978, was the first baby produced from fertilization done in vitro—outside the mother's body. *In vitro fertilization* (IVF) actually occurs in a glass dish, not a test tube, where eggs from the mother's ovary are combined with the father's sperm (in a salt solution). Fertilization should occur within 24 hours and when cell division begins, these fertilized eggs are placed in the mother's womb (or possibly another woman's womb).

What is **lithotripsy**?

Lithotripsy is the use of ultrasonic or shock waves to pulverize kidney stones (calculi), allowing the small particles to be excreted or removed from the body. There are two different methods: *extracorporeal shock wave lithotripsy* (ESWL) and percutaneous lithotripsy. The ESWL method, used on smaller stones, breaks up the stones with external shock waves from a machine called a lithotripter. This technique has eliminated the need for more invasive stone surgery in many cases. For larger stones, a type of endoscope, called a nephroscope, is inserted into the kidney through a small incision. The ultrasonic waves from the nephroscope shatter the stones, and the fragments are removed through the nephroscope.

Prior to the condom, what was the main **contraceptive** practice?

Contraceptive devices have been used throughout recorded history. The most traditional of such devices was a sponge soaked in vinegar. The condom was named for its English inventor, the personal physician to Charles II (1630–1685), who used a sheath of stretched, oiled sheep intestine to protect the king from syphilis. Previously penile sheaths were used, such as the linen one made by Italian anatomist, Gabriel Fallopius (1523–1562), but they were too heavy to be successful.

What is **synthetic skin**?

The material consists of a very porous collagen fiber bonded to a sugar polymer (*glycoaminoglycan*) obtained from shark cartilage. It is covered by a sheet of silicon rubber. Synthetic skin was developed by Ioannis V. Yannas and his colleagues at Massachusetts Institute of Technology around 1895.

How is a **mustard plaster** made?

Mix 1 tablespoon of powdered mustard with 4 tablespoons of flour. Add enough warm water to make a runny paste. Put the mixture in a folded cloth and apply to the chest. Rub olive oil on the chest first if the patient has delicate skin.

What is a **negative ion generator**?

A negative ion generator is an electrostatic air cleaner that sprays a continuous fountain of negatively charged ions into the air. Some researchers say that the presence of these ions causes a feeling of well-being, increased mental and physical energy, and relief from some of the symptoms of allergies, asthma, and chronic headaches. It may also aid in healing of burns and peptic ulcers.

What is **reflexology**?

Reflexology—an ancient Chinese and Egyptian therapy—is the application of specific pressures to reflex points in the hands and feet. The reflex points relate to every organ and every part of the body. Massaging of the reflex points is done to prevent or cure diseases. Dr. William Fitzgerald and Eunice D. Ingham introduced reflexology into the United States earlier this century.

Who developed **psychodrama**?

Psychodrama was developed by Jacob L. Moreno, a psychiatrist born in Romania, who lived and practiced in Vienna until he came to the United States in 1925. He soon began to conduct psychodrama sessions and founded a psychodrama institute in Beacon, New York, in 1934. Largely because of Moreno's efforts it is practiced worldwide.

Psychodrama is defined as a method of group psychotherapy in which personality makeup, interpersonal relationships, conflicts, and emotional problems are explored by means of special dramatic methods.

WEIGHTS, MEASURES, TIME, TOOLS, AND WEAPONS

WEIGHTS, MEASURES, AND MEASUREMENT

See also: Physics and Chemistry—Measurement, Methodology, etc.;
Space—Observation and Measurement; Earth—Observation and Measurement;
Energy—Measures and Measurement; Health and Medicine—Drugs, Medicines, etc.

How much does the **Biblical shekel** weigh in modern units?

A shekel is equal to 0.497 ounces (14.1 grams). Below are listed some ancient measurements with some modern equivalents given.

Biblical
   Volume

omer	=	4.188 quarts (modern) or 0.45 peck (modern) or 3.964 liters (modern)
9.4 omers	=	1 bath
10 omers	=	1 ephah

   Weight

shekel	=	0.497 ounces (modern) or 14.1 grams (modern)

   Length

cubit	=	21.8 inches (modern)

Egyptian
   Weight

60 grams	=	1 shekel
60 shekels	=	1 great mina
60 great minas	=	1 talent

Greek
  Length

cubit	=	18.3 inches (modern)
stadion	=	607.2 or 622 feet (modern)

  Weight

obol or obolos	=	715.38 milligrams (modern) or. 0.04 ounces (modern)
drachma	=	4.2923 grams (modern) or 6 obols
mina	=	0.9463 pounds (modern) or 96 drachmas
talent	=	60 mina

Roman
  Length

cubit	=	17.5 inches (modern)
stadium	=	202 yards (modern) or 415.5 cubits

  Weight

denarius	=	0.17 ounces (modern)

  Volume

amphora	=	6.84 gallons (modern)

What is the **SI system** of measurement?

French scientists as far back as the seventeenth and eighteenth centuries questioned the hodgepodge of the many illogical and imprecise standards used for measurement, and began a crusade to make a comprehensive, logical, precise, and universal measurement system called Système Internationale d'Unités, or SI for short. It uses the metric system as its base. Since all the units are in multiples of tens, calculations are simplified. Today all countries except the United States, Burma, South Yemen, and Tonga use this system. However, elements within American society use SI—scientists, exporting/importing industries, and federal agencies (as of November 30, 1992).

The SI is developed from seven fundamental standards: the meter (for length), the kilogram (for mass), the second (for time), the ampere (for electric current), the kelvin (for temperature), the candela (for luminous intensity), and the mole (for amount of substance). In addition, two supplementary units, the radian (plane angle) and steradian (solid angle) and a large number of derived units compose the current system, which is still evolving. Some derived units are the hertz, newton, pascal, joule, watt, coulomb, volt, farad, ohm, siemens, weber, tesla, henry, lumen, lux, becquerel, gray, and sievert. SI's unit of volume or capacity is the cubic decimeter, but many still use "liter" in its place. Very large or very small dimensions are expressed through a series of prefixes, which increase or decrease in multiples of tens. For example, a decimeter is 1/10 meter, a centimeter is 1/100 of a meter, and a millimeter is 1/1000 of a meter. A dekameter is 10 meters, a hectometer is 100 meters, and a kilometer is 1,000 meters. The use of these prefixes enable the system to express these units in an orderly way, and avoid inventing new names and new relationships.

How was the length of a **meter** originally determined?

It was originally intended that the meter should represent one ten-millionth of the distance along the meridian running from the North Pole to the equator through Dunkirk, France, and Barcelona, Spain. French scientists determined this distance, working nearly six years and completing the task in November, 1798. They decided to use a platinum-iridium bar as the physical replica of the meter. Although the surveyors made an error of about two miles, the error was not discovered until much later. Rather than changing the length of the meter to conform to the actual distance, scientists in 1889 chose the platinum-iridium bar as the international prototype. It was used until 1960. Numerous copies of it are in other parts of the world, including the United States National Bureau of Standards.

How is the **length of a meter** currently determined?

The meter is equal to 39.37 inches. It is currently defined as the distance traveled by light in a vacuum during 1/299,792,458 of a second. From 1960 to 1983, the length of a meter had been defined as 1,650,763.73 times the wavelength of the orange light emitted when a gas consisting of the pure krypton isotope of mass number 86 is excited in an electrical discharge.

How did the measure the **yard** originate?

In early times the natural way to measure lengths was to use various portions of the body (the foot, the thumb, the forearm, etc.). According to tradition, a yard measured on King Henry I (1068–1135) of England became the standard yard still used today. It was the distance from his nose to the middle fingertip of his extended arm.

Other measures derived from physical activity such as a pace, a league (distance that equalled an hour's walking), an acre (amount plowed in a day), and a furlong (length of a plowed ditch). But obviously these units were unreliable. The ell, based on the distance between the elbow and index fingertip, was used to measure out cloth. It ranged from 0.513 to 2.322 meters depending on the locality where it was used and even on the type of goods measured.

Below are listed some linear measurements that evolved from this old reckoning into U.S. customary measures:

U.S. customery linear measures

1 hand	=	4 inches
1 foot	=	12 inches
1 yard	=	3 feet
1 rod (pole or perch)	=	16.5 feet
1 fathom	=	6 feet

U.S. customery linear measures

1 furlong	=	220 yards or 660 feet or 40 rods
1 (statute) mile	=	1,760 yards or 5,280 feet or 8 furlongs
1 league	=	5,280 yards or 15,840 feet or 3 miles
1 international nautical mile	=	6,076.1 feet

Conversion to metric

1 inch	=	2.54 centimeters
1 foot	=	0.304 meters
1 yard	=	0.9144 meters
1 fathom	=	1.83 meters
1 rod	=	5.029 meters
1 furlong	=	201.168 meters
1 league	=	4.828 kilometers
1 mile	=	1.609 kilometers
1 international nautical mile	=	1.852 kilometers

How are U.S. customary measures **converted** to metric measures and vice versa?

Listed below is the conversion process for common units of measure:

To convert from	To	Multiply by
acres	meters, square	4,046.856
centimeters	inches	0.394
centimeters	feet	0.328
centimeters, cubic	inches, cubic	0.06
centimeters, square	inches, square	0.155
feet	meters	0.305
feet, square	meters, square	0.093
gallons, U.S.	liters	3.785
grams	ounces (avoirdupois)	0.035
hectares	kilometers, square	0.01
hectares	miles, square	0.004
inches	centimeters	2.54
inches	millimeters	25.4
inches, cubic	centimeters, cubic	16.387
inches, cubic	liters	0.016387
inches, cubic	meters, cubic	0.0000164
inches, square	centimeters, square	6.4516
inches, square	meters, square	0.0006452
kilograms	ounces, troy	32.15075

To convert from	To	Multiply by
kilograms	pounds (avoirdupois)	2.205
kilograms	tons, metric	0.001
kilometers	feet	3,280.8
kilometers	miles	0.621
kilometers, square	hectares	100
knots	miles per hour	1.151
liters	fluid ounces	33.815
liters	gallons	0.264
liters	pints	2.113
liters	quarts	1.057
meters	feet	3.281
meters	yards	1.094
meters, cubic	yards, cubic	1.308
meters, cubic	feet, cubic	35.315
meters, square	feet, square	10.764
meters, square	yards, square	1.196
miles, nautical	kilometers	1.852
miles, square	hectares	258.999
miles, square	kilometers, square	2.59
miles (statute)	meters	1,609.344
miles (statute)	kilometers	1.609344
ounces (avoirdupois)	grams	28.35
ounces (avoirdupois)	kilograms	0.0283495
ounces, fluid	liters	0.03
pints, liquid	liters	0.473
pounds (avoirdupois)	grams	453.592
pounds (avoirdupois)	kilograms	0.454
quarts	liters	0.946
tons (short/U.S.)	tonne	0.907
ton, long	tonne	1.016
tonne	ton (short/U.S.)	1.102
tonne	ton, long	0.984
yards	meters	0.914
yards, square	meters, square	0.836
yards, cubic	meters, cubic	0.765

What are the equivalents for **dry and liquid measures?**

U.S. customary dry measures

1 pint	=	33.6 cubic inches
1 quart	=	2 pints or 67.2006 cubic inches

U.S. customary dry measures

1 peck	=	8 quarts or 16 pints or 537.605 cubic inches
1 bushel	=	4 pecks or 2,150.42 cubic inches or 32 quarts
1 barrel	=	105 quarts or 7,056 cubic inches
1 pint, dry	=	0.551 liter
1 quart, dry	=	1.101 liters
1 bushel	=	35.239 liters

U.S. customary liquid measures

1 tablespoon	=	4 fluid drams or 0.5 fluid ounce
1 cup	=	0.5 pint or 8 fluid ounces
1 gill	=	4 fluid ounces
4 gills	=	1 pint or 28.875 cubic inches
1 pint	=	2 cups or 16 fluid ounces
2 pints	=	1 quart or 57.75 cubic ounces
1 quart	=	2 pints or 4 cups or 32 fluid ounces
4 quarts	=	1 gallon or 231 cubic inches or 8 pints or 32 gills or 0.833 British quart
1 gallon	=	16 cups or 231 cubic inches or 128 fluid ounces
1 bushel	=	8 gallons or 32 quarts

Conversion to metric

1 fluid ounce	=	29.57 milliliters or 0.029 liter
1 gill	=	0.118 liter
1 cup	=	0.236 liter
1 pint	=	0.473 liter
1 U.S. quart	=	0.833 British quart or 0.946 liter
1 U.S. gallon	=	0.833 British gallon or 3.785 liters

What is the difference between a **short ton**, a **long ton**, and a **metric ton**?

A short ton or U.S. ton or net ton (sometimes just "ton") equals 2,000 pounds; a long ton or an avoirdupois ton is 2,240 pounds; and a metric ton or tonne is 2,204.62 pounds. Some weights for comparison are given below:

U.S. customary measure

1 ounce	=	16 drams or 437.5 grains
1 pound	=	16 ounces or 7,000 grains or 256 drams
1 (short) hundredweight	=	100 pounds
1 long hundredweight	=	112 pounds
1 (short) ton	=	20 hundredweights or 2,000 pounds
1 long ton	=	20 long hundredweights or 2,240 pounds

Conversion to metric

1 grain	=	65 milligrams
1 dram	=	1.77 grams
1 ounce	=	28.3 grams
1 pound	=	453.5 grams
1 metric ton (tonne)	=	2,204.6 pounds

What are the equivalent measurements for **area**?

U.S. customary area measures

1 square foot	=	144 square inches
1 square yard	=	9 square feet or 1,296 square miles
1 square rod or pole or perch	=	30.25 square yards or 272.5 square feet
1 rood	=	40 square rods
1 acre	=	160 square rods or 4,840 square yards or 43,460 square feet
1 section	=	1 mile square or 640 acres
1 township	=	6 miles square or 36 square miles or 36 sections
1 square mile	=	640 acres or 4 roods, or 1 section

International area measures

1 square millimeter	=	1,000,000 square microns
1 square centimeter	=	100 square millimeters
1 square meter	=	10,000 square centimeters
1 are	=	100 square meters
1 hectare	=	100 ares or 10,000 square meters
1 square kilometer	=	100 hectares or 1,000,000 square meters

How much does **water** weigh?

U.S. customary measures

1 gallon	=	4 quarts
1 gallon	=	231 cubic inches
1 gallon	=	8.34 pounds
1 gallon	=	0.134 cubic foot
1 cubic foot	=	7.48 gallons
1 cubic inch	=	.0360 pound
12 cubic inches	=	.433 pound

1 liter	=	1 kilogram
1 cubic meter	=	1 tonne (metric ton)
1 imperial gallon	=	10.022 pounds
1 imperial gallon salt water	=	10.3 pounds

What is a bench mark?

A bench mark is a permanent, recognizable point that lies at a known elevation. It may be an existing object, such as the top of a fire hydrant, or it may be a brass plate placed on top of a concrete post. Surveyors and engineers use bench marks to find the elevation of objects by reading, through a level telescope, the distance a point lies above some already established bench mark.

How are **avoirdupois** measurements converted to **troy measurements** and how do the measures differ?

Troy weight is a system of mass units used primarily to measure gold and silver. An ounce is 480 grains or 31.1 grams. Avoirdupois weight is a system of units that is used for measurement of mass except for precious metals, precious stones, and drugs. It is based on the pound, which is approximately 454 grams. In both systems, the weight of a grain is the same—65 milligrams. However, the two systems do not contain the same weights for other units, even though they use the same name for the unit.

Troy
    1 grain = 65 milligrams
    1 ounce = 480 grains = 31.1 grams
    1 pound = 12 ounce = 5760 grains = 373 grams
Avoirdupois
    1 grain = 65 milligrams
    1 ounce = 437.5 grains = 28.3 grams
    1 pound = 16 ounces   = 7000 grains = 454 grams

To Convert From	To	Multiply By
pounds avdp	ounces troy	14.583
pounds avdp	pounds troy	1.215

To Convert From	To	Multiply By
pounds troy	ounces avdp	1.097
pounds troy	pounds avdp	0.069
ounces avdp	ounces troy	0.911
ounces troy	ounces avdp	1.097

What is a **theodolite**?

This optical surveying instrument used to measure angles and directions is mounted on an adjustable tripod and relies on a spirit level to show when it is horizontal. Similar to the more commonly used transit, the theodolite gives more precise readings; angles can be read to fractions of a degree. It is comprised of a telescope that sights the main target, a horizontal plate to provide readings around the horizon, and a vertical plate and scale for vertical readings. The surveyor uses geometry of triangles to calculate the distance from the angles measured by the theodolite. Such triangulation is used in road- and tunnel-building, and other civil engineering work. One of the earliest forms of this surveying instrument was described by Englishman Leonard Digges in his 1571 treatise, *A Geometrical Treatise Named Pantometria*.

What is meant by **DIN standards**?

They are German standards for products and procedures, covering a wide range of engineering and scientific fields. The German Institute for Standardization (*Deutsches Institut für Normung*) publishes these standards. DIN also represents Germany in the International Organization for Standardization (ISO).

TIME

See also: Biology—Classification, Measurement, and Terms

How is **time** measured?

The passage of time can be measured in three ways. Rotational time is based on the unit of the mean solar day (the average length of time it takes the Earth to complete one rotation around its axis).

Dynamic time, the second way to measure time, uses the motion of the moon and planets to determine time, and avoids the problem of the Earth's varying rotation. The first dynamic time-scale was Ephemeris Time, proposed in 1896 and modified in 1960.

Atomic time is a third way to measure time. This method, using an atomic clock, is based on the "extremely regular oscillations" that occur within atoms. In 1967, the **345** atomic second, the length of time in which 9,192,631,770 vibrations are emitted by a hot cesium atom, was adopted as the basic unit of time. Atomic clocks are now used as international time standards.

But time has also been measured in other, less scientific, terms. Listed below are some other "timely" expressions.

Twilight—The first soft glow of sunlight; the sun is still below the horizon; also the last glow of sunlight after sunset.

Midnight—12 A.M., the point of time when one day becomes the next day and night becomes morning.

Daybreak—The first appearance of the sun.

Dawn—A gradual increase of sunlight.

Noon—12 P.M.; the point of time when morning becomes afternoon.

Dusk—The gradual dimming of sunlight.

Sunset—The last diffused glow of sunlight; the sun is below the horizon.

Evening—A term with wide meaning, evening is generally the period between sunset and bedtime.

Night—The period of darkness, lasting from sunset to midnight or dawn.

The day (24 hours, known as *nychthemeron*) groups into a week (seven days or a *sennight*) and two weeks become a fortnight. A year multiplies into a biennium (two years), a triennium (three years), a quadrennium or an olympiad (four years), and then into a lustrum (five years). Ten years is a decade, or *decennium*, twenty years become a score, or a *vicennium*, five score is a century, and ten centuries become a millennium.

What is the exact length of a **calendar year**?

The calendar year is defined as the time between two successive crossings of the celestial equator by the sun at the vernal equinox. It is exactly 365 days, 5 hours, 48 minutes, 46 seconds. The fact that the year is not a whole number of days has affected the development of calendars, which over time generate an accumulative error. The current calendar used, the Gregorian calendar, named after Pope Gregory XII (1502–1585), attempts to compensate by adding an extra day to the month of February every 4 years. A "leap year" is a year with the extra day added.

When does a **century begin**?

A century has 100 consecutive calendar years. The 1st century consisted of years 01 through 100. The 20th century started with 1901 and will end with 2000. The 21st century will begin January 1, 2001.

When did **January 1** become the first day of the new year?

When Julius Caesar (100 B.C.E.–44 B.C.E.) reorganized the Roman calendar and made it solar rather than lunar in the year 45 B.C.E., he moved the beginning of the year to January 1. When the Gregorian calendar was introduced in 1582, January 1 continued to be recognized as the first day of the year in most places. However, in England and the American colonies, March 25, intended to represent the spring equinox, was the beginning of their year. Under this system, March 24, 1700, was followed by March 25, 1701. In 1752, the British government changed the beginning date of the year to January 1.

Besides the Gregorian calendar, what **other kinds of calendars** have been used?

Babylonian calendar—A lunar calendar composed of alternating 29-day and 30-day months to equal roughly 354 lunar days. When the calendar became too misaligned with astronomical events, an extra month was added. In addition, 3 extra months were added every eight years to coordinate this calendar with the solar year.

Chinese calendar—A lunar month calendar of 12 periods having either 29 or 30 days (to compensate for the 29.5 days from new moon to new moon). The new year begins on the first new moon over China after the sun enters Aquarius (between January 21st and February 19th). Each year has both a number and a name (for example, the year 1992, or 4629 in the Chinese era, was the year of the monkey). The calendar is synchronized with the solar year by the addition of extra months at fixed intervals.

Muslim calendar—A lunar 12-month calendar with 30 and 29 days alternating every month and totalling 354 days. The calendar has a 30-year cycle with designated leap years of 355 days (one day added to last month) in the 30-year period. The Islamic year does not attempt to relate to the solar year (relate to the season). The dating of the beginning of the calendar is 622 C.E. (the date of Mohammed's flight from Mecca to Medina).

Jewish calendar—A blend of the solar and lunar calendar, this calendar adds an extra month (Adar Sheni or the second "Adar" or Veadar) to keep the lunar and solar years in alignment. This occurs seven times during the 19-year cycle. When the extra 29-day month is inserted, the month Adar has 30 days instead of 29. In a usual year the 12 months alternate in a 30-day and 29-day repetition.

Egyptian calendar—The ancient Egyptians were the first to use a solar calendar (about 4236 B.C.E. or 4242 B.C.E.), but their year started with the rising of Sirius, the brightest star in the sky. The year, composed of 365 days, was 0.25 days short of the true solar year, so eventually the Egyptian calendar did not coincide with the seasons. It used twelve 30-day months with five-day weeks and five days of festival.

347

Coptic calendar—Still used in areas of Egypt and Ethiopia, it has a similar cycle to the Egyptian calendar; 12 months of 30 days followed by five complementary days. When a leap year occurs, usually preceding the Julian calendar leap year, the complementary days increase to six.

Roman calendar—Borrowing from the ancient Greek calendar, which had a four-year cycle based on the Olympic Games, the earliest Roman calendar (about 738 B.C.E.) had 304 days with 10 months. Every second year a short month of 22 or 23 days was added to coincide with the solar year. Eventually two more months were added at the end of the year (Januarius and Februarius) to increase the year to 354 days. The Roman republican calendar replaced this calendar during the reign of Tarquinius Priscus (616–579 B.C.E.). This new lunar calendar had 355 days, with the month of February having 28 days. The other months had either 29 or 31 days. To keep the calendar aligned with the seasons, an extra month was added every two years. By the time the Julian calendar replaced this one, the calendar was three months ahead of the season schedule.

Julian calendar—Julius Caesar (100 B.C.E.–44 B.C.E.) in 46 B.C.E., wishing to have one calendar in use for all the empire, had the astronomer Sosigenes develop a uniform solar calendar establishing the year 365 days, with one day ("leap day") added every fourth year to compensate for the true solar year of 365.25 days. The year had 12 months with 30 or 31 days except for February, which had 28 days (or 29 days in a leap year). The first of the year was moved from March 1 to January 1.

Gregorian calendar—Pope Gregory XIII (1502–1585) instituted calendrical reform in 1582 to realign the church celebration of Easter with the vernal equinox (the first day of spring). To better align this solar calendar with the seasons, the new calendar would not have leap year in the century years that were not divisible by 400. Because the solar year is shortening, today a one-second adjustment (usually on December 31st at midnight) is made, when appropriate, to compensate for this effect.

Japanese calendar—It has the same structure as the Gregorian calendar in years, months, and weeks. But the years are enumerated in terms of the reigns of emperors as epochs. The last epoch (for Emperor Akihito) is Epoch Heisei, started January 8, 1989.

Hindu calendars—The principal Indian calendars reckon their epochs from historical events, such as rulers' accessions or death dates, or a religious founder's dates. The Vikrama era (originally from Northern India and still used in western India) dates from February 23rd, 57 B.C.E. in the Gregorian calendar. The Saka era dates from March 3, 78 C.E. in the Gregorian calendar and is based on the solar year with 12 months of 365 days and 366 days in leap years. The first five months have 31 days; the last seven have 30 days. In leap years, the first six months have 31 days and the last six have 30. The Saka era is the national calendar of India (as of 1957). The Buddhist era starts with 543 B.C.E. (believed date of Buddha's death).

Three other secular calendars of note are the Julian Day calendar (a calendar astronomers use that counts days within a 7,980-year period and must be used with a table), the perpetual calendar (which gives the days of the week for the Julian and Gregorian calendars as well), and the World calendar, which is similar to the perpetual calendar, having 12 months of 30 or 31-days a year-day at the end of each year, and a leap-year day before July 1 every four years.

There have been attempts to reform and simplify the calendar. One such example is the Thirteen-Month or International Fixed calendar which would have 13 months of four weeks each. The month *Sol* would come before July; there would be a year-day at the end of each year and a leap-year-day every four years just before July 1. A radical reform was done in France when the French republican calendar (1793–1806) replaced the Gregorian calendar after the French Revolution. It had 12 months of 30 days and 5 supplementary days at the end of the year (6 in a leap year), and weeks were replaced with 10-day decades.

Do calendars other than the Gregorian calendar use the same **twelve-month cycle?**

Some calendars are organized on varying cycle of months, with their first month falling at a different time. Listed below are some variations, with the beginning month of the year in italics:

Months of the year

Gregorian	Hebrew	Hindu
January	Shebat	Magha
February	Adar	Phalguna
March	Nisan	*Caitra*
April	Iyar	Vaisakha
May	Sivan	Jyaistha
June	Tammuz	Asadhe
July	Ab	Sravana
August	Elul	Bhadrapada
September	*Tishri*	Asvina
October	Heshvan	Karttika
November	Kislev	Margasivsa
December	Tebet	Pansa
	Ve-Adar (13th month every three years)	

Muslim	Chinese	
Muharram	*Li Chun*	Li Qui
Safar	Yu Shui	Chu Shu
Rabi I	Jing Zhe	Bai Lu
Rabi II	Chun Fen	Qui Fen

Muslim	Chinese	
Jumada I	Qing Ming	Han Lu
Jumada II	Gu Yu	Shuang Jiang
Rajab	Li Xia	Li Dong
Sha'ban	Xiao Man	Xiao Xue
Ramadan	Mang Zhong	Da Xue
Shawwal	Xia Zhi	Dong Zhi
Dhu'l-Qa'da	Xiao Shu	Xiao Han
Dhu'l-hijja	Da Shu	Da Han

Which animal designations have been given to the Chinese years?

Chinese years are labelled with 12 different names, always used in the same sequence: Rat, Ox, Tiger, Hare (Rabbit), Dragon, Snake, Horse, Sheep (Goat), Monkey, Rooster, Dog, and Pig. The following table may be used.

Chinese year cycle

Rat	Ox	Tiger	Hare (Rabbit)	Dragon	Snake
1900	1901	1902	1903	1904	1905
1912	1913	1914	1915	1916	1917
1924	1925	1926	1927	1928	1929
1936	1937	1938	1939	1940	1941
1948	1949	1950	1951	1952	1953
1960	1961	1962	1963	1964	1965
1972	1973	1974	1975	1976	1977
1984	1985	1986	1987	1988	1989
1996	1997	1998	1999	2000	2001

Horse	Sheep (Goat)	Monkey	Rooster	Dog	Pig
1906	1907	1908	1909	1910	1911
1918	1919	1920	1921	1922	1923
1930	1931	1932	1933	1934	1935
1942	1943	1944	1945	1946	1947
1954	1955	1956	1957	1958	1959
1966	1967	1968	1969	1970	1971
1978	1979	1980	1981	1982	1983
1990	1991	1992	1993	1994	1995
2002	2003	2004	2005	2006	2007

The Chinese year of 354 days begins three to seven weeks into the western 365-day year, so the animal designation changes at that time, rather than on January 1.

What is the **Julian Day Count**?

This system of counting days rather than years was developed by Joseph Justus Scaliger (1540–1609) in 1583. Still used by astronomers today, the Julian Day Count (named after Scaliger's father, Julius Caesar Scaliger) designates Julian Day (JD) 1 as January 1, 4113 B.C.E. On this date the Julian calendar, the ancient Roman tax calendar, and the lunar calendar all coincided. This event would not occur again until 7,980 years later. So each day within the 7,980 year period is numbered. December 31, 1991, at noon is the beginning of JD 2,448,622. The figure reflects the number of days that have passed since the count's inception. To convert Gregorian calendar dates into Julian Day, simple JD conversion tables have been devised for astronomers.

What were the **longest** and **shortest years** on record?

The longest year was 46 B.C.E. when Julius Caesar (100 B.C.E.–44 B.C.E.) introduced his calendar, called the Julian calendar, which was used until 1582. He added 2 extra months and 23 extra days to February to make up for accumulated slippage in the Egyptian Calendar. Thus, 46 B.C.E. was 455 days long. The shortest year was 1582, when Pope Gregory XIII (1502–1585) introduced his calendar, the Gregorian calendar. He decreed that October 5 would be October 15, eliminating 10 days, to make up for the accumulated error in the Julian calendar. Not everyone changed over to this new calendar at once. Catholic Europe adopted it within 2 years of its inception. Many Protestant continental countries did so in 1699–1700; England imposed it on its colonies in 1752 and Sweden in 1753. Many non-European countries adopted it in the 19th century, with China doing so in 1912, Turkey in 1917, and Russia in 1918. To change from the Julian to the Gregorian calendar, 10 days are added to dates October 5, 1582, through February 28, 1700; after that date add 11 days through February 28, 1800; 12 days through February 28, 1900; and 13 days through February 28, 2100.

When does **leap year** occur?

A leap year occurs when the year is exactly divisible by four except centenary years. A centenary year must be divisible by 400 to be a leap year. There was no February 29 in 1900, which was not a leap year. The next centenary year, 2000, will have February 29.

What is a **leap second**?

The Earth's rotation is slowing down, and to compensate for this lagging motion, June 1992 was one second longer than normal. This leap second, announced in February, **351**

1992, by the International Earth Rotation Service, kept calendar time in close alignment with international atomic time. To accomplish this change, 23h 59m 59s universal time on June 30, 1992, was followed by 23h 59m 60s, and this in turn was followed by 0h 0m 0s on July 1.

Where do the names of the **days of the week come from?**

The English days of the week are named for a mixture of figures in Anglo-Saxon and Roman mythology.

Day	Named after
Sunday	The sun
Monday	The moon
Tuesday	Tiu (the Anglo-Saxon god of war, equivalent to the Norse Tyr or the Roman Mars)
Wednesday	Woden (the Anglo-Saxon equivalent of Odin, the chief Norse god)
Thursday	Thor (the Norse god of thunder)
Friday	Frigg (the Norse god of love and fertility, the equivalent of the Roman Venus)
Saturday	Saturn (the Roman god of agriculture)

What is the **origin** of the **week?**

The week originated in the Babylonian calendar, in which one day out of seven was devoted to rest.

How were the **months of the year** named?

The English names of the months of the current (Gregorian) calendar come from the Romans, who tended to honor their gods, specific events, etc., by designating them as month names:

January (*Januarius* in Latin) is named after Janus, a Roman god with two faces, one looking into the past, the other into the future.

February (*Februarium*) from the Latin word *Februare*, meaning to cleanse. At the time of year corresponding to our February, the Romans performed religious rites to purge themselves of sin.

March (*Martius*) is named in honor of Mars, the god of war.

April (*Aprilis*), after the Latin word *Aperio*, meaning to open, because plants begin to grow in this month.

May (*Maius*), after the Roman goddess Maia, as well as from the Latin word *Maiores*, meaning elders, who were celebrated during this month.

June (*Junius*), after the goddess Juno and Latin word *iuniores*, meaning young people.

July (*Iulius*) was, at first, known as *Quintilis,* from the Latin word meaning five, since it was the fifth month in the early Roman calendar. Its name was changed to July, as an honor to Julius Caesar (100 B.C.E.–44 B.C.E.).

August (*Augustus*) is named in honor of the Emperor Octavian (63 B.C.E.–14 B.C.E.), first Roman emperor, known as Augustus Caesar. Originally the month was known as *Sextilis* (sixth month of early Roman calendar).

September (*September*) was the seventh month, and accordingly took its name from *septem*, meaning seven.

October (*October*) takes its name from *octo* (eight), because originally it was the eighth month.

November (*November*) from *novem*, meaning nine, from the ninth month of the early Roman calendar.

December (*December*) from *decem*, meaning ten, from the tenth month of the early Roman calendar.

What are the **beginning dates of the seasons** for 1994 to 2000?

The four seasons in the Northern Hemisphere coincide with astronomical events. The spring season starts with the vernal equinox (around March 21) and lasts until the summer solstice (June 21 or 22); summer is from the summer solstice to the autumnal equinox (about September 21); autumn, from the autumnal equinox to the winter solstice (December 21 or 22); and winter from the winter solstice to the vernal equinox. In the Southern Hemisphere the seasons are reversed; autumn corresponds to spring; winter to summer. The seasons are caused by the tilt of the Earth's axis, which changes the apparent position of the sun in the sky (or angle of declination). In winter the sun is at its lowest position in the sky; in summer, it is at its highest position.

Year	Spring	Summer	Fall	Winter
1994	March 20	June 21	September 23	December 22
1995	March 21	June 21	September 23	December 22
1996	March 20	June 21	September 22	December 21
1997	March 20	June 21	September 22	December 21
1998	March 20	June 21	September 23	December 22
1999	March 21	June 21	September 23	December 22
2000	March 20	June 21	September 22	December 21

How is the date for **Passover** determined?

Passover (*Pesah*) begins at sundown on the evening before the 15th day of the Hebrew month of Nisan, which falls in March or April. Celebrated as a public holiday in Israel, the Passover commemorates the exodus of the Israelites from Egypt in 1290 B.C.E. The term "Passover" is a reference to the last of the 10 plagues that "passed over" the Israelite homes at the end of their captivity in Egypt.

How is the date for **Easter** determined?

The rule for establishing the date in Christian churches is that Easter always falls on the first Sunday after the first full moon that occurs on or just after the vernal equinox. The vernal equinox is the first day of spring in the Northern Hemisphere. Because the full moon can occur on any one of many dates after the vernal equinox, the date of Easter can be as early as March 22 and as late as April 25. For the period 1994–2000 Easter will fall on the following days:

Year	Date
1994	April 3
1995	April 16
1996	April 7
1997	March 30
1998	April 12
1999	April 4
2000	April 23

When is **Daylight Savings Time** observed in the United States?

In 1967, all states and possessions of the United States were to begin observing Daylight Savings Time at 2 A.M. on the first Sunday in April of every year. The clock advances one hour at that time until 2 A.M. of the last Sunday of October when the clock is turned back one hour. The phrase "fall back, spring forward" indicates the direction that the clock setting is to undergo at these appointed dates. A 1972 amendment allowed some areas to be exempt. Arizona, Hawaii, Puerto Rico, the Virgin Islands, American Samoa, and part of Indiana do not follow DST. In the intervening years, the length of the Daylight Savings Time period had been changed, but on July 8, 1986, the starting date of Daylight Saving Time was reinstated as 2 A.M. of the first Sunday in April, with the same ending October date.

Daylight Savings Time was enacted to provide more light in the evening hours during the summer months. Other countries have adopted DST as well. For instance, in western Europe the period is from the last Sunday in March to the last Sunday in

September, with the United Kingdom extending the range to the last Sunday in October. Many countries in the Southern Hemisphere generally maintain DST from October to March; countries near the equator maintain standard time.

How many **time zones** are there in the world?

There are 24 standard time zones that serially cover the Earth's surface at coincident intervals of 15 degrees longitude and 60 minutes Universal Time (UT), as agreed at the Washington Meridian Conference of 1884, thus accounting, respectively, for each 24 hours of the calendar day.

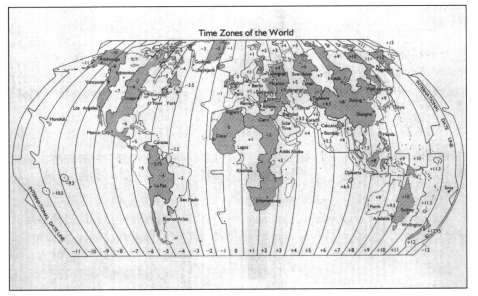

Time Zones of the World

What is meant by **universal time**?

On January 1, 1972, Universal Time (UT) replaced Greenwich Mean Time (GMT) as the time reference coordinate for scientific work. Time is taken from an atomic clock, and is seen as the logical development of the adoption of the atomic second in 1968. An advantage of UT is that the time at which an event takes place can be determined very readily without recourse to the time-consuming astronomical observations and calculations that were necessary before the advent of atomic clocks. Universal Time is also referred to as International Atomic Time. GMT is based on when the sun crosses the Greenwich Meridian (0 degrees longitude, which passes through the Greenwich Observatory).

Who establishes the **correct time** in the United States?

The United States National Institute of Standards and Technology uses an atomic clock accurate to 3 millionths of a second per year to establish the correct time. The atomic clock was constructed at the United States Bureau of Measures in Washington, D.C., in 1948. American physicist William F. Libby formulated the principle of this extremely accurate clock.

What is the **United States Time Standard signal**?

The most-used high-frequency reference signals are the United States Time Standard signals from WWV, Fort Collins, Colorado, and WWVH, Kauai, Hawaii. These stations transmit standard "tic" pulses each second, with minute announcements and special bulletins periodically. They transmit 1.5, 5, 10, 15, and 20 megahertz with very precise time and frequency stability. The first radio station to transmit time signals regularly was the Eiffel Tower Radio Station in Paris in 1913.

What do the initials A.M. and P.M. mean?

A.M. means *ante meridiem,* Latin for "before noon."

P.M. means *post meridiem,* Latin for "after noon."

Why do **clocks and watches** with the **Roman numerals** on their faces use "IIII" for the number 4, rather than "IV"?

A clock with a "IV" has a somewhat unbalanced appearance. Thus tradition has favored the strictly incorrect "IIII," which more nearly balances the equally heavy "VIII."

What is a **floral clock**?

In the Middle Ages it was believed that it was possible to tell the hour of day by the use of flowers, which were supposed to open and close at specific times. These would be planted in flower dials. The first hour belonged to the budding rose, the fourth to hyacinths and the twelfth to pansies. The unreliability of this method of timekeeping must have soon become apparent, but floral clocks can still be seen planted in parks and gardens today. The world's largest clock face is that of the floral clock located inside the Rose Building in Hokkaido, Japan. The diameter of the clock is almost 69 feet (21 meters) and the large hand of the clock is almost 28 feet (8.5 meters) in length.

Where did the term **"grandfather clock"** come from?

The weight-and-pendulum clock was invented by Dutch scientist Christian Huygens (1629–1695) around the year 1656. In the United States, Pennsylvania German settlers considered such clocks, often called long-case clocks, to be a status symbol. In 1876 American songwriter Henry Clay Work (1832–1884) referred to long-case clocks in his song "My Grandfather's Clock" The nickname stuck.

How is **time** denoted at sea?

The maritime day is divided into watches and bells. A watch equals four hours, except for the time period between 4 P.M. and 8 P.M., which has two short watches. From midnight to 4 A.M. is the middle watch; from 4 A.M. to 8 A.M., the morning watch; 8 A.M. to noon, the forenoon watch; noon to 4 P.M., afternoon watch; 4 P.M. to 6 P.M., first dog watch; 6 P.M. to 8 P.M., second or last dog watch; 8 P.M. to midnight, first watch. Within each watch there are eight bells—one stroke for each half hour, so that each watch ends on eight bells except the dog watches, which end at four bells. New Year's Day is marked with 16 bells.

What is **military time**?

Military time divides the day into one set of twenty-four hours, counting from 1:00 A.M. (0100) to midnight (2400). This is expressed without punctuation.

    1:00 A.M. becomes 0100 (pronounced oh-one hundred)
    2:10 A.M. becomes 0210
    Noon becomes 1200
    6:00 P.M. becomes 1800

9:45 P.M. becomes 2145

Midnight becomes 2400 (pronounced twenty-four hundred)

To translate 24-hour time into familiar time, the A.M. times are obvious. For P.M. times, subtract 1200 from numbers larger than 1200; e.g., 1900 − 1200 = 7 P.M.

Who invented the alarm clock?

Levi Hutchins of Concord, New Hampshire, invented an alarm clock in 1787. His alarm clock, however, rang at only one time—4 A.M. He invented this device so that he would never sleep past his usual waking time. He never patented or manufactured it. The first modern alarm clock was made by Antoine Redier (1817–1892) in 1847. It was a mechanical device; the electric alarm clock was not invented until around 1890. The earliest mechanical clock was made in 725 C.E. in China by Yi Xing and Liang Lingzan.

Doomsday clock reset

Who set a **doomsday clock** for nuclear annihilation?

The doomsday clock first appeared on the cover of the magazine *Bulletin of the Atomic Scientists* in 1947 and was set at 11:53 P.M. The clock, created by the magazine's board of directors, represents the threat of nuclear annihilation, with midnight as the time of destruction. It has been recently moved back to 11:43 P.M., the farthest from midnight the clock has ever been. In 1953, just after the United States tested the hydrogen bomb, it was set at 11:58 P.M., the closest to midnight ever.

TOOLS, MACHINES, AND PROCESSES

Which **tools** were used by **Neanderthal man**?

Neanderthal tool kits are termed *Mousterian,* from finds at LeMoustier in France (traditionally dated to the early Fourth Glacial Period, around 40,000 B.C.E.). Using fine-grained glassy stone like flint and obsidian, Neanderthals improved on the already old, established *Levallois* technique for striking one or two big flakes of predetermined shape from a prepared core. They made each core yield many small, thin sharp-edged flakes, which were then trimmed to produce hide scrapers, points, backed knives, stick sharpeners, tiny saws, and borers. These could have served for killing, cutting, and skinning prey, and for making wooden tools and clothing.

What are the six **simple machines**?

All machines and mechanical devices, no matter how complicated, can be reduced to some combinations of six basic, or simple, machines. The lever, the wheel and axle, the pulley, the inclined plane, the wedge, and the screw were all known to the ancient Greeks, who learned that a machine works because an "effort," which is exerted over an "effort distance," is magnified through "mechanical advantage" to overcome a "resistance" over a "resistance distance." Some consider there to be only five simple machines, and regard the wedge as a moving inclined plane.

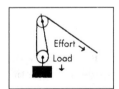

Pulley

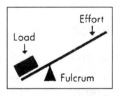

Lever

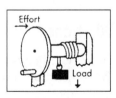

Wheel/Axle

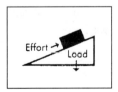

Inclined plane

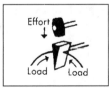

Wedge

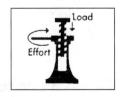

Screw

How does a **four-stroke** differ from a **two-stroke** engine?

A four-stroke engine functions by going through four cycles: 1) the intake stroke draws a fuel-air mixture in on a down stroke; 2) the mixture is compressed on an upward stroke; 3) the mixture is ignited causing a down stroke; and 4) the mixture is exhausted on an up stroke. A two-stroke engine combines the intake and compression strokes (1 and 2) and the power and exhaust strokes (3 and 4) by covering and uncovering ports and valves in the cylinder wall. Two-stroke engines are typically used in small displacement applications, such as chain saws, some motorcycles, and lawnmowers.

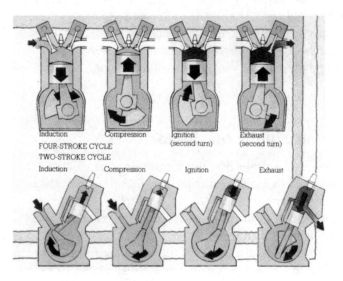

What does "cc" mean in **engine sizes**?

Cubic centimeters (cc), as applied to an internal combustion engine, is a measure of the combustion space in the cylinders. The size of an engine is measured by theoretically removing the top of a cylinder, pushing the piston all the way down and then filling the cylinder with liquid. The cubic centimeters of liquid displaced (spilled out) when the piston is returned to its high position is the measure of the combustion volume of the cylinder. If a motorcycle has four cylinders, each displacing 200 ccs, it has an 800 cc engine. Automobile engines are sized essentially the same way.

What is a **donkey engine**?

A donkey engine is a small auxiliary engine which is usually portable or semi–portable. It is powered by steam, compressed air, or other means. It is often used to power a windlass or lift cargo on shipboard.

What is a **power take-off**?

The standard power take-off is a connection that will turn a shaft inserted through the rear wall of a gear case. It is used to power accessories such as a cable control unit, a winch, or a hydraulic pump. Farmers use the mechanism to pump water, grind feed, or saw wood. A power take-off drives the moving parts of mowing machines, hay balers, combines, and potato diggers.

Who invented the **electron microscope**?

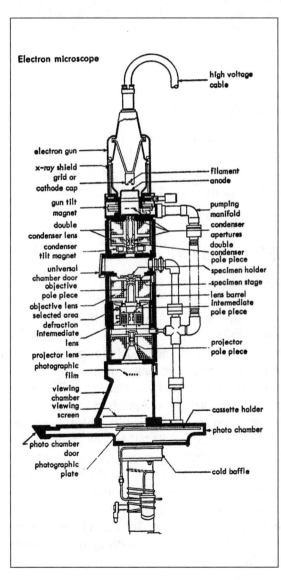

Electron microscope

high voltage cable

electron gun
x-ray shield
grid or cathode cap
gun tilt magnet
double condenser lens
condenser tilt magnet
universal chamber door objective pole piece
objective lens
selected area defraction intermediate lens
projector lens
photographic film
viewing chamber
viewing screen

filament anode
pumping manifold
condenser apertures
double condenser pole piece
specimen holder
specimen stage
lens barrel intermediate pole piece
projector pole piece

cassette holder
photo chamber
photo chamber door
photographic plate
cold baffle

The theoretical and practical limits to the use of the optical microscope were set by the wavelength of light. When the oscilloscope was developed, it was realized that cathode-ray beams could be used to resolve much finer detail because their wavelength was so much shorter than that of light. In 1928, Ernst Ruska (1906–1988) and Max Knoll, using magnetic fields to "focus" electrons in a cathode-ray beam, produced a crude instrument that gave a magnification of 17 and by 1932 they developed an electron microscope having a magnification of 400. By 1937 James Hillier (b. 1915) advanced this magnification to 7,000. The 1939 instrument Vladimir Zworykin (1889–1982) developed gave 50 times more detail than any optical microscope ever could, with a magnification up to two million. The electron microscope revolutionized biological research, especially in the areas of cell structures, proteins, and viruses, where for the first time scientists could "see" molecules of these minute particles.

Who invented the **compound microscope**?

The principle of the compound microscope, in which two or more lenses are arranged to form an enlarged image of an object, occurred independently, at about the same time, to more than one person. Certainly many opticians were active at the end of the 16th century, especially in Holland, in the construction of telescopes, so that it is likely that the idea of the microscope may have occurred to several of them independently. In all probability, the date may be placed within the period 1590–1609, and the credit should go to three spectacle makers in Holland. Hans Janssen, his son Zacharias (1580–1638), and Hans Lippershey (1570–1619) have all been cited at various times as deserving chief credit. An Englishman, Robert Hooke (1635–1703) was the first to make the best use of a compound microscope, and his book, *Micrographia,* published in 1665, contains some of the most beautiful drawings of microscopic observations ever made.

How did Frank B. Gilbreth use his **cyclograph**?

American engineer and efficiency expert Frank Bunker Gilbreth (1868–1924) used his cyclograph, or "motion recorder," to refine his study of time and motion. An ordinary camera and a small electric bulb were all he needed to show the path of movement. The light patterns reveal all hesitation or poor habits interfering with a worker's dexterity.

What does the acronym **"LASER"** stand for?

A "LASER" (Light Amplification by Stimulated Emission of Radiation) is a device that produces a narrow, uniform, high-intensity beam of light made up of one very pure color (one wave length) that can be directed over small and large distances. This light beam is so intense that it can vaporize the hardest and most heat-resistant materials. In 1960 Theodore Maiman (b. 1927) demonstrated a working model of a laser, based on the principles of the maser (a device utilizing single-wave length microwaves, developed by Charles Townes [b. 1915] in 1953) using the wave frequencies of visible light. Ordinary light consists of relatively short wave packets; laser light is made up of long stretches of waves of even amplitude (height) and frequency (width), called "coherent light" because the wave packets or photons seem to stick together. The power of the laser lies in this concentration of photons, which strike the target surface simultaneously. Maiman used a ruby crystal to produce photons by exciting the chromium atoms in the ruby from a low-energy to a high-energy state. After a few thousandths of a second, the atoms revert back to their normal state and emit an identical photon. This photon can stimulate the chromium atoms to emit a photon. The beam is built up by billions of these paired photons, which are reflected back and forth by mirrors until they emerge at one end in bursts of red light.

LASER Applications

Semiconductor industry (Ultraviolet lasers) boards	Micromachining of components and circuit
Telecommunications (diode laser)	Development of optical fiber communications in the long-distance telephone networks
Retailing	Bar code scanners
Printing	Desktop printers, type-setting machines and color scanners
Military	Target designators
Medical	Diagnostic and therapeutics

What is **holography**?

Hungarian–born scientist, Dennis Gabor, invented the technique of holography (image in the round) in 1947, but it was not until 1961 when Emmet Leith and Juris Upatnieks produced the modern hologram using a laser, which gave the hologram the strong, pure light it needed. Three dimensions are seen around an object because light waves are reflected from all around it, overlapping and interfering with each other. This inter-action of these collections of waves, called wave fronts, give an object its light, shade, and depth. A camera cannot capture all the information in these wave fronts, so it pro-duces two dimensional objects. Holography captures the depth of an object by measur-ing the distance light has travelled from the object.

A simple hologram is made by splitting a laser light into two beams through a sil-vered mirror. One beam, called the object beam, lights up the subject of the hologram. These light waves are reflected onto a photographic plate. The other beam, called a ref-erence beam, is reflected directly onto the plate itself. The two beams coincide to cre-ate, on the plate, an "interference pattern." After the plate is developed, a laser light is projected through this developed hologram at the same angle as the original reference beam, but from the opposite direction. The pattern scatters the light to create a pro-jected, three-dimensional, ghostlike image of the original object in space.

In which industries are **robots** being used?

A robot is a device that can execute a wide range of maneuvers under the direction of a computer. Reacting to feedback from sensors or by reprogramming, a robot can alter its maneuvers to fit a changed task or situation. The worldwide population of robots is about 250,000, with approximately 65% being used in Japan and 14% being used in the United States (the second largest user).

Robots are used to do welding, painting, drilling, sanding, cutting, and moving tasks in manufacturing plants. Automobile factories account for more than 50% of robot use in the United States. Robots can work in environments that are extremely threaten-ing to humans or in difficult physical environments. They clean up radioactive areas, **363** extinguish fires, disarm bombs, and load and unload explosives and toxic chemicals. Robots can process light-sensitive materials, such as photographic films that require near darkness—a difficult task for human workers. They can perform a variety of tasks in underwater exploration and in mines (a hazardous, low-light environment). In the military and security fields, robots are used to sense targets or as surveillance devices. In the printing industry, robots perform miscellaneous tasks, such as sorting and tying bundles of output material, delivering paper to the presses, and applying book covers. In research laboratories, small desktop robots prepare samples and mix compounds.

Who was the father of cybernetics?

Norbert Weiner (1894–1964) is considered the creator of cybernetics. Derived from the Greek word *kubernetes,* meaning steerman or helmsman, cybernetics is concerned with the common factors of control and communication in living organisms, automatic machines, or organizations. These factors are exemplified by the skill used in steering a boat, in which the helmsman uses continual judgement to maintain control. The principles of cybernetics are used today in control theory, automation theory, and computer programs to reduce many time-consuming computations and decision-making processes formerly done by people.

What is the **lost wax process**?

This process is used for making valve parts, small gears, magnets, surgical tools, and jewelry. It involves the use of a wax pattern between a two–layered mold. The wax is removed by melting and replaced with molten metal; hence the name "lost wax."

What is **sintering**?

Sintering is the bonding together of compacted powder particles at temperatures below the melting point. This bonding produces larger forms such as cakes and pellets. Sintering is used in powder metallurgy technology, which is defined as the production of useful artifacts from metal powder without passing through the molten state. The heating process in which the powder particles are welded together is called sintering, and the components produced are referred to as sintered parts. These parts are normally quite small, and some typical examples made today are shock absorber pistons, belt pulleys, small helical gears, drive gears for chainsaws, and automotive pump gears.

Because they are molded, sintered parts can have extremely complex shapes, and do not require machining. The toughness and high strength properties of sintered parts make them especially good for today's high technology systems.

WEAPONS

See also: Boats, Trains, Cars, and Planes—Military Vehicles

Who invented the **Bowie knife**?

A popular weapon of the American West, the Bowie knife was named after Jim Bowie (1796–1836), who was killed at the Alamo. According to most reliable sources, his brother Rezin Bowie might have been the actual inventor. The knife's blade measured up to 2 inches (5 centimeters) wide and its length varied from 9 inches to 15 inches (23 to 38 centimeters).

Who invented **mine barrage**?

In 1777, David Bushnell (1742?–1824) conceived the idea of floating kegs containing explosives, which would ignite upon contact with ships.

Which **warfare innovations** were introduced during the American **Civil War**?

Barbed wire, trench warfare, hand grenades, land mines, armored trains, ironclad ships, aerial reconnaissance, submarine vessels, machine guns, and even a primitive flamethrower were products of the American Civil War, 1861–1865.

When was the **Colt revolver** patented?

The celebrated six-shooter of the American West was named after its inventor, Samuel Colt (1814–1862). Although he did not invent the revolver, he perfected the design, which he first patented in 1835 in England and then in the United States in the following year. Colt had hoped to mass-produce this weapon, but failed to raise enough backing to acquire the necessary machinery. Consequently, the guns, made by hand, were expensive and attracted only limited orders. In 1847, the Texas Rangers ordered one thousand Colt pistols, finally enabling Colt to set up assembly-line manufacture in a plant in Hartford, Connecticut.

Who invented the **machine gun**?

The first successful machine gun, invented by Richard J. Gatling (1818–1903), was patented in 1862 during the Civil War. Its six barrels were revolved by gears operated by a hand crank, and it fired 1,200 rounds per minute. Although there had been several partially successful attempts at building a multi-firing weapon, none were able to overcome the many engineering difficulties until Gatling. In his gear-driven machine, cocking and firing were performed by cam action. The United States Army officially adopted the gun on August 24, 1866.

The first automatic machine gun was a highly original design by Hiram S. Maxim (1840–1915). In 1884, Maxim designed a portable, single-barreled automatic weapon that made use of the recoil energy of a fired bullet to eject the spent cartridge and load the next.

The original "tommy gun" was the Thompson Model 1928 SMG. This 45-caliber machine gun, designed in 1918 by General John Taliaferro Thompson (1860–1940), was to be used in close-quarter combat. World War I ended before it went into production however, and Thompson's Auto Ordnance Corporation did not do well until the gun was adopted by American gangsters during Prohibition. The image of a reckless criminal spraying his enemies with bullets from his hand-held "tommy gun" became a symbol of the depression years. The gun was modified several times and was much used during World War II.

How did the **bazooka** get its name?

The term was coined by American comedian Bob Burns (1893–1956), but not for a weapon. As a prop in his act he used a unique musical instrument that was long and cylindrical and resembled an oboe. When United States Army soldiers in World War II were first issued hollow-tube rocket launchers, they named them bazookas because of their similarity to Burns' instrument.

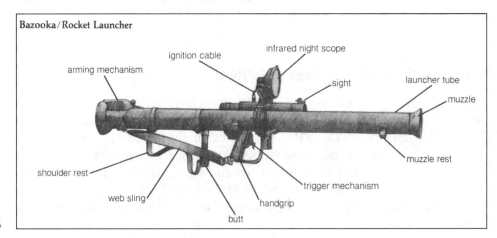

Bazooka/Rocket Launcher

Who was known as the **Cannon King**?

Alfred Krupp (1812–1887), whose father Friedrich (1787–1826) established the family's cast-steel factory in 1811, began manufacturing guns in 1856. Krupp supplied large weapons to so many nations that he became known as the "Cannon King." Prussia's victory in the Franco-German War of 1870-1871 was largely the result of Krupp's field guns. In 1933 when Hitler came to power, this family business began manufacturing a wide range of artillery. Alfred Krupp (1907–1967), the great-grandson of Alfred, supported the Nazis in power and accrued staggering wealth for the company. The firm seized property in occupied countries and used slave labor in its factories. After the war, Alfred was imprisoned for twelve years and had to forfeit all his property. Granted amnesty in 1951, he restored the business to its former position by the early 1960s. On his death in 1967, however, the firm became a corporation and the Krupp family dynasty ended.

How did **Big Bertha** get its name?

This was the popular name first applied to the 42-centimeter (about 16.5 inches) howitzers used by the Germans and Austrians in 1914, the year World War I started. Subsequently the term included other types of huge artillery pieces used during World Wars I and II. The large gun was built by the German arms manufacturer Friedrich A. Krupp (1854–1902), and was named after his only child, Bertha Krupp (1886–1957).

These large guns were used to destroy the concrete and steel forts defending Belgium in 1914; their shells weighed 205 pounds (930 kilograms) and were nearly as tall as a man. Since it was slow and difficult to move such guns, the use of them was practical only in static warfare. In World War II, bomber aircraft took away the role of long-range bombardment from these cumbersome guns.

What was the **Manhattan Project**?

The Manhattan Engineer District was the formal code name for the United States government project to develop an atomic bomb during World War II. It soon became known as the Manhattan Project—a name taken from the location of the office of Colonel James C. Marshall, who had been selected by the United States Army Corp of Engineers to build and run the bomb's production facilities. When the project was activated by the United States War Department in June, 1942, it came under the direction of Colonel Leslie R. Groves (1896–1970).

The first major accomplishment of the project's scientists was the successful initiation of the first self-sustaining nuclear chain reaction, done at a University of Chicago laboratory on December 2, 1942. The project tested the first experimental detonation of an atomic bomb in a desert area near Alamogordo, New Mexico, on July 16, 1945. The test site was called Trinity, and the bomb generated an explosive power equivalent to between 15,000 and 20,000 tons (15,240 and 20,320 tonnes) of TNT. Two of the project's bombs were dropped on Japan the following month (Hiroshima on August 6 and Nagasaki on August 9, 1945), resulting in the Japanese surrender to end World War II.

How far can a **Pershing missile** travel?

The Pershing surface-to-surface nuclear missile, which is 34.5 feet long (10.5 meters) and weighs 10,000 pounds (4,536 kilograms), has a range of about 1,120 miles (1,800 kilometers). It was developed by the United States Army in 1972. Other surface-to-surface missiles are the Polaris, with a range of 2,860 miles (4,600 kilometers); the Minuteman, with a range of 1,120 miles (1,800 kilometers); the Tomahawk, 2,300 miles (3,700 kilometers); the Trident, 4,600 miles (7,400 kilometers); and the Peacemaker, 6,200 miles (10,000 kilometers).

What is a **nuclear winter**?

The term *nuclear winter* was coined by American physicist Richard P. Turco in a 1983 article for the journal *Science,* in which he describes a hypothetical post-nuclear war scenario bringing severe worldwide climatic changes: prolonged periods of darkness, below-freezing temperatures, violent windstorms, and persistent radioactive fallout. This would be caused by billions of tons of dust, soot, and ash being tossed into the atmosphere, accompanied by smoke and poisonous fumes from firestorms. In the case of a severe nuclear war, within a few days the entire Northern Hemisphere would be under a blanket so thick that as little as $\frac{1}{10}$ of 1% of available sunlight would reach the Earth. Without sunlight, temperatures would drop well below freezing for a year or longer, causing dire consequences for all plant and animal life on Earth.

Reaction to this doomsday prediction lead critics to coin the term *nuclear autumn,* which downplayed such climatic effects and casualties. In January, 1990, the release of *Climate and Smoke: An Appraisal of Nuclear Winter,* based on five years of laboratory studies and field experiments, reinforced the original 1983 conclusions.

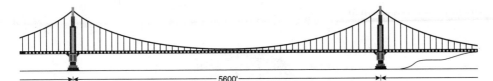

5600'

BUILDINGS, BRIDGES, AND OTHER STRUCTURES

BUILDINGS AND BUILDING PARTS

How many acres of **trees** are used in the construction of a **single-family home**?

On the average, an acre of softwood forest is used for a typical 2,000 square-foot house.

How does a **chimney** differ from a **flue**?

A chimney is a brick and masonry construction that contains one or more flues. A flue is a passage within a chimney through which smoke, fumes, and gases ascend. A flue is lined with clay or steel to contain the combustion wastes. By channeling the warm, rising gases, a flue creates a draft that pulls the air over the fire and up the flue. Each heat source needs its own flue, but one chimney can house several flues.

What is a **clerestory**?

In architecture, a clerestory is a windowed wall that extends above the ceiling level of a room or hall, providing lighting and sometimes ventilation. The most highly developed of the many types of clerestories are found in Gothic churches. The earliest known example of the use of the clerestory is at the great Temple of Amon-Ra at Karnak, **369**

Egypt, in its hypostyle hall (a hall in which the roof is supported only by columns, without vaults or domes).

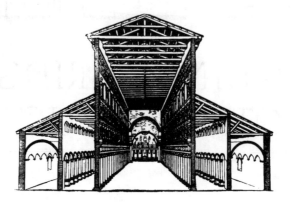

Clerestory level above roofline

Which part of a door is the **doorjamb**?

The doorjamb is not part of the door, but is the surrounding case into and out of which the door opens and closes. It consists of two upright pieces, called side jambs, and a horizontal head jamb.

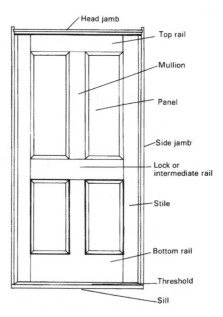

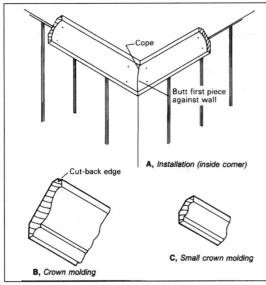

A, *Installation (inside corner)*

B, *Crown molding*

C, *Small crown molding*

Cope

Butt first piece against wall

Cut-back edge

What is **crown molding**?

Crown molding is a wood, metal, or plaster finishing strip placed on a wall where it intersects the ceiling. If the molding has a concave face, it is called a cove molding. In inside corners it must be cope-jointed to insure a tight joint.

What name is given to the the covering at the **top of a concrete wall**?

A coping is the protective covering over any vertical construction, such as a wall or chimney. A coping may be made of masonry, metal, or wood and is usually sloping or beveled to shed water in such a way that it does not run down the vertical face of the wall.

What is **R-value**?

An R-value, or resistance to the flow of heat, is a special measurement of insulation that represents the difficulty with which heat flows through an insulating material. The higher the R-value, the greater the insulating value of the material. A wall's component R-values can be added up to get its total R-value:

Wall component	R-value
Inside Air Film	0.7
½" Gypsum Wallboard	0.5
R-13 Insulation	13.0
½" Wood Fiber Sheathing	1.3
Wood Siding	0.8
Outside Air Film	0.2
Total R-value	16.5

Building component	R-value
Standard attic ceiling	19
Standard 4-inch-thick insulated wall	11
Typical single-pane glass window	1
Double-glazed window	2

Building component	R-value
"Superwindows" (inner surface of a pane coated with an infrared radiation reflector material such as tin oxide and argon gas filling in the space between the panes of a double-glazed window.)	4

What is an **STC rating** and what does it mean?

The STC (Sound Transmission Class) tells how well a wall or floor assembly inhibits airborne sound. The higher the number, the greater the sound barrier. Typical STC ratings are as follows:

STC number	What it means
25	Normal speech can be easily understood
30	Loud speech can be understood
35	Loud speech is audible, but not intelligible
42	Loud speech audible as a murmur
45	Must strain to hear loud speech
48	Some loud speech barely audible
50	Loud speech not audible

The degree to which sound barriers work depends greatly on the elimination of air gaps under doors, through electrical outlets, and around heating ducts.

What is the **BOCA code**?

The *Building Officials and Code Administrators* (BOCA) International is a service organization that issues a series of model regulatory construction codes for the protection of public health, safety, and welfare. The codes are published in sections, such as the National Building Code, National Plumbing Code, National Fire Prevention Code, etc. They are designed for adoption by state or local governments and may be amended or modified to accomplish desired local requirements.

Why are **Phillips screws** used?

The recessed head and cross-shaped slots are self-centering and allow a closer, tighter fit than conventional screws. Straight-slotted screws can cause the screwdriver to slip out of the groove, ruining the wood.

Screws were used in carpentry as far back as the 16th century, but slotted screws with a tapering point were made at the beginning of the 19th century. The great advantage of screws over nails is that they are extremely resistant to longitudinal tension. Larger-size screws that require considerable force to insert have square heads that can be tightened with a wrench. Screws provide more holding power than nails and can be withdrawn without damaging the material. Types of screws include wood screws, lag screws (longer and heavier than wood screws), expansion anchors (for masonry usually), and sheet metal screws. The screw sizes vary from ¼ to 6 inches.

Why is the term **"penny"** used in **nail sizes**?

The term "penny" is a measurement relating to the length of nails, originating in England. One explanation is that it refers to cost, with the cost of 100 nails of a certain size being 10 pence or 10d ("d" being the British symbol for a penny). Another explanation suggests that the term refers to the weight of 1,000 nails, with "d" at one time being used as an abbreviation for a pound in weight.

Mankind has been using nails for the last 5,000 years or so. Nails are known to have been used in Ur (ancient Iraq) to fasten together sheet metal. Before 1500, nails were made by hand, by drawing small pieces of metal through a succession of graded holes in a metal plate. In 1741, 60,000 people were employed in England making nails.

The first nail-making machine was invented by the American Ezekiel Reed. In 1851, Adolphe F. Brown of New York invented a wire-nail-making machine. This enabled nails to be mass-produced cheaply.

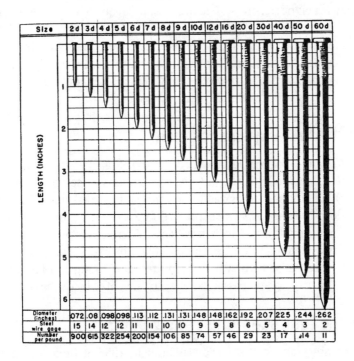

Size	2 d	3d	4 d	5 d	6 d	7 d	8 d	9 d	10d	12 d	16 d	20 d	30d	40 d	50 d	60d
Diameter (inches)	072	.08	.098	098	.113	.112	.131	.131	.148	.148	.162	.192	.207	225	.244	.262
Steel wire gage	15	14	12	12	11	11	10	10	9	9	8	6	5	4	3	2
Number per pound	900	615	322	254	200	154	106	85	74	57	46	29	23	17	.14	11

Sizes of penny-nails

What is a yurt?

Originally a Mongolian hut, the yurt has been adapted in the United States as a low-cost structure that can be used as a dwelling. The foundation is built of wood on a hexagonal frame. The wooden lattice-work side walls have a tension cable sandwiched between the wall pieces at the top to keep the walls from collapsing. The walls are insulated and covered with boards, log slabs, canvas, or aluminum siding. A shingle roof, electricity, plumbing, and a small heating stove may be installed. The interior can be finished as desired with shelves, room dividers, and interior siding. The yurt is beautiful, practical, and relatively inexpensive to erect.

How much does the **leaning tower of Pisa** lean?

The leaning tower of Pisa, 184.5 feet (56.23 meters) tall, is about 17 feet (5.18 meters) out of perpendicular, increasing by about $\frac{1}{20}$ inch (1.25 millimeters) a year. The Romanesque-style tower was started by Bonanno Pisano in 1173 as a campanile, or bell tower, for the nearby Baptistry, but was not completed until 1372. Built entirely of white marble, with eight tiers of arched arcades, it began to lean during construction. Although the foundation was dug down to 10 feet (3.04 meters) the builders did not reach bedrock for a firm footing. Ingenious attempts were made to compensate for the tilt by straightening up the subsequent stories and making the pillars higher on the south side than on the north. During the 1960s, cement was added to the foundation to strengthen it, but it is still threatened with collapse.

When was the **first skyscraper** built?

Designed by William Le Baron Jenney (1832–1907), the first skyscraper, the 10-story Home Insurance Company Building in Chicago, Illinois, was completed in 1885. A skyscraper—a very tall building supported by an internal frame (skeleton) of iron and steel rather than by load-bearing walls—maximizes floor space on limited land. Three technological developments made skyscrapers feasible: a better understanding of how materials behave under stress and load (from engineering and bridge design); the use of steel or iron framing to create a structure, with the outer skin "hung" on the frame; and the introduction of the first "safety" passenger elevator, invented by Elisha Otis (1811–1861).

When was the **first shopping center** built?

The first shopping center in the world was built in 1896 at Roland Park, Baltimore, Maryland. One of the world's largest shopping centers is the West Edmonton Mall in

Alberta, Canada, which covers 5.2 million square feet (480,000 square meters) on a 121-acre (49-hectare) site. It features 828 stores and services, with parking for 20,000 vehicles.

What is the ground area of the **Pentagon Building** in Washington, D.C.?

The Pentagon, the headquarters of the United States Department of Defense, is the world's largest office building in terms of ground space. Its construction was completed on January 15, 1943—in just over 17 months. With a gross floor area of over 6.5 million square feet, this five-story, five-sided building has three times the floor space of the Empire State Building and is 1½ times larger than the Sears Tower in Chicago. The World Trade Center complex in New York City, completed in 1973, is larger, with over 9 million square feet (836,000 square meters), but it is basically two structures (towers). Each of the Pentagon's five sides is 921 feet (281 meters) long, with a perimeter of 4,610 feet (1,405 meters). The Secretary of Defense, the Secretaries of the three military departments, and the military heads of the Army, Navy, and Air Force are all located in the Pentagon. The National Military Command Center, which is the nation's military communications hub, is located in the area of the Joint Chiefs of Staff. It is commonly called the "war room."

The largest commercial building in the world under one roof is the flower auction building of the cooperative VBA in Aalsmeer, Netherlands. In 1986, the floor plan was extended to 91.05 acres (36.85 hectares). It measures 2,546 x 2,070 feet (776 x 639 meters). The largest building in the United States, and the largest assembly plant in the world, is the Boeing 747 assembly plant in Everett, Washington, with a capacity of 200 million cubic feet (5.5 million cubic meters) and covering 47 acres (19 hectares).

Who invented the **geodesic dome**?

A geodesic line is the shortest distance between two points across a surface. If that surface is curved, a geodesic line across it will usually be curved as well. A geodesic line on the surface of a sphere will be part of a great circle. Buckminster Fuller (1895–1983) realized that the surface of a sphere could be divided into triangles by a network of geodesic lines and that structures could be designed so that their main elements either followed those lines or were joined along them.

This is the basis of his very successful geodesic dome: a structure of generally spherical form, constructed of many light, straight structural elements in tension, arranged in a framework of triangles to reduce stress and weight. These contiguous tetrahedrons are made from lightweight alloys with high tensile strength.

An early example of a geodesic structure is Britain's Dome of Discovery, built in 1951. It was the first dome ever built with principal framing members intentionally aligned along great circle arcs. The ASM (*American Society for Metals*) dome built east of Cleveland, Ohio, in 1959–1960 is an open lattice-work geodesic dome. Built in 1965, the Houston Astrodome forms a giant geodesic dome.

Geodesic dome

Which building has the **largest non-air-supported clear span roof?**

The Suncoast Dome in St. Petersburg, Florida, completed in 1990, has a clear span of 688 feet (209.7 meters). Covering an area of 372,000 square feet (34,570.1 square meters), the fabric–covered dome is a cable roof structure—the newest structural system for trussed domes. Its structural behavior is exactly opposite that of the traditional dome: the base ring is in compression rather than tension, and the ring at the crown is in tension rather than compression. Also, the space enclosed is not totally free and unobstructed, but includes structural, top-to-bottom members. To retain its characteristic lightness, the roofing surface employs flexible fabric membranes. These must be flexible since cable structures undergo major structural distortions, and they actually change shape under different load conditions. The largest air-supported building is the 80,600-capacity octagonal Pontiac Silverdome Stadium in Pontiac, Michigan. 522 feet (159.1 meters) wide and 722 feet (220.1 meters) long, it is supported by a 10 acre (1.62 hectare) translucent fiberglass roof.

What are the **tallest buildings** in the United States?

The following buildings are the tallest United States buildings and the tallest in the

world:

Building	Location	Year built	Height		
			Feet	Meters	Stories
Sears Tower	Chicago, Illinois	1974	1454	443	110
plus two T.V. towers			1707	520.9	
World Trade Center	New York, New York				
North tower		1972	1368	417	110
South tower		1973	1362	415	110
Empire State Bldg	New York, New York	1931	1250	381	102
plus two T.V. towers			1414	431	

The tallest self-supporting structure in the world is the CN Tower in Toronto, Canada, at 1815 feet (533 meters), built in 1975. The tallest structure in the world was the Guyed Warszawa Radio mast at Konstantynow, Poland, at 2,121 feet (646.4 meters). On August 10, 1991, however, it fell during renovation.

What is the "topping out" party performed by iron workers?

When the last beam is placed on a new bridge, skyscraper, or building, ironworkers hoist up an evergreen tree, attach a flag or a handkerchief, and brightly paint the final beam and autograph it.

This custom of raising an evergreen tree goes back to Scandinavia around 700 C.E., when attaching the tree to the building's ridge pole signalled to all who helped, that the celebration of its completion would begin.

When did Sears, Roebuck and Company sell homes by mail order?

Sears sold mail-order homes from 1909 to 1937. All construction materials and detailed plans for electric wiring were included. The price for the precut homes ranged from $595 to $5,000.

ROADS, BRIDGES, AND TUNNELS

Who is known as the father of civil engineering?

Thomas Telford (1757–1834), the first President of the Institute of Civil Engineers, is the father figure of the British civil engineering profession. Telford established the professional ethos and tradition of the civil engineer—a tradition followed by all engineers today. He built bridges, roads, harbors, and canals. His greatest works include the Menai Strait Suspension Bridge and Pont y Cysyllte Aqueduct, the Gotha Canal in Sweden, the Caledonian Canal, and many Scottish roads. He was the first and greatest master of the iron bridge.

Which state has the **most miles of roads**?

Texas lays claim to the most road mileage, with 305,692 miles (491,858 kilometers) lacing through the Lone Star State. California is a distant second, with 164,298 miles (264,355 kilometers) of roads. On the other end of the scale, only three states have less than 10,000 miles (16,090 kilometers) of roads: Delaware, Hawaii, and Rhode Island. The total road milage count for the United States is 3,876,501 miles (6,237,290 kilometers).

When was the **first U.S. coast-to-coast highway** built?

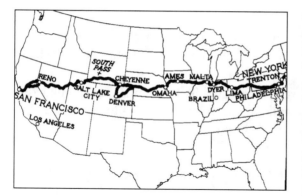

Lincoln Highway

Completed in 1923, after ten years of planning and construction, the Lincoln Highway was the first transcontinental highway connecting the Atlantic coast (New York) to the Pacific (California). Its original length of 3,389 miles (5,453 kilometers) was later shortened by relocations and improvements to 3,143 miles (5,057 kilometers). Crossing 13 states—New York, New Jersey, Pennsylvania, Ohio, Indiana, Illinois, Iowa Nebraska, Colorado, Wyoming, Utah, Nevada, and California—the Lincoln Highway became, for most of its length, U.S. Route 30 in 1925.

How are U.S. **highways numbered**?

The main north–south interstate highways always have odd numbers of one or two digits. The system, beginning with Interstate 5 on the West Coast, increases in number as it moves eastward, and ends with Interstate 95 on the East Coast.

The east–west interstate highways have even numbers. The lowest numbered highway begins in Florida with Interstate 4, increasing in number as it moves northward, and ends with Interstate 96. Coast-to-coast east–west interstates, such as Routes 10, 40, and 80, end in zero. An interstate with three digits is either a beltway or a spur route.

U.S. routes follow the same numbering system as the interstate system, but increase in number from east to west and from north to south. U.S. Route 1, for example, runs along the East Coast; U.S. Route 2 runs along the Canadian border. U.S. route numbers may have anywhere from one to three digits.

Which city had the **first traffic light**?

On December 10, 1868, the first traffic light was erected on a 22-foot-high (6.7-meter) cast-iron pillar at the corner of Bridge Street and New Palace Yard off Parliament Square in London, England. Invented by J.P. Knight, a railway signalling engineer, the light was a revolving lantern illuminated by gas, with red and green signals. It was turned by hand using a lever at the base of the pole.

Cleveland, Ohio, installed an electric traffic signal at Euclid Avenue and 105th Street on August 5, 1914. It had red and green lights with a warning buzzer as the color changed.

Around 1913, Detroit, Michigan, used a system of manually operated semaphores. Eventually the semaphores were fitted with colored lanterns for night traffic. New York City installed the first three-color light signals in 1918; these signals were also operated manually.

Why are **manhole covers** round?

Circular covers are almost universally used on sewer manholes because they cannot drop through the opening. The circular cover rests on a lip that is smaller than the cover. Any other shape—square, rectangle, etc.—could slip into the manhole opening.

What is the world's **longest road tunnel**?

The two-lane road tunnel at St. Gotthard Pass between Göschenen, Switzerland, and Airolo, Italy, constructed between 1970 and 1980, is 10.1 miles (16.3 kilometers) long. The Mont Blanc tunnel, completed in 1965, is the next longest road tunnel. Its two lanes connect the 7.2 miles (11.6 kilometers) between Chamonix, France, and the Valle d'Aosta, Italy. The longest road tunnel in the United Kingdom is the four-lane Mersey Tunnel, joining Liverpool and Birkenhead, Merseyside. It is 2.13 miles (3.43 kilometers) long. In the United States, the longest road tunnel is the 2.5 mile (4.02 kilometer) Lincoln tunnel, connecting New York City and New Jersey, and constructed under the Hudson River in 1937.

Where is the **longest bridge-tunnel** in the world located?

Completed in 1964 after 42 months' work, costing $200 million and spanning a great distance of open sea, the Chesapeake Bay Bridge-Tunnel is a 17.5-mile (28.2 kilometer) combination of trestles, bridges, and tunnels that connect Norfolk with Cape Charles in Virginia. Its only rival for crossing so much open and deep water is the Zuider Zee Dam in Holland—a road-carrying structure of similar length, but without the same water depth and same length of open sea.

What are the various **types of bridge structures**?

There are four basic types of structures that can be used to bridge a stream or similar obstacle: rigid beam, cantilever, arch, and suspension systems.

The rigid beam bridge, the simplest and most common form of bridging, has straight slabs or girders carrying the roadbed. The span is relatively short and its load rests on its supports or piers.

The arch bridge is in compression, and thrusts outward on its bearings at each end.

In the suspension bridge, the roadway hangs on steel cables, with the bulk of the load carried on cables anchored to the banks. It can span a great distance without intermediate piers.

Each arm of a cantilever bridge is, or could be, free-standing, with the load of the short central truss span pushing down through the piers of the outer arms and pulling up at each end. The outer arms are usually anchored at the abutments and project into the central truss.

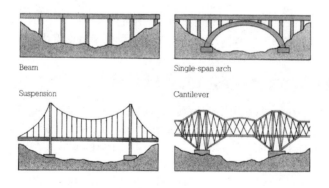

Beam

Single-span arch

Suspension

Cantilever

Bridge types

What is a **"kissing bridge"**?

Covered bridges with roofs and wooden sides are called "kissing bridges," because people inside the bridge could not be seen from outside. Such bridges can be traced back to the early nineteenth century. Contrary to folk wisdom, they were not designed to produce rural "lovers' lanes," but were covered to protect the structures from deterioration.

How many **covered bridges** are there in the United States?

There were more than 10,000 covered bridges built across the United States between 1805 (when the first was erected in Philadelphia) and the early 20th century. As of January, 1980, only 893 of these covered bridges remained—231 in Pennsylvania, 157 in Ohio, 103 in Indiana, 100 in Vermont, 54 in Oregon, and 52 in New Hampshire. Three interstate covered bridges link New Hampshire and Vermont. The remainder are scattered throughout the country. Non-authentic covered bridges—built or covered for visual effect—appear in each state.

What are the **world's longest bridge spans**?

The world's longest main-span in a suspension bridge is the Humber Estuary Bridge in England, at 4,626 feet (1,410 meters). Designed by Freeman Fox and Partners and begun on July 27, 1972, it was opened on July 18, 1981. The Humber Bridge is longer than the Verrazano–Narrows Bridge in New York, but it will be surpassed by two bridges now under construction. The double-deck Akashi-Kaikyo Bridge linking Honshū and Shikoku, Japan, will have a main span of 6,529 feet (1,990 meters). The Messing Bridge that will link Sicily with Calabria on the Italian mainland will become the longest by far when it is completed, with a single span of 10,892 feet (3,320 meters).

The world's longest cable-stayed bridge span is the 1,608-foot (490 meter) Ikuchi Bridge between Honshū and Shikoku, Japan, completed in 1991.

The world's longest cantilever bridge is the Quebec Bridge over the St. Lawrence River in Canada, opened in 1917, which has a span of 1,800 feet (548.6 meters) between piers, with an overall length of 3,239 feet (987.3 meters).

Completed in 1977, the world's longest steel arch bridge is the New River Gorge Bridge near Fayetteville, West Virginia, with a span of 1,700 feet (518.2 meters).

The world's longest concrete arch bridge is the Jesse H. Jones Memorial Bridge, which spans Houston Ship Canal in Texas. Completed in 1982, the bridge measures 1,500 feet (457.2 meters).

The world's longest stone arch bridge is the 3,810-foot–long (1161-meter) Rockville Bridge, completed in 1901, north of Harrisburg, Pennsylvania, with 48 spans containing 216.051 tons (219.52 tonnes) of stone.

Where is the **longest suspension bridge** in the United States?

Spanning New York (City) Harbor, the Verrazano–Narrows Bridge is the longest suspension bridge in the United States. With a span of 4,260 feet (1,298.5 meters) from tower to tower, its total length is 7,200 feet (2,194.6 meters). Named after Giovanni da Verrazano (1485–1528), the Italian explorer who discovered New York Harbor in April 1524, it was erected under the direction of Othmar H. Ammann (1879–1965), and completed in 1964. To avoid impeding navigation in and out of the harbor, it provides a clearance of 216 feet (65.8 meters) between the water level and the bottom of the bridge deck. Like other suspension bridges, the bulk of the load of the Verrazano–Narrows Bridge is carried on cables anchored to the banks.

Which United States place has the **most bridges**?

With the possible exception of Venice, Italy, Pittsburgh, Pennsylvania is the bridge capital of the world. There are more than 1,700 bridges in Allegheny County's 731 square miles. The county has 2.3 bridges per square mile or one bridge for every mile of highway. Nationwide, there are approximately 564,000 bridges.

Are there any **floating bridges** in the United States?

The three floating pontoon bridges in the United States are all located in the state of Washington. The Evergreen Point Bridge (1963) and the Third Lake Washington Bridge (1989) in Seattle are 7,518 feet (2,291.5 meters) and 6,130 feet (1868.5 meters) long respectively. The Hood Canal Bridge (1961) in Port Gamble is 6,471 feet (1972.4 meters) long.

How was the safety of the **first bridge across the Mississippi River** at St. Louis in 1874 proved?

According to a historical appraisal of the bridge by Howard Miller:

> Progressively heavier trains shuttled back and forth across the bridge as its engineer, James B. Eads, took meticulous measurements. However, the general public was probably more reassured by a nonscientific test. Everyone knew that elephants had canny instincts and would not set foot on an unsound bridge. The crowd cheered as a great beast from a local menagerie mounted the approach without hesitation and lumbered placidly across to the Illinois side.

Who built the **Brooklyn Bridge**?

John A. Roebling (1806–1869), a German-born American engineer, constructed the first truly modern suspension bridge in 1855. Towers supporting massive cables, tension anchorage for stays, a roadway suspended from the main cables, and a stiffening deck below or beside the road deck to prevent oscillation are all characteristics of Roebling's suspension bridge. In 1867, Roebling was given the ambitious task of constructing the Brooklyn Bridge. In his design he proposed the revolutionary idea of using steel wire for cables rather than the less-resilient iron. Just as construction began, Roebling died of tetanus when his foot was crushed in an accident; his son, Washington A. Roebling (1837–1926), assumed responsibility for the bridge's construction. Fourteen years later, in 1883, the bridge was completed. At that time, it was the longest suspension bridge in the world, spanning the East River and connecting New York's Manhattan with Brooklyn. The bridge has a central span of 1,595.5 feet (486.3 meters), with its masonry towers rising 276 feet (84 meters) above high water. Today, the Brooklyn Bridge is among the best known of all American civil engineering accomplishments.

Who designed the **Golden Gate Bridge**?

Joseph B. Strauss (1870–1938), formally named chief engineer for the Golden Gate project in 1929, was assisted by Charles Ellis and Leon Moisseiff in the design. An engineering masterpiece that opened to traffic in May, 1937, this suspension bridge spans San Francisco Bay, linking San Francisco with Marin County, California. It has a central span of 4,200 feet (1280 meters), with towers rising 746 feet (227.4 meters).

MISCELLANEOUS STRUCTURES

Will building a **seawall** protect a beach?

It may for a while, but during a storm, the sand cannot follow its natural pattern of allowing waves to draw the sand across the lower beach, making the beach flatter. With a seawall, the waves carry off more sand, dropping it into deeper water. A better alternative to the seawall is the revetment. This is a wall of boulders, rubble, or concrete block, tilted back away from the waves. It imitates the way a natural beach flattens out under wave attack.

Where is the **highest dam** in the United States and what is its capacity?

Oroville, the highest dam in the United States, is an earth-fill dam that rises 754 feet (229.8 meters) and extends more than a mile across the Feather River, near Oroville, California. Built in 1968, it forms a reservoir containing about 3.5 million acre-feet (4.3 million cubic meters) of water. The next highest dam in the United States is the Hoover Dam on the Colorado River, on the Nevada–Arizona border. It is 726 feet (221.3 meters) high, and was, for 22 years, the world's highest. Currently, nine dams are higher than the Oroville—the highest is the 984-foot (299.9-meter) Nurek Dam on the Vakhsh River in Tadzhikistan near the Afghanistan border. But it will be surpassed by the 1,098-foot (334.7-meter) Rogun(skaya) earth-fill dam, currently under construction, which also crosses the Vaksh River in Tadzhikistan.

How big is the **Hoover Dam**?

Formerly called Boulder Dam, the Hoover Dam is located between Nevada and Arizona on the Colorado River. The highest concrete arch dam in the United States, the dam is 1,244 feet (379 meters) long and 726 feet (221 meters) high. It has a base thickness of 660 feet (201 meters) and a crest thickness of 45 feet (13.7 meters). It stores 21.25 million acre-feet of water in the 115-mile-long (185-kilometer) Lake Mead reservoir.

The dam was built because the Southwest was faced with constantly recurring cycles of flood and drought. Uncontrolled, the Colorado River had limited value, but regulated, the flow assures a stabilized, year-round water supply and low-lying valleys are protected against floods. On December 21, 1928, the Boulder Canyon Project Act became law, and the project was completed on September 30, 1935—two years ahead of schedule. For 22 years, Hoover Dam was the highest dam in the world.

How tall is the figure on the **Statue of Liberty**, and how much does it weigh?

The Statue of Liberty, conceived and designed by the French sculptor Fréderic-Auguste Bartholdi (1834–1904), was given to the United States to commemorate its first centennial. Called (in his patent, U.S. Design Patent no. 11,023, issued February 18, 1879) "Liberty Enlightening the World," it is 152 feet (46.3 meters) high, weighs 225 tons (204 metric tons), and stands on a pedestal and base that are 151 feet (46 meters) tall. Its flowing robes are made from more than 300 sheets of hand-hammered copper over a steel frame. Constructed and finished in France in 1884, the statue's exterior and interior were taken apart piece by piece, packed into 200 mammoth wooden crates, and shipped to the United States in May, 1885. The statue was placed by Bartholdi on Bedloe's Island, at the mouth of New York City Harbor. On October 28, 1886, ten years after the centennial had passed, the inauguration celebration was held.

It was not until 1903 that the inscription "Give me your tired, your poor, Your huddled masses yearning to breathe free ..." was added. The verse was taken from *The New Colossus* composed by New York City poet Emma Lazarus in 1883. The statue, the tallest in the United States, and the tallest metal statue in the world, was recently refurbished for its own centennial, at a cost of $698 million, reopening on July 4, 1986. One visible difference is the flame of her torch is now 24-karat gold-leaf, just as in the original design. In 1916 the flame was redone into a lantern of amber glass. Concealed within the rim of the statue's crown is the observation deck that can be reached by climbing 171 steps or by taking the newly installed hydraulic elevator.

What are the dimensions of the **Eiffel Tower**?

Eiffel Tower

Now France's most instantly recognizable landmark, the Eiffel Tower was built in Paris during 1887–1889 by Gustave Eiffel (1832–1923), a French engineer, for the 100th anniversary of the French Revolution. For this momentous occasion, for which a huge exhibition, Exposition Universelle, was planned, Eiffel wished to erect something extraordinary made from modern materials. His design, a structure of wrought-iron ribs, held together by rivets and resting on a solid masonry foundation, was initially denounced by some as a "tragic lamppost," but its elegance has come to be recognized by most.

It stands 985 feet 11 inches (300.5 meters) high on a square base measuring 328 feet (100 meters) on each side. The first-story platform is at 189 feet (57.6 meters), with sides 232 feet (70.7 meters) long. The second-story platform is at 379 feet (115.5 meters), with sides 134 feet (40.8 meters) long. There is an intermediate platform at 643 feet (196 meters)—this is only a point to change elevators. The third-story platform is at 905 feet (275.8 meters). Later a 66-foot (20-meter) television antenna was attached, which extended the tower's total height to 1052 feet (320.7 meters).

Who invented the **Ferris Wheel** and when?

Originally called "pleasure wheels," the first such rides were described by English traveller Peter Mundy in 1620. In Turkey he saw a ride for children consisting of two vertical wheels, 20 feet (6.1 meters) across, supported by a large post on each side. Such rides were called "ups-and-downs" at the St. Bartholomew Fair of 1728 in England, and in 1860, a French pleasure wheel was turned by hand and carried sixteen passengers. They were also in use in the United States by then, with a larger, wooden wheel operating at Walton Spring, Georgia.

Wanting a spectacular attraction to rival that of the 1889 Paris Centennial celebration—the Eiffel Tower—the directors of the 1893 Columbian Exposition held a design competition. The prize was won by the American bridge builder George Washington Gale Ferris (1859–1896). In 1893 he designed and erected a gigantic revolving steel wheel whose top reached 264 feet (80.5 meters) above ground. The wheel—825 feet (251.5 meters) in circumference, 250 feet (76.2 meters) in diameter, and 30 feet (9.1 meters) wide—was supported by two 140-foot (42.7 meters) towers. Attached to the wheel were 36 cars, each able to carry 60 passengers. Opening on June 21, 1893, at the exposition in Chicago, Illinois, it was extremely successful. Thousands lined up to pay 50 cents for a 20-minute ride—a large sum in those days, considering that a merry-go-round ride cost only 4 cents. In 1904 it was moved to St. Louis, Missouri, for the Louisiana Purchase Exposition. It was eventually sold for scrap. The largest-diameter wheel currently operating is the Cosmoclock 21 at Yokohama City, Japan. It is 344.48 feet (105 meters) high and 328 feet (100 meters) in diameter.

How many **roller coasters** are there in the United States, and how many people ride them?

There were 164 roller coasters in the United States in 1989 and an estimated 214 million people rode them during that year. Roller coasters have had a long history in thrill-giving. During the 15th and 16th centuries, the first known gravity rides were built in St. Petersburg and were called "Russian Mountains." A wheeled roller coaster, called the "Switchback," was used as early as 1784 in Russia. By 1817, the first roller coaster whose cars were locked to the tracks was in operation in France. The first United States roller coaster patent was granted to J.G. Taylor in 1872, and LaMarcus Thompson built the first known roller coaster in the United States at Coney Island, Brooklyn, New York, in 1884. Until recently, the longest roller coaster in the world was "The Beast" at Kings Island, Ohio, which has a run of 1.4 miles (2.25 kilometers) with 800 feet (243.84 meters) of tunnels. The current record-holder is the "Ultimate" at Lightwater Valley, Ripon, Great Britain, with a run of 1.42 miles (2.28 kilometers).

BOATS, TRAINS, CARS, AND PLANES

BOATS AND SHIPS

What is **dead reckoning**?

Dead reckoning is the determination of a craft's current latitude and longitude by advancing its previous position to the new one on the basis of assumed distance and direction traveled. The influences of current and wind as well as compass errors are taken into account in this calculation, all done without the aid of any celestial or physical observation. This is a real test of a navigator's skill.

Nuclear-powered submarines, which must retain secrecy of movement and cannot ascend to the surface, use the SINS system (Ship's *I*nertial *N*avigation *S*ystem) developed by the United States Navy. It is a fully self-contained system, which requires no receiving or transmitting apparatus and thus involves no detectable signals. It consists of accelerometers, gyroscopes, and a computer. Together they produce inertial navigation, which is a sophisticated form of dead reckoning.

Why is the right side of a ship called **starboard**?

In the time of the Vikings, ships were steered by long paddles or boards placed over the right side. They were known in Old English as *steorbords,* evolving into the word *starboard.* The left side of a ship, looking forward, is called *port.* Formerly the left side was called *larboard,* originating perhaps from the fact that early merchant ships were always loaded from the left side. Its etymology is Scandinavian, being *lade* (load) and *bord* (side). The British Admiralty ordered port to be used in place of *larboard* to prevent confusion with *starboard.*

Where does the term "mark twain" originate?

Mark twain is a riverboat term meaning 2 fathoms (a depth of 12 feet, or 3.65 meters). A hand lead is used for determining the depth of water where there is less than 20 fathoms. The lead consists of a lead weight of 7 to 14 pounds (3 to 6 kilograms) and a line of hemp or braided cotton, 25 fathoms (150 feet, or 46 meters) in length. The line is marked at 2, 3, 5, 7, 10, 15, 17, and 20 fathoms. The soundings are taken by a leadsman who calls out the depths while standing on a platform projecting from the side of the ship, called "the chains." The number of fathoms always forms the last part of the call. When the depth corresponds to any mark on the lead line, it is reported as "By the mark 7," "By the mark 10," etc. When the depth corresponds to a fathom between the marks on the line, it is reported as "By the deep 6," etc. When the line is a fraction greater than a mark, it is reported as "And a half 7," "And a quarter 5"; a fraction less than a mark is "Half less 7," "Quarter less 10," etc. If bottom is not reached, the call is "No bottom at 20 fathoms."

"Mark Twain" was also the pseudonym chosen by American humorist Samuel L. Clemens. Supposedly, he chose the name because of its suggestive meaning, since it was a riverman's term for water that was just barely safe for navigation. One implication of this "barely safe water" meaning was, as his character Huck Finn would later remark, "Mr. Mark Twain ... he told the truth, mostly." Another implication was that "barely safe water" usually made people nervous, or at least uncomfortable.

How is a ship's tonnage calculated?

Tonnage of a ship is not necessarily the number of tons that the ship weighs. There are at least six different methods of rating ships; the most common are:

Displacement tonnage—used especially for warships and U.S. merchant ships—is the weight of the water displaced by a ship. Since a ton of sea water occupies 35 cubic feet, the weight of water displaced by a ship can be determined by dividing the cubic footage of the submerged area of the ship by 35. The result is converted to long tons (2,240 pounds). Loaded displacement tonnage is the weight of the water displaced when a ship is carrying its normal load of fuel, cargo, and crew. Light displacement tonnage is the weight of water displaced by the unloaded ship.

Gross tonnage (GRST) or *gross registered tonnage* (GRT)—used to rate merchant shipping and passenger ships—is a measure of the enclosed capacity of a vessel. It is the sum in cubic feet of the vessel's enclosed space divided by 100 (100 such cubic feet is considered one ton). The result is gross (registered) tonnage. For example, the old *Queen Elizabeth* did not weigh 83,673 tons, but had a capacity of 8,367,300 cubic feet.

Deadweight tonnage (DWT)—used for freighters and tankers—is the total weight in long tons (2,240 pounds) of everything a ship can carry when fully loaded. It represents the amount of cargo, stores, bunkers, and passengers that are required to bring a ship down to her loadline, i.e. the carrying capacity of a ship.

Net registered tonnage (NRT)—used in merchant shipping—is the gross registered tonnage minus the space that cannot be utilized for paying passengers or cargo (crew space, ballast, engine room, etc.).

What does the term **loadline** mean in shipping?

A loadline or load waterline is an immersion mark on the hull of a merchant ship. This indicates her safe load limit. The lines vary in height for different seasons of the year and areas of the world. Also called the *Plimsoll Line* or *Plimsoll Mark,* it was accepted as law by the British Parliament in the Merchant Shipping Act of 1875, primarily at the instigation of Samuel Plimsoll (1824–1898). This law prevented unscrupulous owners from sending out unseaworthy and overloaded, but heavily insured, vessels (so-called "coffin ships"), which risked the crew's lives.

Who created the **Liberty ship**?

The Liberty ship of World War II was the brainchild of Henry J. Kaiser (1882–1967), an American industrialist who had never run a shipyard before 1941. The huge loss of merchant tonnage during the war created an urgent need for protection of merchant vessels transporting weapons and supplies, and the Liberty ship was born. They were standard merchant ships with a deadweight tonnage of 10,500 long tons and a service speed of 11 knots. They were built to spartan standards and production was on a massive scale. Simplicity of construction and operations, rapidity of building, and large cargo carrying capacity were assets. To these, Kaiser added prefabrication, and welding instead of riveting. The ships were a deciding factor on the side of the Allies. In four years 2,770 ships with a deadweight tonnage of 29,292,000 long tons were produced.

When was the first **hospital ship** built?

It is believed that the Spanish Armada fleet of 1587–1588 included hospital ships. England's first recorded hospital ship was the *Goodwill* in 1608, but it was not until after 1660 that the Royal Navy made it a regular practice to set aside ships for hospital use. The United States Government outfitted six hospital ships, some of which were permanently attached to the fleet, during the Spanish–American War of 1898. Congress authorized the construction of the USS *Relief* in August, 1916. It was launched in 1919 and delivered to the Navy in December, 1920.

Why did the **Titanic** sink?

On its maiden voyage from Southampton, England, to New York City, the British luxury liner *Titanic* sideswiped an iceberg at 11:40 P.M. on Sunday, April 14, 1912, and was badly damaged. The 882-foot–long (268.8-meter) liner, whose eight decks rose to the height of an 11–story building, sank two hours and forty minutes later. Of the 2,227 passengers and crew, 705 escaped in 20 lifeboats and rafts; 1,522 drowned.

Famous as the greatest disaster in transatlantic shipping history, circumstances made the loss of life in the sinking of the *Titanic* exceptionally high. Although Capt. E.J. Smith was warned of icebergs in shipping lanes, he maintained his speed of 22 knots, and did not post additional lookouts. Later inquiries revealed that the liner *Californian* was only 20 miles (32.18 kilometers) away and could have helped, had its radio operator been on duty. The *Titanic* had an insufficient number of lifeboats, and those available for use were badly managed, with some leaving the boat only half-full. The only ship responding to distress signals was the ancient *Carpathia,* which saved 705 people.

Contrary to a long-held belief, the *Titanic* had not been sliced open by the iceberg. When Dr. Robert Ballard (b. 1942) from Woods Hole Oceanographic Institution descended to the site of the sunken vessel in the research vessel *Alvin* in July, 1986, he found that the ship's starboard bow plates had buckled under the impact of the collision. This caused the ship to be opened up to the sea.

Ballard found the bow and the stern more than 600 yards apart on the ocean's floor, and speculated on what happened after the collision with the iceberg. "Water entered six forward compartments after the ship struck the iceberg. As the liner nosed down, water flooded compartments one after another, and the ship's stern rose even higher out of the water, until the stress amidships was more then she could bear. She broke apart." and the stern soon sank by itself.

What is the world's **largest ship?**

Among passenger liners, the cruise ship *Sovereign of the Seas,* built in France for the Norwegian Royal Caribbean Cruise Lines, was the world's largest until 1990, with a gross registered tonnage of 73,192. Completed in 1988, its dimensions are 874 x 105 feet (266.4 x 32 meters) with a depth of 49 feet (14.94 meters). It can accommodate 2,763 passengers plus crew (over 3,000 people total). In the autumn of 1990, the cruise ship *Norway,* built in 1961 as the *France* and renamed in 1979, increased her gross registered tonnage from 70,202 to 76,000 to become not only the longest liner, at 1035.5 feet (315.66 meters), but also the largest. However, the largest passenger liner ever built was Cunard's *Queen Elizabeth,* which was 1031 feet (314.25 meters) long and 118.5 feet (36.12 meters) wide with a final gross registered tonnage of 82,998 (originally 83,673). With her last voyage ending on November 15, 1968, the Queen Elizabeth in 1970 became a floating marine university, *Seaside University,* in Hong Kong, and then burned on January 9, 1972.

The largest ship of any kind currently in service is the tanker *Hellas Fos,* owned by the Bilinder Marine Corporation of Athens, Greece. Built in 1979, this steam turbine tanker has a deadweight tonnage of 611,832.5 (gross registered tonnage of 254,583). The largest ship in the world is the *Happy Giant,* formerly the *Seawise Giant,* with a deadweight tonnage of 622,571.5, a length of 1,504 feet (458.42 meters), and a beam of

226 feet (68.88 meters). Attacked and damaged by Iraqi Mirage jets in the Persian Gulf on December 22, 1987, and again on May 14, 1988, the tanker is in the process of being refitted. The steam turbines are being replaced with a diesel engine, which will decrease her deadweight tonnage to about 420,000 long tons.

When were the first **nuclear-powered vessels** launched?

A controlled nuclear reaction generates tremendous heat, which turns water into steam for running turbine engines. The USS *Nautilus* was the first submarine to be propelled by nuclear power, making her first sea run on January 17, 1955. It has been called the first true submarine, since it can remain underwater for an indefinite period of time; as a nuclear reactor requires no oxygen to operate, only the crew's need for oxygen limits the submersion time. The *Nautilus,* 324 feet (98.75 meters) long, has a range of 2,500 miles (4,022.5 kilometers) submerged, a diving depth of 700 feet (213.36 meters), and can travel submerged at 20 knots.

The first nuclear warship was the 14,000-ton cruiser USS *Long Beach,* launched on July 14, 1959. The USS *Enterprise* was the first nuclear-powered aircraft carrier. Launched on September 24, 1960, the *Enterprise* was 1,101.5 feet (335.74 meters) long, and designed to carry 100 aircraft.

The first nuclear-powered merchant ship was the *Savannah,* a 20,000-ton vessel, launched in 1962. The United States built it largely as an experiment and it was never operated commercially. In 1969, Germany built the *Otto Hahn,* a nuclear-powered ore carrier. The most successful use of nuclear propulsion in non-naval ships has been as icebreakers. The first nuclear-powered icebreaker was the Soviet Union's *Lenin,* commissioned in 1959.

TRAINS AND TROLLEYS

What is a **standard gauge** railroad?

The first successful railroads in England used steam locomotives built by George Stephenson (1781–1848) to operate on tracks with a gauge of 4 feet 8.5 inches (1.41 meters), probably because that was the wheel spacing common on the wagons and tramways of the time. Stephenson, a self-taught inventor and engineer, had developed in 1814 the steam-blast engine that made steam locomotives practical. His railroad rival, Isambard K. Brunel (1806–1859), laid out the line for the Great Western Railway at 7 feet 0.25 inches (2.14 meters), and the famous "Battle of the gauges" began. A commission appointed by the British Parliament decided in favor of Stephenson's narrower gauge, and the Gauge Act of 1846 prohibited using other gauges. This width eventually became accepted by the rest of the world. The distance is measured between the inner sides of the heads of the two rails of the track at a distance of ⅝ inch (1.58 centimeters) below the top of the rails.

What is the **world's longest railway?**

The Trans-Siberian Railway, from Moscow to Vladivostok, is 5,777 miles (9,297 kilometers) long. If the spur to Nakhodka is included, the distance becomes 5,865 miles (9,436 kilometers). It was opened in sections, and the first goods train reached Irkutsk on August 27, 1898. The Baikal-Amur Northern Main Line, begun in 1938, shortens the distance by about 310 miles (500 kilometers). The journey takes approximately seven days, two hours, and crosses seven time zones. There are nine tunnels, 139 large bridges or viaducts, and 3,762 smaller bridges or culverts on the whole route. Nearly the entire line is electrified.

In comparison, the first American transcontinental railroad, completed on May 10, 1869, is 1,780 miles (2,864 kilometers) long. The Central Pacific Railroad built eastward from Sacramento, California, and the Union Pacific Railroad was built westward to Promontory Point, Utah, where the two lines met to connect the entire route.

What is the **fastest train?**

The world's fastest train is France's TGV (Train à Grande Vitesse or train of great speed) with a top speed of 186 miles (299 kilometers) per hour and an average speed of 132 miles (212.5 kilometers) per hour. However, soon it will be outdistanced by the MAGLEV (*mag*netic *lev*itation) trains of Japan and Germany, which can travel between 250 to 300 miles (402 to 483 kilometers) per hour. These trains run on a bed of air produced from the repulsion or attraction of powerful magnetic fields (based on the principle that like poles of magnets repel and unlike poles attract). The German Transrapid uses conventional magnets to levitate the train. The principle of attraction in magnetism, the employment of wing-like flaps extending under the train to fold under a T-shaped guideway, and the use of electromagnets on board (that are attracted to the non–energized magnetic surface) are the guiding components. Interaction between the train's electromagnets and those built on top of the T-shaped track lifts the vehicle 3/8 inch (1 centimeter) off the guideway. Another set of magnets along the rail sides provides lateral guidance. The train rides on electromagnetic waves. Alternating current in the magnet sets in the guideway changes their polarity to alternately push and pull the train along. Braking is done by reversing the direction of the magnetic field (caused by reversing the magnetic poles). To increase train speed, the frequency of current is raised.

The Japanese MLV002 uses the same propulsion system; the difference is in the levitation design in which the train rests on wheels until it reaches a speed of 100 miles (161 kilometers) per hour. Then it levitates 4 inches (10 centimeters) above the guideway. The levitation depends on superconducting magnets and a repulsion system (rather than the attraction system that the German system uses).

When was the **first railroad in the United States** chartered?

The first American railroad charter was obtained on February 6, 1815, by Colonel John Stevens (1749–1838) of Hoboken, New Jersey, to build and operate a railroad between the Delaware and Raritan rivers near Trenton and New Brunswick. However, lack of financial backing prevented its construction. The Granite Railway built by Gridley Bryant was chartered on October 7, 1826. It ran from Quincy, Massachusetts, to the Neponset River—a distance of three miles (4.8 kilometers). The main cargo was granite blocks used in building the Bunker Hill Monument.

What was the **railroad velocipede**?

Velocipede

In the nineteenth century, railroad track maintenance workers used a three-wheeled handcar to speed their way along the track. The handcar was used for interstation express and package deliveries and for delivery of urgent messages between stations that could not wait until the next train. Also called an "Irish Mail," this 150-pound (68 kilogram) three-wheeler resembled a bicycle with a sidecar. The operator sat in the middle of the two-wheel section and pushed a crank back and forth, which propelled the triangle-shaped vehicle down the tracks. This manually powered handcar was replaced after World War I by a gasoline-powered track vehicle. This, in turn, was replaced by a conventional pickup truck fitted with an auxiliary set of flanged wheels.

What is a **Johnson bar** on a railroad?

It is the reverse lever of a locomotive. "To put the Johnson bar against the running board" means to get speed.

What does the term **gandy dancer** mean?

This name for a track laborer derived from the special tools used for track work made by the Gandy Manufacturing Company of Chicago, Illinois. These tools were used during the 19th century almost universally by section gangs.

What was the route of the **Orient Express**?

This luxury train service was inaugurated in June, 1883, to provide through connection between France and Turkey. It was not until 1889 that the complete journey could be made by train. The route left Paris and traveled via Chalons, Nancy, and Strasbourg into Germany (via Karlsruhe, Stuttgart, and Munich), then into Austria (via Salzburg, Linz, and Vienna), into Hungary (through Gyor and Budapest), south to Belgrade, Yugoslavia, through Sofia, Bulgaria, and finally to Istanbul (Constantinople), Turkey. It ceased operation in May, 1977. In 1982, part of the line, the *Venice–Simplon–Orient Express* went into operation.

How does a **cable car,** like those in San Francisco, move?

A cable runs continuously in a channel, between the tracks located just below the street. The cable is controlled from a central station, and usually moves about 9 miles (14.5 kilometers) per hour. Each cable car has an attachment, on the underside of the car, called a grip. When the car operator pulls the lever, the grip latches onto the moving cable and is pulled along by the moving cable. When the operator releases the lever, the grip disconnects from the cable and comes to a halt when the operator applies the brakes. Also called an endless ropeway, it was invented by Andrew S. Hallide (1836–1900), who first operated his system in San Francisco in 1873.

MOTOR VEHICLES

See also: Energy—Consumption and Conservation
Buildings, Bridges, and Other Structures—Roads, Bridges, and Tunnels

How did the term **horsepower** originate?

Horsepower is the unit of energy needed to lift 550 pounds the distance of 1 foot in 1 second. Near the end of the 18th century Scottish engineer James Watt (1736–1819) made improvements in the steam engine and wished to determine how its rate of pumping water out of coal mines compared with that of horses, which had previously been used to operate the pumps. In order to define a horsepower, he tested horses and concluded that a strong horse could lift 150 pounds 220 feet in 1 minute. Therefore, 1 horsepower was equal to 150 x 220/1 or 33,000 foot pounds per minute (also expressed as 745.2 joules per second, 7,452 million ergs per second, or 745.2 watts).

The term horsepower was frequently used in the early days of the automobile because the "horseless carriage" was generally compared to the horse-drawn carriage. Today this inconvenient unit is still used routinely to express the power of motors and engines, particularly of automobiles and aircraft. A typical automobile requires about 20 horsepower to propel it at 50 miles (80.5 kilometers) per hour.

Who **invented the automobile?**

Although the idea of self-propelled road transportation originated long before, Karl Benz (1844–1929) and Gottlieb Daimler (1834–1900), are both credited with the invention of the gasoline-powered automobile, because they were the first to make their automotive machines commercially practicable. Benz and Daimler worked independently, unaware of each other's endeavors. Both built compact, internal-combustion engines to power their vehicles. Benz built his three-wheeler in 1885; it was steered by a tiller. Daimler's four-wheeled vehicle was produced in 1887.

Earlier self-propelled road vehicles include a steam-driven contraption invented by Nicolas-Joseph Cugnot (1725–1804), who rode the Paris streets at 2.5 miles (4 kilometers) per hour in 1769. Richard Trevithick (1771–1833) also produced a steam-driven vehicle that could carry eight passengers. It first ran on December 24, 1801, in Camborne, England. Londoner Samuel Brown built the first practical four-horsepower gasoline-powered vehicle in 1826. The Belgian engineer J.J. Etienne Lenoire (1822–1900) built a vehicle with an internal combustion engine that ran on liquid hydrocarbon fuel in 1862, but he did not test it on the road until September, 1863, when it travelled a distance of 12 miles (19.3 kilometers) in 3 hours. The Austrian inventor Siegfried Marcus (1831–1898) invented a four-wheeled gasoline-powered handcart in 1864 and a full-size car in 1875; the Viennese police objected to the noise that the car made, and Marcus did not continue its development. Edouard Delamare-Deboutteville invented an eight-horsepower vehicle in 1883 that was not durable enough for road conditions.

Is the **electric automobile** an idea of the last twenty years or so?

During the last decade of the 19th century, the electric vehicle became especially popular in cities. People had grown familiar with electric trolleys and railways, and technology had produced motors and batteries in a wide variety of sizes. The Edison Cell, a nickel-iron battery, became the leader in electric vehicle use. By 1900, electric vehicles nearly dominated the pleasure car field. In that year, 4,200 automobiles were sold in the United States. Of these 38% were powered by electricity, 22% by gasoline, and 40% by steam. By 1911, the automobile starter motor did away with hand-cranking gasoline cars, and Henry Ford had just begun to mass-produce his Model T's. By 1924, not a single electric vehicle was exhibited at the National Automobile Show, and the Stanley Steamer was scrapped the same year.

Because of the energy crises of the 1970s and the 1990s concern for the environment (as well as "Clean Air" legislation), the large American auto manufacturers, along with the Japanese, are working toward producing a practical electric automobile. Chrysler and Ford have built electric versions of their minivans, and Nissan, Mitsubishi, and Daihatsu are competing, having announced aggressive programs of their own. In Europe—where electric vehicles (EVs) such as the General Motors one-ton van the Griffon have been used for years for local deliveries and in service fleets—VW, BMW, Audi, Fiat, and Peugeot are also developing products. But the odds-on favorite for the first electric car to win a mass market is General Motors' Impact, due out in 1995.

The basic problem, a technological one, still remains—the limitations of battery technology. The electric car remains a poor performer in speed and range, and the cost of battery pack replacement is expensive. Two new types of batteries are under development: the lithium-sulfide battery and the sodium-sulfur unit. Also, manufacturers are experimenting with an alternating-current (AC) engine to replace the present heavier direct-current (DC) power trains.

Who started the **first American automobile company**?

Charles Duryea (1861–1938), a cycle manufacturer from Peoria, Illinois, and his brother, Frank (1869–1967), founded America's first auto-manufacturing firm and became the first to build cars for sale in the United States. The Duryea Motor Wagon company, set up in Springfield, Massachusetts, in 1895, built gasoline-powered, horseless carriages similar to those built by Benz in Germany.

However, the Duryea brothers did not build the first automobile factory in the United States. Ransom Eli Olds (1864–1950) built it in 1899 in Detroit, Michigan, to manufacture his Oldsmobile. Over 10 vehicles a week were produced there by April, 1901 (433 in 1901; 2,500 in 1902; and 5,508 in 1904). In 1906, there were 125 companies making automobiles in the United States. In 1908, the American engineer Henry Ford (1863–1947) initiated assembly line techniques that made automotive manufacture quick and cheap; his company sold 10,660 vehicles that year.

When did the first **used-car dealership** open?

The first used-car dealership in Britain, the Motor Car Company of London, England, opened in September, 1897, with 17 second-hand vehicles.

When was the **Michelin tire** introduced?

The first pneumatic (air-filled) tire for automobiles was produced in France by André (1853–1931) and Edouard Michelin (1859–1940) in 1885. The first radial-ply tire, the Michelin X, was made and sold in 1948. In radial construction, layers of cord materials called plies are laid across the circumference of the tire from bead to bead (perpendicular to the direction of the tread centerline). The plies can be made of steel wires or belts which circle the tire. Radial tires are said to give longer tread life, better handling, and a softer ride at medium and high speeds than bias or belted bias tires (both of which have plies laid diagonally). Radials give a firm, almost hard, ride at low speeds.

When were **tubeless automobile tires** first manufactured?

In Akron, Ohio, the B.F. Goodrich Company announced the manufacture of tubeless tires on May 11, 1947. Dunlop was the first British firm to make tubeless tires, in 1953.

What do the **numbers** mean **on automobile tires**?

The numbers and letters associated with tire sizes and types are complicated and confusing. The "Metric P" system of numbering is probably the most useful method of indicating tire sizes. For example, if the tire had P185/75R-14, then "P" means the tire is for a passenger car. The number 185 is the width of the tire in millimeters. 75 is the aspect ratio, i.e., the height of the tire from the rim to the road is 75% of the width. R indicates that it is a radial tire. 14 is the wheel diameter. 13 and 15 inches are also common sizes.

Which vehicle had the **first modern automobile air conditioner**?

The first air-conditioned automobile was manufactured by the Packard Motor Car Company in Detroit, Michigan, and was exhibited publicly November 4–12, 1939, at the 40th Automobile Show in Chicago, Illinois. Air in the car was cooled to the temperature desired, dehumidified, filtered, and circulated. The first fully automatic air conditioning system was Cadillac's "Climate Control," introduced in 1964.

What was the **first** car manufactured with an **automatic transmission**?

The first of the modern generation of automatic transmissions was General Motors' Hydramatic, first offered as an option on the Oldsmobile during the 1940 season. Between 1934 and 1936, a handful of 18 horsepower Austins were fitted with the American-designed Hayes infinitely variable gear. The direct ancestor of the modern automatic gearbox was patented in 1898.

Has there ever been a **nuclear-powered automobile**?

In the 1950s, Ford automotive designers envisioned the Ford Nucleon, which was to be propelled by a small atomic reactor core, located under a circular cover at the rear of the car. It was to be recharged with nuclear fuel. The car was never built.

Where was the first **automobile license plate** issued?

Leon Serpollet of Paris, France, obtained the first license plate in 1889. They were first required in the United States by New York State in 1901. Registration was required within 30 days. Owners had to provide their names and addresses as well as a description of their vehicles. The fee was one dollar. The plates bore the owner's initials and were required to be over 3 inches high. Permanent plates made of aluminum were first issued in Connecticut in 1937.

What information is available from the **vehicle identification number (VIN), body number plate,** and **engine** on a car?

These coded numbers reveal the model and make, model year, type of transmission, plant of manufacture, and sometimes even the date and day of the week a car was made. The form and content of these codes is not standardized and often changes from one year to the next for the same manufacturer. Various components of a car may be made in different plants so a location listed on a VIN may differ from one on the engine number. The official shop manual lists the codes for a particular make of car.

How many **motor vehicles** are registered **in the United States**?

Total United States registration of motor vehicles in 1991 was estimated to be 190,741 thousand. Of the total, 145 million were automobiles and 45 million were trucks and buses. Worldwide, there were 556,031,486 total motor vehicles registered in 1989 (approximately 80% of which were passenger cars).

How much does it **cost to operate an automobile** in the United States?

Below is listed the average cost per mile to operate an automobile in the United States in cents per mile. Figures are given for suburban driving conditions:

	Large	Intermediate	Compact	Sub-compact	Passenger van
Depreciation	9.6	8.6	7.3	5.9	10.7
Maintenance	6.0	5.2	4.6	5.1	6.9
Gas and oil	7.0	5.7	4.6	4.4	9.1
Parking and tolls	0.9	0.9	0.9	0.9	0.9
Insurance	4.9	5.6	4.3	5.0	8.9
Taxes	2.2	1.8	1.6	1.4	2.7
Total costs	30.6	27.8	23.3	22.7	39.2

Large (weight not more than 3,500 pounds)

Intermediate (weight less than 3,500 pounds)

Compact (weight less than 3,000 pounds)

Subcompact (weight less than 2,500 pounds)

Passenger van (weight less than 5,000 pounds)

How many new passenger cars in the United States are imported cars?

In 1990, imported cars accounted for 25.8% of the total passenger car sales. On total sales of 9,301,324 cars, 6,896,888 were domestic and 2,404,416 were imports. Japanese car sales (1,719,839) accounted for 18.5% of total car sales, and 71.5% of total import car sales. In 1990, the top new passenger car models sold in the United States were the Honda Accord (417,179), Ford Taurus (313,274), Chevrolet Cavalier (295,123), Ford Escort (288,727), and Toyota Camry (264,595).

Which states require regular automobile safety inspections?

Twenty-six states require regular inspection of automobiles to rid the highways of dangerous vehicles with bald tires, wobbly suspensions, smoke-filled exhausts, and defective brakes and lights. These states are:

Arkansas

California

Delaware

District of Columbia

Hawaii

Louisiana

Maine

Maryland

Massachusetts

Michigan

Missouri

New Hampshire

New Jersey

New York

North Carolina

Ohio

Oklahoma

Pennsylvania

Rhode Island

South Carolina

Tennessee

Texas

Utah

Vermont

Virginia

West Virginia

What is the braking distance for an automobile at different speeds?

Average stopping distance is directly related to vehicle speed. On a dry, level concrete surface, the minimum stopping distances are as follows (including driver reaction time to apply brakes):

Speed		Reaction time distance		Braking distance		Total distance	
Mph	Kph	Feet	Meters	Feet	Meters	Feet	Meters
10	16	11	3.4	9	2.7	20	6.1
20	32	22	6.7	23	7.0	45	13.7
30	48	33	10.1	45	13.7	78	23.8
40	64	44	13.4	81	24.7	125	38.1
50	80	55	16.8	133	40.5	188	57.3
60	97	66	20.1	206	62.8	272	82.9
70	113	77	23.5	304	92.7	381	116.1

When was a **speed trap** first employed to apprehend speeding automobile drivers?

In 1905, William McAdoo, police commissioner of New York City, was stopped for travelling at 12 miles per hour in an 8 mile per hour zone in rural New England. The speed detection device consisted of two lookout posts, camouflaged as dead tree trunks, spaced one mile apart. A deputy with a stopwatch and a telephone kept watch for speeders. When a car appeared to be travelling too fast, the deputy pressed his stopwatch and telephoned ahead to his confederate who immediately consulted a speed-mileage chart and phoned ahead to another constable manning a road block to apprehend the speeder. McAdoo invited the New England constable to set up a similar device in New York City.

One of the most famous speed traps was in the Alabama town of Fruithurst, on the Alabama–Georgia border. In one year, this town of 250 people collected over $200,000 in fines and forfeitures from unwary "speeders."

Which colors of cars are the safest?

Tests at the University of California concluded that either blue or yellow is the best color for car safety. Blue shows up best during daylight and fog; yellow is best at night. The worst color from the visibility standpoint is gray. In another study by Mercedes-Benz in Germany, white ranked the highest in all-around visibility, except in situations of completely snow-covered roads or white sand. In such extremes, bright yellow and bright orange ranked second and third respectively in visibility. The least visible car color in the Mercedes-Benz test was dark green.

How does **VASCAR** work?

Invented in 1965, VASCAR (*Visual Average Speed Computer and Recorder*) is a calculator that determines a car's speed from two simple measurements of time and distance. No radar is involved. VASCAR can be used at rest or while moving to clock traffic in both directions. A patrol car employing VASCAR can be behind, ahead of, or even perpendicular to the target vehicle. The device measures the length of a speed trap and then determines how long it takes the target car to cover that distance. An internal calculator does the math and displays the average speed on an LED readout. Most police departments now use several forms of moving radar which are less detectable and more accurate.

How does **police radar** work?

Austrian physicist Christian Doppler (1803–1853) discovered that the reflected radio waves bouncing off a moving object are returned at a different frequency (shorter or longer waves, cycles, or vibrations). This phenomenon, called the *Doppler effect,* is the basis of police radar. Directional radio waves are transmitted from the radar device. The waves bounce off the targeted vehicle and are received by a recorder. The recorder compares the difference between the sent and received waves, and translates the information into miles per hour and displays the speed on a dial.

Which states do not allow **radar detection devices** in motor vehicles?

Only Connecticut, Virginia, and the District of Columbia do not allow drivers to install radar detectors.

How does an **air bag work** to prevent injury in an automobile crash?

An air bag, stored in the steering wheel or dashboard, is activated during a frontal collision involving a force roughly equivalent to hitting a solid barrier head-on at about 12 miles (19.31 kilometers) per hour. Sensors trigger the release of chemicals (sodium azide and nitrogen) that inflate the bag fully, in about 1/30 of a second after impact, to create a protective cushion. Immediately the air bag begins to deflate and the harmless nitrogen gas escapes through holes in the back, so that the bag is out of the way almost at once.

When was the **air bag invented**?

Patented ideas on air bag safety devices began appearing in the early 1950s. U.S. patent 2,649,311 was granted on August 18, 1953, to John W. Hetrick for an inflated safety **401** cushion to be used in automotive vehicles. The Ford Motor Company studied the use of air bags around 1957, and other undocumented work was carried out by Assen Jordanoff before 1956. There are other earlier uses of an air bag concept, and it is rumored that some World War II pilots inflated their life vests before a crash.

In the mid-70s, General Motors geared up to sell 100,000 air bag–equipped cars a year in a pilot program to offer them as a discounted option on luxury models. GM dropped the option after only 8,000 buyers ordered air bags in three years. As of September 1, 1989, all new passenger cars produced for sale in the United States are required to be equipped with passive restraints (either automatic seatbelts or air bags). Currently, a major consumer shift toward air bags is taking place, and in 1991, 5 million United States cars were equipped with air bags.

How many motor vehicles are **stolen each year**?

Automobile theft has been occurring since an automobile mechanic stole Baron de Zuylen's Peugeot in Paris in June, 1896. Motor vehicle thefts in 1989 in the United States totaled 1,564,800—averaging 816 thefts for every 100,000 vehicles and increasing 9.2% from 1988 figures. Thefts declined in 12 states and the District of Columbia, but were up significantly in several others. Arizona led these states with a 53.5% increase. Other states with a vehicle theft increase of 20% minimum included Arkansas, Delaware, Kansas, Missouri, and Wisconsin.

Which automobile is the one **most often stolen**?

The 1986 Camaro was the most stolen car in 1991, followed by the 1984 Oldsmobile Cutlass Supreme and the 1987 Camaro. The three most stolen cars of 1991 were also the top three in 1990. On April 24, 1986, the United States Congress ruled that certain parts on "high-theft" cars must be marked by the manufacturer with an identification number. Cars affected by this were the Buick Riviera, Toyota Celica Supra, Cadillac Eldorado, Chevrolet Corvette, Pontiac Firebird, Mazda RX-7, Chevrolet Camaro, Porsche 911, Pontiac Grand Prix, and Oldsmobile Toronado.

What is the difference between a **medium truck** and a **heavy truck**?

Medium trucks weigh 14,001 to 33,000 pounds (6,351 to 14,969 kilograms). They span a wide range of sizes and uses from step-van route trucks to truck tractors. Common examples include beverage trucks, city cargo vans, and garbage trucks. Heavy trucks weigh 33,001 pounds (14,969 kilograms) or greater. Heavy trucks include over-the-road 18-wheelers, dump trucks, concrete mixers, and fire trucks. These trucks have come a long way from the first carrying truck, built in 1870 by John Yule, which moved at a rate of 0.75 miles (12 kilometers) per hour.

What is the origin of the term **taxicab**?

The term *taxicab* is derived from two words—*taximeter* and *cabriolet*. The taximeter, an instrument invented by Wilhelm Bruhn in 1891, automatically recorded the distance traveled and/or the time consumed. This enabled the fare to be accurately measured. The cabriolet is a two-wheeled, one-horse carriage that was often rented.

The first taxicabs for hire were two Benz-Kraftdroschkes operated by "Droschkenbesitzer" Dütz in the spring of 1896 in Stuttgart, Germany. In May, 1897, a rival service was started by Friedrich Greiner. In a literal sense, Greiner's cabs were the first "true" taxis because they were the first motor cabs fitted with taximeters.

AIRCRAFT

Who made the **first nonstop transatlantic flight**?

The first nonstop flight across the Atlantic Ocean, from Newfoundland, Canada, to Ireland, was made by two British aviators, Capt. John W. Alcock (1892–1919) and Lt. Arthur W. Brown (1886–1948), on June 14–15, 1919. The aircraft, a converted twin-engined Vickers Vimy bomber, took 16 hours 27 minutes to fly 1,890 miles (3,032 kilometers). Later Charles A. Lindbergh (1902–1974) made the first solo crossing flight on May 20–21, 1927, in the single-engined Ryan monoplane *Spirit of St. Louis*. His flight from New York to Paris covered a distance of 3,609 miles (5,089 kilometers) and lasted 33½ hours. The first woman to fly solo across the Atlantic was Amelia Earhart (1897-1937), who flew from Newfoundland to Ireland May 20–21, 1932.

Who made the first **supersonic flight**?

Supersonic flight is flight at or above the speed of sound. The speed of sound is 760 miles (1,223 kilometers) per hour in warm air at sea level. At a height of about 37,000 feet (11,278 kilometers), its speed is only 660 miles (1,062 kilometers) per hour. The first person credited with reaching the speed of sound (Mach 1) was Major Charles E. (Chuck) Yeager (b. 1923), of the United States Air Force. In 1947, he attained Mach 1.45 at 60,000 feet (18,288 meters) while flying the Bell *X-1* rocket research plane. This plane had been carried aloft by a B-29 and released at 30,000 feet (9,144 meters). In 1949, the Douglas *Skyrocket* became the first supersonic jet-powered aircraft to reach Mach 1 when Gene May flew at Mach 1.03 at 26,000 feet (7,925 meters).

When was the **first non-stop, unrefueled airplane flight around the world?**

Dick Rutan (b. 1943) and Jeana Yeager (b. 1952) flew the *Voyager,* a trimaran monoplane, in a closed circuit loop westbound and back to Edwards Air Force Base, California, December 14–23, 1986. The flight lasted 9 days, 3 minutes, 44 seconds and covered 24,986.7 miles (40,203.6 kilometers). The first successful round-the-world flight was made by two Douglas World Cruisers between April 6 and September 28, 1924. Four aircraft originally left Seattle, Washington, and two went down. The two successful planes completed 27,553 miles (44,332.8 kilometers) in 175 days—with 371 hours 11 minutes being their actual flying time. Between June 23 and July 1, 1931, Wiley Post (1900–1935) and Harold Gatty (1903–1957) flew around the world, starting from New York, in their Lockheed Vega, *Winnie Mae.*

What is **avionics?**

Avionics, a term derived by combining aviation and electronics, describes all of the electronic navigational, communications, and flight management aids with which airplanes are equipped today. In military aircraft this term also covers electronically controlled weapons, reconnaissance, and detection systems. Until the 1940s, the systems involved in operating aircraft were purely mechanical, electric, or magnetic, with radio apparatus being the most sophisticated instrumentation. The advent of radar and the great advance made in airborne detection during World War II led to the general adoption of electronic distance-measuring and navigational aids. In military aircraft such devices improve weapon delivery accuracy and in commercial aircraft they provide greater safety in operation.

Where is the **black box** carried on an airplane?

Actually painted bright orange to make it more visible in an aircraft's wreckage, the black box is a tough metal and plastic case containing two recorders. Installed in the rear of the aircraft—the area most likely to survive a crash—the case has two shells of stainless steel with a heat-protective material between the shells. The case must be able to withstand a temperature of 2,000°F (1,100°C) for 30 minutes. Inside it, mounted in a shockproof base, are the aircraft's flight data and cockpit voice recorders. The flight data recorder provides information about airspeed, direction, altitude, acceleration, engine thrust, and rudder and spoiler positions from sensors that are located around the aircraft. The data is recorded as electronic pulses on stainless steel tape that is about as thick as aluminum foil. When the tape is played back, it generates a computer printout. The cockpit voice recorder records the previous 30 minutes of the flight crew's conversation and radio transmission on a continuous tape loop. If a crash does not stop the recorder from recording, vital information can be lost as the loop continues to record in the aftermath.

What is the maximum **seating capacity** in a Boeing 747?

The seating capacity of the 747 and that of some other jets servicing cities in the United States are listed below.

Airplane	Maximum seating capacity
Boeing 707	179
Boeing 707-320, 707-420	189
Boeing 720	149
Boeing 727	125
Boeing 747	498
Boeing 757	196
Boeing 767	289
Boeing 777	375
Concorde (SST)	110
Lockheed L-1011 TriStar	345
McDonnell Douglas DC-8	189
McDonnell Douglas DC-9	
Series 20	119
Series 30 & 40	125
Series 50	139
McDonnell Douglas DC-10	380
Tupolev Tu-144 (Soviet SST)	140

When was the first **full-scale wind tunnel** for testing airplanes used?

It began operations on May 27, 1931, at the Langley Research Center of the National Advisory Committee for Aeronautics, Langley Field, Virginia. This tunnel, still in use, is 30 feet (9.14 meters) high and 60 feet (18.29 meters) wide. A wind tunnel is used to simulate air flow for aerodynamic measurement; it consists essentially of a closed tube, large enough to hold the airplane or other craft being tested, through which air is circulated by powerful fans.

Who designed the **Spruce Goose**?

Howard Hughes (1905–1976) designed and built the all-wood H-4 Hercules flying boat, nicknamed the *Spruce Goose*. The aircraft had the greatest wingspan ever built and **405** was powered by eight engines. It was only flown once—covering a distance of less than one mile at Los Angeles harbor—on November 2, 1947, lifting only 33 feet (10.6 meters) off the surface of the water.

After the Japanese attack on Pearl Harbor on December 7, 1941, and the subsequent entry of the United States into World War II, the United States government needed a large, cargo-carrying airplane that could be made from non-critical wartime materials, such as wood. Henry J. Kaiser (1882–1967), whose shipyards were producing Liberty ships at the rate of one per day, hired Howard Hughes to build such a plane. Hughes eventually produced a plane that weighed 400,000 pounds (181,440 kilograms) and had a wingspan of 320 feet (97.5 meters). Unfortunately, the plane was so complicated that it was not finished by the end of the war. In 1947, Hughes flew the boat himself during its only time off the ground—supposedly just to prove that something that big could fly. Today, the plane is on public exhibition in Long Beach, California.

What is the difference between an **amphibian plane** and a **seaplane**?

The primary difference is that an amphibian plane has retractable wheels that enable it to operate from land as well as water, while a seaplane is limited to water take-offs and landings, having only pontoons without wheels. Because its landing gear cannot retract, a seaplane is less aerodynamically efficient than an amphibian.

What is the world record for the fastest kite?

On May 17, 1987, Troy Vickstrom piloted a speeding 10-foot (3.05-meter) Flexifoil across the beach in Lincoln City, Oregon. Documentation of the record came from the local police, who issued Vickstrom a traffic citation for exceeding the beach's posted speed limit of 20 miles (32.18 kilometers) per hour. The fastest speed attained by a flying kite was 120 miles (193 kilometers) per hour on September 22, 1989, by Pete DiGiacomo at Ocean City, Maryland.

Why did the dirigible Hindenburg **explode**?

Despite the official United States and German investigations into the explosion, it still remains a mystery today. The most plausible explanations are structural failure, St. Elmo's Fire, static electricity, or sabotage. The *Hindenburg,* built following the great initial success of the *Graf Zeppelin,* was intended to exceed all other airships in size, speed, safety, comfort, and economy. At 803 feet (244.75 meters) long, it was 80% as long as the liner *Queen Mary,* and it was 135 feet (41.1 meters) in diameter.

In 1935, the German Air Ministry virtually took over the Zeppelin Company to use it to spread Nazi propaganda. After its first flight in 1936, the airship was very popular with the flying public. No other form of transport could carry passengers so swiftly, reliably, and comfortably between continents. During 1936, 1,006 passengers flew over the North Atlantic Ocean in the *Hindenburg.* On May 6, 1937, while landing at Lakehurst, New Jersey, its hydrogen burst into flames, and the airship was completely destroyed. Of the 97 people aboard, 62 survived.

MILITARY VEHICLES

Where did the **military tank** get its name?

During World War I, when the British were developing the tank, they called these first armored fighting vehicles "water tanks" to keep their real purpose a secret. This code word has remained in spite of early efforts to call them "combat cars" or "assault carriages."

Who invented the **culin device** on a tank?

In World War II, American tank man Sergeant Curtis G. Culin devised a crossbar welded across the front of the tank with four protruding metal tusks. This device made it possible to break through the German hedgerow defenses and save thousands of lives. In the hedgerow country of Normandy, France, countless rows or stands of bushes or trees surrounded the fields, limiting tank movement. The culin device—also known as the "Rhinoceros" because its steel angled teeth formed a tusk-like structure—cut into the base of the hedgerow and pushed a complete section ahead of it into the next field, burying any enemy troops dug in on the opposite side.

What is a **Sopwith Camel** and why is it so called?

The most successful British fighter of World War I, the Camel was a development of the earlier Sopwith Pup, with a much larger rotary engine. Its name "camel" was derived from the humped shape of the covering of its twin synchronized machine guns. The highly maneuverable Camel, credited with 1,294 enemy aircraft destroyed, proved far superior to all German types as dogfighters, until the introduction of the Fokker D. VII in 1918. Altogether, 5,490 Camels were built by Sopwith Aircraft. The plane's top speed was 118 miles (189 kilometers) per hour and it had a ceiling of 24,000 feet (7,300 meters).

Who was the "Red Baron"?

Manfred von Richthofen (1892–1918), a German fighter pilot during World War I, was nicknamed "Red Baron" by the Allies because he flew a red-painted Albatros fighter. He was the top ace of the war, shooting down 80 Allied planes. He was killed on April 21, 1918, when he was attacked both by Roy Brown, a Canadian ace, and by Australian ground machine-gunners. Both parties claimed responsibility for his death.

When was the B-17 Flying Fortress introduced?

A Fortress prototype first flew on July 28, 1935, and the first Y1B-17 was delivered to the Air Corps in March, 1937, followed by an experimental Y1B-17A fitted with turbo-supercharged engines in January, 1939. An order for 39 planes was placed for this model under the designation B-17B. In addition to its bombing function, the B-17 was used for many experimental duties, including serving as a launching platform in the U.S.A.A.F. guided missile program and in radar and radio-control experiments. It was called a "Flying Fortress" because it was the best defended bomber of World War II. Altogether, it carried thirteen 50 caliber Browning M-2 machine guns, each having about 700 pounds (317.5 kilograms) of armor-piercing ammunition. Ironically, the weight of all its defensive armament and manpower severely restricted the plane's bomb-carrying capacity.

Who were the "Flying Tigers"?

They were members of the American Volunteer Group who were recruited early in 1941 by Major General Claire Lee Chennault (1890–1958) to serve in China as mercenaries. Some 90 veteran United States pilots and 150 support personnel served from December, 1941, until June, 1942, during World War II. The airplanes they flew were P-40 Warhawks, which had the mouths of tiger sharks painted on the planes' noses. These painted-on images inspired the group nickname the "Flying Tigers."

Why was the designation MiG chosen for the Soviet fighter plane used in World War II?

The *MiG* designation, formed from the initials of the plane's designers, Artem I. Mikoyan and Mikhail I. Gurevich, sometimes is listed as the Mikoyan-Gurevich MiG. Appearing in 1940 with a maximum speed of 400 miles (643.6 kilometers) per hour, the MiG-3, a piston-engined fighter, was one of the few Soviet planes whose performance was comparable with Western types during World War II. One of the best-known fighters, the MiG-15, first flown in December, 1977, was powered by a Soviet version of a Rolls-Royce turbo jet engine. This high performer saw action during the Korean

Conflict (1950–1953). In 1955 the MiG-19 became the first Soviet fighter capable of supersonic speed in level flight.

What is the name of the airplane that carried the **first atomic bomb**?

During World War II, the *Enola Gay,* a modified Boeing B-29 bomber, dropped the first atomic bomb on Hiroshima, Japan, at 8:15 A.M. on August 6, 1945. It was piloted by Col. Paul W. Tibbets, Jr. (b.1915) of Miami, Florida. The bombardier was Maj. Thomas W. Ferebee of Mocksville, North Carolina. Bomb designer Capt. William S. Parsons was aboard as an observer.

Three days later, another B-29 called *Bock's Car* dropped a second bomb on Nagasaki, Japan. The Japanese surrendered unconditionally on August 15, which confirmed the American belief that a costly and bloody invasion of Japan could be avoided at Japanese expense.

The *Enola Gay* is currently undergoing restoration at the National Air and Space Museum's Silver Hill, Maryland, facility. *Bock's Car* is on display at the U.S. Air Force Museum, Wright-Patterson Air Force Base, Dayton, Ohio.

COMMUNICATIONS

SYMBOLS, WRITING, AND CODES

Which **animals** other than horses have been **used to deliver the mail**?

During the 19th century, cows hauled mail wagons in some German towns. In Texas, New Mexico, and Arizona, camels were used. In Russia and Scandinavia, reindeer pulled mail sleighs. The Belgian city of Liége even tried cats, but they proved to be unreliable.

What is the **fog index** formula?

This is a system that has been developed to judge the reading level of a text. It is calculated in three steps:

1. From several randomly selected samples of about 100 words, divide the number of words in the total sample by the number of complete thought sentences, to get average sentence average. Count every independent clause (separated by colons, commas, and semicolons) as a thought sentence.

2. Count the number of "hard" words having three of more syllables. Do not count as "hard" words the following: proper nouns, compound words made from shorter words (e.g., basketball), or verb forms whose third syllable ends in "ed" or "es" (e.g., "decided"). Count the syllables as they are generally pronounced, not as the dictionary divides them. If the same word is repeated, count it again. Divide the number of hard words by the total number of words in the sample. The figure should be computed to four decimal places. (In order to avoid a distorted result, terms such as "nuclear magnetic resonance" should be counted as "NMR.")

3. Add the results from 1 and 2 together and multiply by 0.4. This will give the grade level of the reading sample.

How much does a **cubic foot of books** weigh?

A cubic foot of books weighs 65 pounds (29.48 kilograms). This does not take into consideration any type of variation in paper quality, binding, etc.

What is the **international phonetic alphabet**?

Letter	Phonetic equivalent
A	Alpha
B	Bravo
C	Charlie
D	Delta
E	Echo
F	Fox Trot
G	Gold
H	Hotel
I	India
J	Juliett
K	Kilo
L	Lima
M	Mike
N	November
O	Oscar
P	Papa
Q	Quebec
R	Romeo
S	Sierra
T	Tango
U	Uniform
V	Victor
W	Whiskey
X	X-ray
Y	Yankee

Who invented the **Braille** alphabet?

The Braille system, used by the blind to read and write, employs combinations of raised
dots that form characters corresponding to the letters of the alphabet, punctuation

marks, and common words such as "and" and "the." Louis Braille (1809–1852), blind himself since the age of three, began working on developing a practical alphabet for the blind shortly after he started a school for the blind in Paris. He experimented with a communication method called night-writing, which the French army used for night-time battlefield missives. With the assistance of an army officer, Captain Charles Barbier, Braille pared the method's 12-dot configurations down to a 6-dot one and devised a code of 63 characters. The system was not widely accepted for several years; even Braille's own Paris school did not adopt the system until 1854 (two years after his death). In 1916, the United States sanctioned Louis Braille's original system of raised dots, and in 1932 a modification called "Standard English Braille, Grade 2" was adopted throughout the English-speaking world. The revised version changed the letter-by-letter codes into common letter combinations, such as "ow," "ing," and "ment," making reading and writing a faster activity.

Before Braille's system, one of the few effective alphabets for the blind was devised by another Frenchman, Valentin Haüy (1745–1822), who was the first to emboss paper to help the blind read. Haüy's letters in relief were actually a punched alphabet, and imitators immediately began to copy and improve on his system. Another letter-by-letter system of nine basic characters was devised by Dr. William Moon (1818–1894) in 1847, but it is less versatile in its applications.

Braille alphabet

What is the **Morse Code**?

The success of any electrical communication system lies in its coding interpretation, for only series of electric impulses can be transmitted from one end to the other. These impulses must be "translated" from/into words, numbers, etc. This problem plagued early telegraphy until American painter-turned-scientist Samuel F.B. Morse (1791–1872) with the help of Alfred Vail (1807–1859) devised in 1835 a code composed of dots and dashes to represent letters, numbers, and punctuation. Telegraphy uses an electromagnet—a device that becomes magnetic when activated and raps against a metal contact. A series of short electrical impulses repeatedly can make and break this magnetism, resulting in a tapped-out message.

Having secured a patent on the code in 1837, Morse and Vail established a communications company on May 24, 1844. The first long-distance telegraphed message was sent by Morse in Washington, D.C., to Vail in Baltimore. This was the same year that Morse took out a patent on telegraphy; Morse never acknowledged the unpatented contributions of Joseph Henry (1797–1878), who invented the first electric motor and working electromagnet in 1829 and the electric telegraph in 1831.

The International Morse Code (shown below) uses sound or a flashing light to send messages. The dot is a very short sound or flash; a dash equals three dots. The pauses between sounds or flashes should equal one dot. An interval of the length of one dash is left between letters; an interval of two dashes is left between words.

A .-	J .——	S ...
B -...	K -.-	T -
C -.-.	L .-..	U ..—
D -..	M —	V ...-
E .	N -.	W .——
F ..-.	O ——	X -..-
G ——.	P .——.	Y -.——
H	Q ——.-	Z ——..
I ..	R .-.	

1 .————	6 -....	Period .-.-.-
2 ..———	7 ——...	Comma ——..——
3 ...——	8 ———..	
4 ....-	9 ————.	
5	0 —————	

What were Enigma and Purple in World War II?

Enigma and *Purple* were the electric rotor cipher machines of the Germans and Japanese, respectively. The Enigma machine, used by the Nazis, was invented in the 1920s. The Japanese machine, called Purple by the Americans, was adapted from Enigma.

Cryptography—the art of sending messages in such a way that the real meaning is hidden from everyone but the sender and the recipient—is done in two ways: code and cipher. A code is like a dictionary in which all the words and phrases are replaced by codewords or codenumbers. A codebook is used to read the code. A cipher works with single letters, rather than complete words or phrases. There are two kinds of ciphers: transposition and substitution. In a transposition cipher, the letters of the ordinary message (or plain text) are jumbled to form the cipher text. In substitution, the plain text letters can be replaced by other letters, numbers, or symbols.

What are the "10-codes"?

Almost as many different codes exist as there are agencies using codes in radio transmission. The following are officially suggested by the Associated Public Safety Communications Officers (APCO):

Ten-1 Cannot understand your message.

Ten-2 Your signal is good.

Ten-3 Stop transmitting.

Ten-4 Message received ("O.K.").

Ten-5 Relay information to _____.

Ten-6 Station is busy.

Ten-7 Out of service.

Ten-8 In service.

Ten-9 Repeat last message.

Ten-10 Negative ("no").

Ten-11 _____ in service.

Ten-12 Stand by.

Ten-13 Report _____ conditions.

Ten-14 Information.

Ten-15 Message delivered.

Ten-16 Reply to message.

Ten-17 Enroute.

Ten-18 Urgent.

Ten-19 Contact _____.

Ten-20 Unit location.

Ten-21 Call _____ by telephone.

Ten-22 Cancel last message.

Ten-23 Arrived at scene.

Ten-24 Assignment completed.

Ten-25 Meet _____.

Ten-26 Estimated time of arrival is _____.

Ten-27 Request for information on license.

Ten-28 Request vehicle registration information.

Ten-29 Check records.

Ten-30 Use caution.

Ten-31 Pick up.

Ten-32 Units requested.

Ten-33 Emergency! Officer needs help.

Ten-34 Correct time.

What do the lines in a **Universal Product Code** mean?

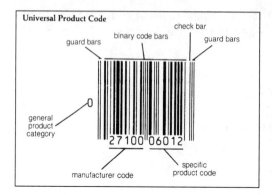

Also called a bar code, a Universal Product Code (UPC) is a product description code designed to be read by a computerized scanner or cash register. It consists of 11 numbers in groups of "0"s (dark strips) and "1"s (white strips). A bar will be thin if it has only one strip or thicker if there are two or more strips side by side.

The first number describes the type of product. Most products begin with a "0"; exceptions are variable weight products such as meat and vegetables (2), health-care products (3), bulk-discounted goods (4), and coupons (5). Since it might be misread as a bar, the number 1 is not used.

The next five numbers describe the product's manufacturer. The following five numbers after that describe the product itself including its color, weight, size, and other distinguishing characteristics. The price of the item is not included; it is coded into the cash register separately.

The last number is a check digit which is used to tell the scanner if there is an error in the other numbers. The preceding numbers, when added, multiplied, and subtracted in a certain way will equal this number. If they do not, a mistake exists somewhere.

What does the code that follows the letters **ISBN** mean?

ISBN, or International Standard Book Number, is an ordering and identifying code for book products. It forms an unique number to identify that particular item. The first number of the series relates to the language the book is published in, for example, zero is designated for the English language. The second set of numbers identifies the publisher and the last set of numbers are unique numbers that are used only to identify the particular item. The very last number is a "check number," which mathematically makes certain that the previous numbers have been entered correctly.

RADIO AND TELEVISION

Who **invented radio?**

Guglielmo Marconi (1874–1937), of Bologna, Italy, was the first to prove that radio signals could be sent over long distances. Radio is the radiation and detection of signals propagated through space as electromagnetic waves to convey information. It was first called wireless telegraphy because it duplicated the effect of telegraphy without using wires. On December 12, 1901, Marconi successfully sent Morse code signals from Newfoundland, Canada, to England.

In 1906, American inventor Lee DeForest (1873–1961) built what he called "the Audion," which became the basis for the radio amplifying vacuum tube. This device made voice radio practical, because it magnified the weak signals without distorting them. The next year, DeForest began regular radio broadcasts from Manhattan, New York. As there were still no home radio receivers, DeForest's only audience was ship wireless operators in New York City Harbor.

What was the **first radio broadcasting station**?

Credit for the first radio broadcasting station has customarily gone to Westinghouse station KDKA in Pittsburgh for its broadcast of the Harding-Cox presidential election returns on November 2, 1920. Although there were other, earlier radio transmissions, KDKA was one of the first stations to use electron tube technology to generate the transmitted signal and hence to provide what could be described as broadcast quality. It was also the first to operate with a well-defined commercial purpose—it was not a hobby or a publicity stunt. It was the first broadcast station to be licensed on a frequency outside the amateur bands. Altogether, it was the direct ancestor of modern broadcasting.

How are the **call letters** beginning with K or W assigned to **radio stations**?

These beginning call letters are assigned on a geographical basis. For the majority of radio stations located east of the Mississippi River, their call letters begin with the letter *W;* if the stations are west of the Mississippi, their first call letter is the letter *K.* However, there are exceptions to this rule. Stations founded before this rule went into effect kept their old letters. So, for example, KDKA in Pittsburgh has retained the first letter K; likewise some western pioneer stations have retained the letter W. Since many AM licensees also operate FM and TV stations, a common practice is to use the AM call letters followed by *-FM* or *-TV.*

Why do **FM radio stations** have a limited broadcast range?

Usually radio waves higher in frequency than approximately 50 to 60 megahertz are not reflected by the Earth's ionosphere, but are lost in space. Television, FM radio, and high frequency communications systems are therefore limited to approximately line-of-sight **417** ranges. The line-of-sight distance depends on the terrain and antenna height, but is usually limited to from 50 to 100 miles (80 to 161 kilometers). FM (frequency-modulation) radio uses a wider band than AM (amplitude-modulation) radio to give broadcasts high fidelity, especially noticeable in music—crystal clarity to high frequencies and rich resonance to bass notes, all with a minimum of static and distortion. Invented by Edwin Howard Armstrong (1891–1954) in 1933, FM receivers became available in 1939.

Can **radio transmissions** between **space shuttles** and ground control be picked up by shortwave radio?

Amateur radio operators at Goddard Space Flight Center, Greenbelt, Maryland, retransmit shuttle space-to-ground radio conversations on shortwave frequencies. These retransmissions can be heard freely around the world. To hear astronauts talking with ground controllers during liftoff, flight, and landing, a shortwave radio capable of receiving single-sideband signals should be tuned to frequencies of 3.860, 7.185, 14.295, and 21.395 megahertz. British physics teacher Geoffrey Perry, at the Kettering Boys School, has taught his students how to obtain telemetry from orbiting Russian satellites. Since the early 1960s Perry's students have been monitoring Russian space signals using a simple taxicab radio and manipulating the data to calculate position and orbits of the spacecraft.

Why do AM stations have a wider broadcast range at night?

This variation is caused by the nature of the ionosphere of the Earth. The ionosphere consists of several different layers of rarefied gases in the upper atmosphere that have become conductive through the bombardment of the atoms of the atmosphere by solar radiation, by electrons and protons emitted by the sun, and by cosmic rays. These layers, sometimes called the Kennelly-Heaviside layer, reflect AM radio signals, enabling AM broadcasts to be received by radios that are great distances from the transmitting antenna. With the coming of night the ionosphere layers partially dissipate and become an excellent reflector of the short waveband AM radio waves. This causes distant AM stations to be heard more clearly at night.

Who was the **father of television**?

The idea of television (or "seeing by electricity," as it was called in 1880) was posed by several people over the years, and many individuals contributed a multiplicity of partial inventions. For example, in 1897 Ferdinand Braun (1850–1918) constructed the first cathode ray oscilloscope, a fundamental component to all television receivers. In 1907, Boris Rosing proposed using Braun's tube to receive images, and in the following year Alan Campbell-Swinton likewise suggested using the tube, now called the cathode-ray tube, for both transmission and receiving. However, the key figure, frequently called the father of television, was Russian-born American, Vladimir K. Zworykin (1889–1982). A former pupil of Rosing, he produced a practical method of amplifying the electron beam so that the light/dark pattern would produce a good image. In 1923, he patented the iconoscope (which would become the television camera) and in 1924 he patented the kinoscope (television tube). Both inventions rely on streams of electrons for both scanning and creating the image on a fluorescent screen. By 1938, after adding new and more sensitive photo cells, Zworykin demonstrated his first practical model.

During the early twentieth century others worked on different approaches to television. The best-known is John L. Baird (1888–1946), who in 1936 used a mechanized scanning device to transmit the first recognizable picture of a human face. Limitations in his designs made any further improvements in the picture quality impossible.

How does **rain affect television reception** from a satellite?

Television's incoming microwave signals are absorbed by rain and moisture, and severe rainstorms can reduce signals by as much as 10 decibels (reduction by a factor of 10). If the installation cannot cope with this level of signal reduction, the picture may be momentarily lost. Even quite moderate rainfall can reduce signals enough to give noisy reception on some receivers. Another problem associated with rain is an increase in noise due to its inherent noise temperature. Any body above the temperature of absolute zero (0°K or -459°F or -273°C) has an inherent noise temperature generated by the release of wave packets from the body's molecular agitation (heat). These wave packets have a wide range of frequencies, some of which will be within the required bandwidth for satellite reception. The warm Earth has a high noise temperature, and consequently rain does as well.

What name is used for a **satellite dish** that picks up **TV broadcasts**?

Earth station is the term used for the complete satellite receiving or transmitting station. It includes the antenna, the electronics, and all associated equipment necessary to receive or transmit satellite signals. It can range from a simple, inexpensive, receive-only Earth station that can be purchased by the individual consumer, to elaborate, two-way communications stations that offer commercial access to the satellite's capacity. Signals are captured and focused by the antenna into a feedhorn and low noise amplifier. These are relayed by cable to a down converter and then into the satellite receiver/modulator.

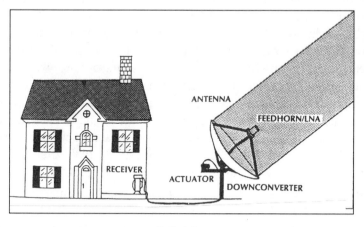

Earth station

What is **high definition television?**

The amount of detail shown in a television picture is limited by the number of lines that make it up and by the number of picture elements on each line. The latter is mostly determined by the width of the electron beam. To obtain pictures closer to the quality associated with 35 millimeter photography, a new television system HDTV (*H*igh *D*efinition *T*elevision) has been developed that has more than twice the number of scan lines with a much smaller picture element. Currently American and Japanese television has 525 scanning lines, while Europe uses 625 scanning lines. HDTV has received wide publicity in recent years, but it is currently in an engineering phase, and not yet commercially available.

The Japanese are generally given credit for pioneering in HDTV, ever since NHK, the Japanese broadcasting company, began research in 1968. In fact, the original pioneer was RCA's Otto Schade, who began his research after the end of World War II. Schade was ahead of his time, and decades passed before television pickup tubes and other components became available to take full advantage of his research.

HDTV cannot be used in the commercial broadcast bands until technical standards are approved by the United States Federal Communications Commission (FCC) or the various foreign regulating agencies. The more immediate problem however, is a technological one—HDTV needs to transmit five times more data than is currently assigned to each television channel. One approach is signal compression—squeezing the 30 megahertz bandwidth signal that HDTV requires into the 6 megahertz bandwidth currently used for television broadcasting. The Japanese and Europeans are exploring analog systems that use wavelike transmission, while the Americans are basing their HDTV development on digital transmission systems.

How do **submerged submarines communicate?**

Using frequencies from very high to extremely low, submarines can communicate by radio when submerged if certain conditions are met, and depending on whether or not detection is important. Submarines seldom transmit on long-range high radio frequencies if detection is important, as in war. However, Super (SHF), *U*ltra (UHF), or *Very H*igh Frequency (VHF) two-way links with cooperating aircraft, surface ships, via satellite, or with the shore are fairly safe with high data rate, though they all require that the boat show an antenna or send a buoy to the surface.

TELECOMMUNICATIONS, RECORDING, ETC.

When was the **first commercial communications satellite** used?

In 1960 *ECHO 1,* the first communications satellite, was launched. Two years later, on July 10, 1962, the first commercially funded satellite, *Telstar 1,* (paid for by American Telephone and Telegraph) was launched into low Earth orbit. It was also the first true communications satellite, being able to relay not only data and voice, but television as well. The first broadcast, which was relayed from the United States to England, showed an American flag flapping in the breeze. The first commercial satellite (in which its operations are conducted like a business) was *Early Bird,* which went into regular service on June 10, 1965, with 240 telephone circuits. *Early Bird* was the first satellite launched for Intelsat (International Telecommunications Satellite Organization). Still in existence, the system is owned by member nations—each nation's contribution to the operating funds are based on its share of the system's annual traffic.

How does a **telefax** work?

Telefacsimile (also telefax or facsimile or fax) transmits graphic and textual information from one location to another through telephone lines. A transmitting machine uses either a digital or analog scanner to convert the black and white representations of the image into electrical signals that are transmitted through the telephone lines to a designated receiving machine. The receiving unit converts the transmission back to **421** an image of the original and prints it. In its broadest definition, a facsimile terminal is simply a copier equipped to transmit and receive graphics images.

What is a **fiber optic cable**?

A fiber optic cable is composed of many very thin strands of coated glass fibers. It can transmit messages or images by directing beams of light inside itself over very short or very long distances (thousands of miles). The pattern of light waves forms a code that carries a message. At the receiving end the light beams are converted back into electric current and decoded. Since light beams are immune to electrical noise and can be carried greater distances before fading, this technology is used heavily in telecommunications. Other applications include the use of medical fiber optic viewers, such as endoscopes and fiberscopes, to see internal organs; fiber optic message devices in aircraft and space vehicles; and fiber optic connections in automotive lighting systems.

What is a Clarke belt?

In 1945 Arthur C. Clarke (b. 1917), the famous scientist and science fiction writer, predicted that an artificial satellite placed at a height of 22,248 miles (35,803 kilometers) directly above the equator would orbit the globe at the same speed with which the Earth was rotating. As a result, the satellite would remain stationary with respect to any point on the Earth's surface. This equatorial belt, rather like one of Saturn's rings, is affectionately known as the Clarke belt.

How does a **car phone** work?

A cellular phone system has three parts: the installation in the car; the cell site; and the mobile telephone switching office. Each cell has at its center a cell site where the fixed radio receiver and transmitter are located. All the cell sites belonging to a particular system are connected together at a mobile telephone switching office, which ties them to the local phone system. As a call passes from one cell site and enters into another, the call is transferred or "handed off" to an adjoining cell without any noticeable interruption.

What is the **Dolby** noise reduction system?

The magnetic action of audiotape produces a background hiss—a drawback in sound reproduction on tape. A noise reduction system known as *Dolby* (named after R.M.

Dolby [b. 1933], its American inventor) is widely used to deal with the hiss. In quiet passages, electronic circuits automatically boost the signals before they reach the recording head, drowning out the hiss. On playback, the signals are reduced to their correct levels. The hiss is reduced at the same time, becoming inaudible.

What is **digital audio tape (DAT)**?

DAT, a new concept in magnetic recording, produces a mathematical value for each sound based on the binary code. When the values are reconstructed during playback, the reconstructed sound is so much like the original that the human ear cannot distinguish the difference. The reproduction is so good that American record companies have lobbied lawmakers to prevent DAT from being sold in the United States, claiming that it could encourage illegal CD copying.

How are **compact discs (CDs)** made?

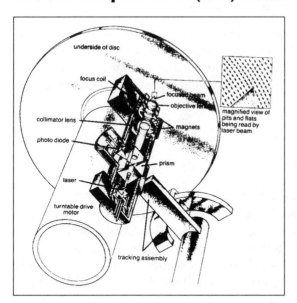

Compact disk player

The master disc for a CD is an optically flat glass disc coated with a resist. The resist is a chemical that is impervious to an etchant that dissolves glass. The master is placed on a turntable. The digital signal to be recorded is fed to a laser, turning the laser off and on in response to a binary on-off signal. When the laser is on, it burns away a small amount of the resist on the disc. While the disc turns, the recording head moves across the disc, leaving a spiral track of elongated "burns" in the resist surface. After the recording is complete, the glass master is placed in the chemical etchant bath. This developing removes the glass only where the resist is burned away. The spiral track now contains a series of small pits of varying length and constant depth. To play a recorded CD, a laser beam scans the three miles (five kilometers) of playing track and converts the "pits" and "lands" of the CD into binary codes. A photodiode converts these into a coded string of electrical impulses. In October, 1982, the first CDs were marketed; they were invented by Phillips (Netherlands) Company and Sony in Japan in 1978. **423**

What is the lifespan of a **CD-ROM** disc?

Although manufacturers claim that a CD-ROM disc will last twenty years, recent statements by the United States National Archives and Records Administration suggest that a lifespan of three to five years is more accurate. The main problem is that the aluminum substratum on which the data is recorded is vulnerable to oxidation.

COMPUTERS

What is an **algorithm**?

An algorithm is a set of clearly defined rules and instructions for the solutions of a problem. It is not necessarily applied only in computers, but can be a step-by-step procedure for solving any particular kind of problem. A nearly 4,000-year-old Babylonian banking calculation inscribed on a tablet is an algorithm, as is a computer program, which consists of the step-by-step procedures for solving a problem.

The term is derived from the name of Muhammad ibn Musa al Kharizmi (ca. 780–ca.850), a Baghdad mathematician who introduced Hindu numerals (including 0) and decimal calculation to the west. When his treatise was translated into Latin in the 12th century, the art of computation with Arabic (Hindu) numerals became known as *algorism.*

Who **invented the computer**?

Computers developed from calculating machines. One of the earliest mechanical devices for calculating, still widely used today, is the abacus—a frame carrying parallel rods on which beads or counters are strung. Herodotus, a Greek historian who lived around 400 B.C.E., mentions the use of the abacus in Egypt. In 1617, John Napier (1550–1617) invented "Napier's Rods"—marked pieces of ivory for multiples of numbers. In the middle of the same century, Blaise Pascal (1623–1662) produced a simple mechanism for adding and subtracting. Multiplication by repeated addition was a feature of a stepped drum or wheel machine of 1694 invented by Gottfried Wilhelm Leibniz (1646–1716). In 1823, English visionary Charles Babbage (1792–1871) persuaded the British government to finance an "analytical engine." This would have been a machine that could undertake any kind of calculation. It would have been driven by steam, but the most important innovation was that the entire program of operations was stored on a punched tape. Babbage's machine was not completed and would not have worked if it had been. The standards required were far beyond the capabilities of

the engineers of the time, and in any case, rods, levers, and cogs move too slowly for really quick calculations. Only electrons, which travel at near the speed of light, are rapid enough. Although he never built a working computer, Babbage thought out many of the basic principles that guide modern computers.

Based on the concepts of British mathematician Alan M. Turing (1912–1954), the earliest programmable electronic computer was the 1,500-valve "Colossus," formulated by Max Newman (1897–1985), built by T.H. Flowers, and used by the British government in 1943 to crack the German codes generated by the coding machine "Enigma."

What was the first major use for **punched cards**?

Punched cards were a way of programming, or giving instructions to, a machine. In 1801, Joseph Marie Jacquard (1752–1834) built a device that could do automated pattern weaving. Cards with holes were used to direct threads in the loom, creating predefined patterns in the cloth. The pattern was determined by the arrangement of holes in the cards, with wire hooks passing through the holes to grab and pull through specific threads to be woven into the cloth.

By the 1880s, Herman Hollerith (1860–1929) was using the idea of punched cards to give machines instructions. He built a punched card tabulator that processed the data gathered for the 1890 United States Census in six weeks (three times the speed of previous compilations). Metal pins in the machine's reader passed through holes punched in cards the size of dollar bills, momentarily closing electric circuits. The resulting pulses advanced counters assigned to details such as income and family size. A sorter could also be programmed to pigeonhole cards according to pattern of holes, an important aid in analyzing census statistics. Later, Hollerith founded Tabulating Machines Co., which in 1924 became IBM. When IBM adopted the 80-column punched card (measuring 7⅜ by 3¼ inches, or 18.7 by 8.25 centimeters, and .007 inches, or .018 centimeters, thick), the de facto industry standard was set, which has endured for decades.

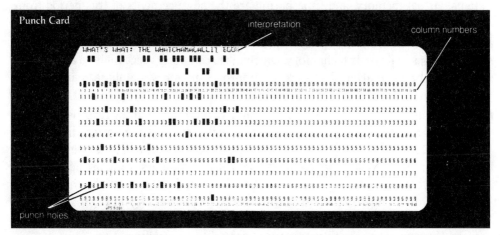

What is the origin of the expression "Do not fold, spindle, or mutilate"?

This is the inscription on an IBM punched card. Frequently, office workers organize papers and forms by stapling or folding them together, or by impaling them on a spindle. Because Hollerith (punched) card readers scan uniform rectangular holes in a precise arrangement, any damage to the physical card makes it unusable. In the 1950s and 1960s, when punched cards became widespread, manufacturers printed a warning on each card; IBM's "Do not fold, spindle, or mutilate" was the best-known. In 1964, the student revolution at the University of California, Berkeley used the phrase as a symbol of authority and regimentation.

What is meant by fifth generation computers? What are the other four generations?

The evolution of computers has advanced so much in the past few decades that "generations" are used to describe these important advances:

First generation computer—a mammoth computer using vacuum tubes, drum memories, and programming in machine code as its basic technology. Univax 1, used in 1951, was one of the earliest of these vacuum-tube based electronic computers. The generation starts at the end of World War II and ends about 1957.

Second generation computer—a computer using discrete transistors as its basic technology. Solid-state components replaced the vacuum tubes during this period from 1958 to 1963. Magnetic core memories store information. This era includes the development of high-level computer languages.

Third generation computer—a computer having integrated circuits, semiconductor memories, and magnetic disk storage. New operating systems, minicomputer systems, virtual memory, and timesharing are the advancements of this period from 1963 to 1971.

Fourth generation computer—a computer using microprocessors and large scale integrated chips as its basic technology, which made computers accessible to a large segment of the population. Networking, improved memory, database management systems, and advanced languages mark the period from 1971 to the end of the 1980s.

Fifth generation computer—a computer that uses inference to draw reasoned conclusions from a knowledge base, and interacts with its users via an intelligent user interface to perform such functions as speech recognition, machine translation of natural languages, and robotic operations. These computers using artificial intelligence have been under development since the early 1980s, especially in Japan, as well as in the United States and Europe. However, in 1991, Japan began a new ten-year initiative to investigate neural networks, which will probably divert resources from development of the fifth generation as traditionally defined.

A lot of people have heard of **ENIAC**, the first large electronic computer. What was **MANIAC**?

MANIAC (*m*athematical *a*nalyzer, *n*umerator, *i*ntegrator, *a*nd *c*omputer) was built at the Los Alamos Scientific Laboratory under the direction of Nicholas C. Metropolis between 1948 and 1952. It was one of several different copies of the high-speed computer built by John von Neumann (1903–1957) for the Institute for Advanced Studies (IAS). It was constructed primarily for use in the development of atomic energy applications, specifically the hydrogen bomb.

It was originated with the work on ENIAC (*e*lectronic *n*umerical *i*ntegrator *a*nd *c*omputer), the first fully operational large-scale electronic digital computer. ENIAC was built at the Moore School of Electrical Engineering at the University of Pennsylvania between 1943 and 1946. Its builders, John Prosper Eckert, Jr., and John William Mauchly (1907–1980), virtually launched the modern era of the computer with ENIAC.

For what purpose was **MADAM** designed?

MADAM (*M*anchester *a*utomatic *d*igital *m*achine) is a chess-playing machine designed by Alan M. Turing (1912–1954) in 1950. Turing was one of the first individuals to program a computer to play chess. His machine was a very poor chess player and made foolish moves. After several moves the machine would be forced to give up. Today it is possible to play a fairly advanced game of chess with a computer. However, no machine has been designed that analyzes every possible strategy corresponding to any move. It would take 10^{108} years to play all the possible games if we had a machine that could play a million games a second.

What is a **silicon chip**?

A silicon chip is an almost pure piece of silicon, usually less than one centimeter square and about half a millimeter thick. It contains hundreds of thousands of microminiature electronic circuit components, mainly transistors, packed and interconnected in layers beneath the surface. These components can perform control, logic, and/or memory functions. There is a grid of thin metallic strips on the surface of the chip; these wires are used for electrical connections to other devices. The silicon chip was developed independently by two researchers: Jack Kilby of Texas Instruments in 1958, and Robert Noyce (b. 1927) of Fairchild Semiconductor in 1959.

While silicon chips are essential to most computer operation today, a myriad of other devices depend on them as well, including calculators, microwave ovens, automobile diagnostic equipment, and VCRs.

What are the **sizes of silicon chips**?

SSI—*s*mall *s*cale *i*ntegration

MSI—*m*edium *s*cale *i*ntegration

LSI—*l*arge *s*cale *i*ntegration

VLSI—*v*ery *l*arge *s*cale *i*ntegration

ULSI—*u*ltra *l*arge *s*cale *i*ntegration

GSI—*g*i*g*a*s*cale *i*ntegration

The number of components packed into a single chip is very loosely defined. ULSI units can pack millions of components on a chip. GSI, a long-term target for the industry, will potentially hold billions of components on a chip.

Are any devices being developed to **replace silicon chips**?

When transistors were introduced in 1948, they demanded less power than fragile, high-temperature vacuum tubes; allowed electronic equipment to become smaller, faster, and more dependable; and generated less heat. These developments made computers much more economical and accessible; they also made portable radios practical. However, the smaller components were harder to wire together, and hand wiring was both expensive and error-prone.

In the early 1960s, circuits on silicon chips allowed manufacturers to build increased power, speed, and memory storage into smaller packages, which required less electricity to operate and generated even less heat. While through most of the 1970s manufacturers could count on doubling the components on a chip every year without increasing the size of the chip, the size limitations of silicon chips are becoming more restrictive. Though components continue to grow smaller, the same rate of shrinking cannot be maintained.

Researchers are investigating different materials to use in making circuit chips. Gallium arsenide is harder to handle in manufacturing, but it has potential for greatly increased switching speed. Organic polymers are potentially cheaper to manufacture, and could be used for liquid-crystal and other flat screen displays, which need to have their electronic circuits spread over a wide area. Unfortunately, organic polymers do not allow electricity to pass through as well as the silicons do. Several researchers are working on hybrid chips, which could combine the benefits of organic polymers with those of silicon. Researchers are also in the initial stages of developing integrated optical chips, which would use light rather than electric current. Optical chips would generate little or no heat, would allow faster switching, and would be immune to electrical noise.

How is a **hard disk** different from a **floppy disk**?

Both types of disk use a magnetic recording surface to record, access, and erase data, in much the same way as magnetic tape records, plays, and erases sound or images. A read/write head, suspended over a spinning disk, is directed by the *c*entral *p*rocessing *u*nit (CPU) to the sector where the requested data is stored, or where the data is to be recorded. A hard disk uses rigid aluminum disks coated with iron oxide for data storage. It has much greater storage capacity than several floppy disks (from 10 to hundreds of megabytes). While most hard disks used in microcomputers are "fixed" (built into the computer), some are removable. Minicomputer and mainframe hard disks include both fixed and removable (in modules called disk packs or disk cartridges). A floppy disk, also called a diskette, is made of plastic film covered with a magnetic coating, which is enclosed in a nonremovable plastic protective envelope. Floppy disks vary in storage capacity from 100,000 bytes to more than 2 megabytes. Floppy disks are generally used in minicomputers and microcomputers.

In addition to storing more data, a hard disk can provide much faster access to storage than a floppy disk. A hard disk rotates from 2,400 to 3,600 *revolutions per minute* (rpm) and is constantly spinning (except in laptops, which conserve battery life by spinning the hard disk only when in use). An ultra-fast hard disk has a separate read/write head over each track on the disk, so that no time is lost in positioning the head over the desired track; accessing the desired sector takes only milliseconds, the time it takes for the disk to spin to the sector. A floppy disk does not spin until a data transfer is requested, and the rotation speed is only about 300 rpm.

How much **data** can a **floppy disk** hold?

The three common floppy disk (diskette) sizes vary widely in storage capacity.

Envelope size (inches)	Storage capacity
8	100,000–500,000 bytes
5.25	100 kilobytes–1.2 megabytes
3.5	400 kilobytes–more than 2 megabytes

An 8-inch or 5-inch diskette is enclosed in a plastic protective envelope, which does not protect the disk from bending or folding; parts of the disk surface are also exposed, and can be contaminated by fingerprints or dust. The casing on a 3.5-inch floppy disk is rigid plastic, and includes a sliding disk guard that protects the disk surface, but allows it to be exposed when the disk is inserted in the disk drive. This protection, along with the increased data storage capacity, makes the 3.5-inch disk currently the most popular.

Why does a computer **floppy disk** have to be **"formatted"**?

A disk must first be organized so that data can be written to it and retrieved from it. The data on a floppy disk or a hard disk is arranged in concentric tracks. Sectors, which can hold blocks of data, occupy arc-shaped segments of the tracks. Most floppy disks are soft-sectored, and formatting is necessary to record sector identification so that data blocks can be labeled for retrieval. Hard-sectored floppy disks use physical marks to identify sectors; these marks cannot be changed, so the disks cannot be reformatted. The way that sectors are organized and labeled dictates system compatibility: disks formatted for DOS computers can only be used in other DOS machines; those formatted for Macintoshes can only be used in other Macintoshes. Formatting erases any pre-existing data on the disk. Hard disk drives are also formatted before being initialized, and should be protected so that they are not reformatted unintentionally.

Who was the **first programmer**?

According to historical accounts, Lord Byron's daughter, Augusta Ada Byron, the Countess of Lovelace, was the first person to write a computer program for Charles Babbage's (1792–1871) "analytical engine." This machine, never built, was to work by means of punched cards that could store partial answers that could later be retrieved for additional operations, and that would print results. Her work with Babbage and the essays she wrote about the possibilities of the "engine" established her as a "patron saint," if not a founding parent, of the art and science of programming. The programming language called "Ada" was named in her honor by the United States Department of Defense. In modern times the honor goes to Commodore Grace Murray Hopper (1906–1992) of the United States Navy. She wrote the first program for the Mark I computer.

Who invented the computer mouse?

A computer "mouse" is a hand-held input device that, when rolled across a flat surface, causes a cursor to move in a corresponding way on a display screen. A prototype mouse was part of an input console demonstrated by Douglas C. Englehart in 1968 at the Fall Joint Computer Conference in San Francisco. Popularized in 1984 by the Macintosh from Apple Computer, the mouse was the result of 15 years devoted to exploring ways to make communicating with computers simpler and more flexible.

The physical appearance of the small four-sided box with the dangling tail-like wire suggested the informal name of "mouse," which quickly superseded the formal name.

What is **Hopper's rule**?

Electricity travels one foot in a nanosecond (a billionth of a second). This is one of a number of rules compiled for the convenience of computer programmers. This is also considered to be a fundamental limitation on the possible speed of a computer—signals in an electrical circuit cannot move any faster.

Is an **assembly language** the same thing as a **machine language**?

While the two terms are often used interchangeably, an assembly language is a more "user friendly" translation of a machine language. A machine language is the collection of patterns of bits recognized by a *c*entral *p*rocessing *u*nit (CPU) as instructions. Each particular CPU design has its own machine language. The machine language of the CPU of a microcomputer generally includes about 75 instructions; the machine language of the CPU of a large mainframe computer may include hundreds of instructions. Each of these instructions is a pattern of 1's and 0's that tells the CPU to perform a specific operation.

An assembly language is a collection of symbolic, mnemonic names for each instruction in the machine language of its CPU. Like the machine language, the assembly language is tied to a particular CPU design. Programming in assembly language requires intimate familiarity with the CPU's architecture, and assembly language programs are difficult to maintain and require extensive documentation.

The computer language C, developed in the late 1980s, is now frequently used instead of assembly language. It is a high-level programming language that can be compiled into machine languages for almost all computers, from microcomputers to mainframes, because of its functional structure.

Who invented the **COBOL** computer language?

COBOL (*c*ommon *b*usiness *o*riented *l*anguage) is a prominent computer language designed specifically for commercial uses, created in 1960 by a team drawn from several computer makers and the Pentagon. The best-known individual associated with COBOL was then-Lieutenant Grace M. Hopper (1906–1992), who made fundamental contributions to the United States Navy standardization of COBOL. COBOL excels at the most common kinds of data processing for business—simple arithmetic operations performed on huge files of data. The language endures because its syntax is very much like English and because a program written in COBOL for one kind of computer can run on many others without alteration.

How is a **byte** defined?

A byte, a common unit of computer storage, holds the equivalent of a single character, such as a letter ("A"), a number ("2"), a symbol ("$"), a decimal point, or a space. It is usually equivalent to eight "data bits" and one "parity bit." A bit (a *bi*nary dig*it*), the **431** smallest unit of information in a digital computer, is equivalent to a single "0" or "1". The parity bit is used to check for errors in the bits making up the byte. Although eight data bits per byte is the most common size, computer manufacturers are free to define a differing number of bits as a byte. Six data bits per byte is another common size.

What does it mean to **"boot"** a computer?

Booting a computer is starting it, in the sense of turning control over to the operating system. The term comes from bootstrap, because bootstraps allow an individual to pull on boots without help from anyone else. Some people prefer to think of the process in terms of using bootstraps to lift oneself off the ground, impossible in the physical sense, but a reasonable image for representing the process of searching for the operating system, loading it, and passing control to it. The commands to do this are embedded in a *read only memory* (ROM) chip that is automatically executed when a microcomputer is turned on or reset. In mainframe or minicomputers, the process usually involves a great deal of operator input. A cold boot powers on the computer and passes control to the operating system; a warm boot resets the operating system without powering off the computer.

What is meant by **default**?

As used in computer, communication, and data processing systems, a default is the response that a system "assumes" in the absence of specific instructions. It can be a value (6), a condition (true), or an action (close file).

Where did the term **bug** originate?

The slang term *bug* is used to describe problems and errors occurring in computer programs. The term may have originated during the early 1940s at Harvard University, when a computer malfunctioned. A dead moth found in the system was thought to be the cause of the system's failure. This famous carcass, taped to a page of notes, is preserved with the trouble log notebook at the Virginia Naval Museum.

What is a **computer "virus"** and how is it spread?

Taken from the obvious analogy with biological viruses, a computer "virus" is a program that searches out other programs and "infects" them by replicating itself in them. When the programs are executed, the embedded virus is executed too, thus propagating the "infection." This normally happens invisibly to the user. A virus cannot infect other computers, however, without assistance. It is spread when users communicate by computer, often when they trade programs. The virus may do nothing but propagate itself and then allow the program to run normally. Usually, however, after propagating silently for a while, it starts interfering in ways that vary from inserting "cute" messages to destroying all of the user's files. Computer "worms" and "logic bombs" are similar to viruses, but they do not replicate themselves within programs as viruses do. A logic bomb does its damage immediately—destroying data, inserting garbage into data files, or reformatting the hard disk; a worm can alter the program and database either immediately or over a period of time.

In the 1990s, viruses, worms, and logic bombs have become such a serious problem, especially among IBM PC and Macintosh users, that the production of special detection and "inocculation" software has become an industry.

What is a **pixel**?

A pixel (*pix* [for picture] *el*ement) is the smallest element on a video display screen. A screen contains thousands of pixels, each of which can be made up of one or more dots or a cluster of dots. On a simple monochrome screen, a pixel is one dot; the two colors of image and background are created when the pixel is switched either on or off. Some monochrome screen pixels can be energized to create different light intensities, to allow a range of shades from light to dark. On color screens, three dot colors are included in each pixel—red, green, and blue. The simplest screens have just one dot of each color, but more elaborate screens have pixels with clusters of each color. These more elaborate displays can show a large number of colors and intensities. On color screens, black is created by leaving all three colors off; white by all three colors on; and a range of grays by equal intensities of all the colors.

The most economical displays are monochrome, with one bit per pixel, and settings limited to on and off. High-resolution color screens, which can use a million pixels, with each color dot using four bytes of memory, need to reserve many megabytes just to display an image.

What does **DOS** stand for?

DOS stands for *d*isk *o*perating *s*ystem, a program that controls the computer's transfer of data to and from a hard or floppy disk. Frequently it is combined with the main operating system. The operating system was originally developed at Seattle Computer Products as SCP-DOS. When IBM decided to build a personal computer and needed an operating system, it chose the SCP-DOS after reaching an agreement with the Microsoft Corporation to produce the actual operating system. Under Microsoft, SCP-DOS became MS-DOS, which IBM referred to as PC-DOS (personal computer), and which everyone eventually simply called DOS.

What is **E mail**?

Electronic mail, also known as E mail or E–mail, uses communication facilities to transmit messages. Many systems use computers as transmitting and receiving inter- **433** faces, but fax communication is also a form of E mail. A user can send a message to a single recipient, or to many. Different systems offer different options for sending, receiving, manipulating text, and addressing. For example a message can be "registered," so that the sender is notified when the recipient looks at the message (though there is no way to tell if the recipient has actually read the message). Many systems allow messages to be forwarded. Usually messages are stored in a simulated "mailbox" in the network server or host computer; some systems announce incoming mail if the recipient is logged onto the system. An organization (such as a corporation, university, or professional organization) can provide electronic mail facilities; there are also national and international networks. In order to use E mail, both sender and receiver must have accounts on the same system or on systems connected by a network.

What is a **hacker**?

A hacker is a skilled computer user. The term originally denoted a skilled programmer, particularly one skilled in machine code and with a good knowledge of the machine and its operating system. The name arose from the fact that a good programmer could always hack an unsatisfactory system around until it worked.

The term later came to denote a user whose main interest is in defeating password systems. The term has thus acquired a pejorative sense, with the meaning of one who deliberately and sometimes criminally interferes with data available through telephone lines. The activities of such hackers have led to considerable efforts to tighten security of transmitted data.

Who coined the term **technobabble**?

John A. Barry used the term *technobabble* to mean the pervasive and indiscriminate use of computer terminology, especially as it is applied to situations that have nothing at all to do with technology. He first used it in the early 1980s.

GENERAL SCIENCE AND TECHNOLOGY

TERMS, AND THEORIES

What is **High Technology** or **High Tech**?

This buzz term used mainly by the lay media (as opposed to scientific, medical, or technological media) appeared in the late 1970s. It was initially used to identify the newest, "hottest" application of technology to fields such as medical research, genetics, automation, communication systems, and computers. It usually implied a distinction between technology to meet the information needs of society and traditional heavy industry which met more material needs. By the mid-1980s, the term had become a catch-all applying primarily to the use of electronics (especially computers) to accomplish everyday tasks.

How is a **therblig** defined?

Frank Bunker Gilbreth (1868–1924), the founder of modern motion study technique, called the fundamental motions of the hands of a worker *therbligs* (Gilbreth spelled backwards). He concluded that any and all operations are made up of series of these 17 divisions. The 17 divisions are search, select, grasp, reach, move, hold, release, position, pre-position, inspect, assemble, disassemble, use, unavoidable delay, avoidable delay, plan, and test to overcome fatigue.

What does the term **ergonomics** mean?

The study of human capability and psychology in relation to the working environment and the equipment operated by the worker is variously known as ergonomics, human **435** engineering, human factors engineering, engineering psychology, or biotechnology. Ergonomics is based on the premise that tools humans use and the environment they work in should be matched with their capabilities and limitations, rather than forcing humans to adapt to the physical environment. Researchers in ergonomics try to determine optimum conditions in communication, cognition, reception of sensory stimuli, physiology, and psychology, and examine the effect of adverse conditions. Specific areas of study include design of work areas (including seats, desks, consoles, and cockpits) in terms of human physical size, comfort, strength, and vision; effects of physiological stresses such as work speed, work load, decision making, fatigue, and demands on memory and perception; and design of visual displays to enhance the quality and speed of interpretation.

What is the new science of **chaos**?

Chaos or chaotic behavior is the behavior of a system whose final state depends very sensitively on the initial conditions. The behavior is unpredictable and cannot be distinguished from a random process, even though it is strictly determinate in a mathematical sense. Chaos studies the complex and irregular behavior of many systems in nature, such as changing weather patterns, flow of turbulent fluids, and swinging pendulums. Scientists once thought they could make exact predictions about such systems, but found that a tiny difference in starting conditions can lead to greatly different results. Chaotic systems do obey certain rules which mathematicians have described with equations, but the science of chaos demonstrates the difficulty of predicting the long-range behavior of chaotic systems.

INVENTION AND DISCOVERY

What are the ten outstanding **engineering achievements** of the last 25 years?

The National Academy of Engineering lists them as follows:

> The Moon Landing
> Application Satellites
> Microprocessors
> Computer-Aided Design and Manufacturing
> CAT Scan
> Advanced Composite Materials
> Jumbo Jets
> Lasers
> Fiber-Optic Communication
> Genetically Engineered Products

Which **scientific discoveries** were **made accidentally**?

Some accidental discoveries include velcro, penicillin, x-rays, dynamite, vulcanization of rubber, synthetic dyes, rayon, saccharin, and the discovery of iodine and helium.

What are the areas of **critical technology**?

There are 22 areas of technological development that the United States considers to be critical to the prosperity and security of the country. They are aeronautics; applied molecular biology; ceramics; composites; computer simulation; data storage; electronic and photonic materials; energy; environment; flexible computer integrated manufacturing; high-definition imaging and displays; high–performance computing/networking; high-performance metals and alloys; intelligent processing equipment; materials processing; medicine; micro- and nano-fabrication; microelectronics and optoelectronics; sensors and signal processing; software; surface transportation; and systems-management.

Which inventions are credited to **Thomas Jefferson,** even though he never received a patent?

Thomas Jefferson (1743–1826), the third United States president, is recognized as having invented the swivel chair, pedometer, shooting stick, a hemp-treating machine, and an improvement in the moldboard of a plow.

What is a **Rube Goldberg** contraption?

A *Rube Goldberg contraption* is an overengineered solution to a simple problem. Ruben Lucius Goldberg (1883–1970) was a skilled engineering draftsman, a stand-up comic, a reporter, and one of the world's greatest cartoonists. His bizarre inventions mixed grand silliness with high seriousness. His devices used balloons, leaking hot water bottles, candle flames, pulleys, strings, levers, and various animals in elaborate schemes for accomplishing simple tasks.

Rube Goldberg juice squeezer

PATENTS AND TRADEMARKS

When was the **first United States patent** received?

The first patent was issued on July 31, 1790, to Samuel Hopkins for "making pot and pearl ashes"—a cleaning formula called potash (a key ingredient for soap-making then). The first really significant patent issued by the U.S. Patent Office was one for a cotton gin given to Eli Whitney (1765–1825) on March 14, 1794. Its importance lies in the fact that with the mechanical separation of cotton seeds from the lint, cotton became a staple crop of the South. This crop was a major factor in retaining the then-declining institution of slavery.

Between Hopkins's first patent in 1790, and Whitney's in 1794, the U.S. Patent Office issued 70 patents to inventors, some of whom have famous names (Oliver Evans, James Rumsey, John Fitch, and John Stevens) for their work with steam propulsion.

How many **United States patents** have been given?

Over five million patents have been given by the United States Patent Office since its inception. Below are listed the total number given from the period 1901–1990.

Year	Total number of patents granted
1901	25,546
1921	37,798
1930	45,226
1940	42,238
1950	43,040
1960	47,170
1965	62,857
1970	64,427
1980	61,800
1985	71,700
1987	83,000
1988	77,900
1989	102,712
1990	90,366

What **cannot be patented?**

Many things are not open to patent protection:

The laws of nature, physical phenomena and abstract ideas.

A new mineral or a new plant found in the wild.

Inventions useful solely in the utilization of special nuclear material or atomic energy for weapons.

A machine that is not *useful* (i.e. does not have a useful purpose or does not operate to perform the intended purpose).

Methods of doing business.

Printed matter.

In the case of mixtures of ingredients, such as medicines, a patent cannot be granted unless the effect of mixture is more than the effect of its components.

Human beings cannot be patented.

Mere substitution of one material for another or changes in size to a previously known useful invention without "novelty."

Must all United States **patent applications** be accompanied by a **drawing**?

It is required by law in most cases that a drawing of the invention accompany the patent application. Composition of matter or some processes *may* be exempt, although in these cases the commissioner of Patents and Trademarks may require a drawing for a process when it is useful. The required drawing must show every feature of the invention specified in the claims, and be in a format specified by the Patent and Trademark Office. Drawings must be in black and white unless waived by the Deputy Assistant Commissioner for Patents. Color drawings are permitted for plant patents where color is a distinctive characteristic.

What **types of patents** are available?

There are three types:

1. Utility patents, granted for new, useful and nonobvious process, machine, manufactured article, composition, or an improvement in any of the above.

2. Design patents, granted for new, original and ornamental designs for an article of manufacture.

3. Plant patents, provided to anyone who has invented or discovered and asexually reproduced any distinct and new variety of plant.

When was the **first patent covering an animal** issued?

On April 12, 1988, patent history was made when a patent was issued covering specifically genetically-engineered mice for cancer research.

Which **patent** was issued to a **United States president?**

On May 22, 1849, 12 years before he became the 16th United States president, Abraham Lincoln (1809–1865), was granted United States patent number 6,469 for a device to help steamboats pass over shoals and sand bars. The device, never tested or manufactured, had a set of adjustable buoyancy chambers (made from metal and waterproof cloth) attached to the ship's sides below the waterline. Bellows could fill the chambers with air to float the vessel over the shoals and sandbars. It was the only patent ever held by a United States president.

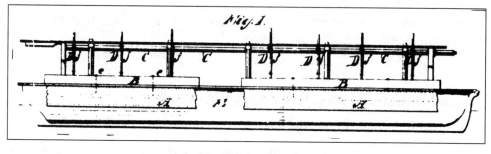

Lincoln's shoal device

Which **patents** were issued to **Mark Twain?**

The American writer Mark Twain (1835–1910), whose real name was Samuel Clemens, received three patents. Patent No. 121,992 issued in 1871 was for suspenders. His second, issued in 1873, was for his famous "Mark Twain's Self-Pasting Scrapbook." The third patent in 1885 was for an educational game that helped players remember important historical dates.

For what is **Percy Lavon Julian** noted?

Dr. Percy Lavon Julian (1899–1975), a grandson of a former slave, is noted for his work on synthesizing hormones and medicinal substances, as well as for developing many industrial applications for plant products. He obtained 130 chemical patents in his lifetime. For instance, in 1935 he synthesized *physostigmine,* a powerful drug for the treatment of the disease glaucoma (build-up of pressure inside the eyeball that gradually destroys the retina, causing blindness). A year later, he isolated and prepared soybean protein to be used for paper coatings, textile sizing, and cold-water paints, and as a chemical fire-extinguishing foam. He also developed methods of synthesizing cortisone (used for treatment of rheumatoid arthritis and other inflammatory diseases) and the hormones progesterone and testosterone.

OFFICE ITEMS

Why are the letters all mixed up on the typewriter keyboard?

Although Christopher Latham Scholes did not invent the typewriter, he made it work faster with his keyboard design. The standard keyboard (QUERTY) as used today was established by Scholes in 1872. He rearranged the keyboard from its alphabetical arrangement, which caused the bars of the machine to jam, to one in which the letter arrangement made the bars hit the inked ribbon from opposite directions, resulting in less jamming. Eventually the Remington Fire Arms Company bought the patent and mass-produced the machine in 1874. American writer Mark Twain (1835–1910) was one of the 400 customers who purchased Remington typewriters in 1874. The first 10 years of production saw disappointing sales. One reason was the use of aniline-inked typewriter ribbons that would eventually fade. With the introduction in 1885 of permanent ink ribbons, the typewriter became an essential office item.

What is a **Wooton Patent Desk**?

With dozens of pigeonholes and compartments, hinged and rotating parts, and elaborate exteriors, Wooton Patent Desks were manufactured in Indiana from 1874 to 1897 by William S. Wooton and Co. (changed in 1880 to Wooton Desk Manufacturing Co.). The design of the desk provided a solution to the businessman's problem of organizing the increasing volume of paperwork that accompanied the rapid expansion of business. These desks served well until the 1890s, when the widespread use of manila folders and file drawers replaced the limited storage capacity of the desk.

Who invented the **Xerox® machine**?

After 20 years of experimentation, American physicist Chester F. Carlson (1906–1968) developed in 1938 a method of copying that used dry powder, electric charge, and light. Because nothing moist was used in the process, the procedure was called *xerography* (meaning dry writing in Greek). But Carlson could not interest the industry of this new method until the Haloid Company of Rochester, New York, acquired his patent. The company, which later became the Xerox Corporation, gave the first public demonstration of this copier at the annual meeting of the Optical Society of America, in Detroit, Michigan, on October 22, 1948.

Carlson's new method used static electricity as the basis of the xerographic process. An electrostatic charge is induced on an insulating photoconductive surface in darkness. Then this surface is exposed to light reflecting from the original (similar to a photographic plate). The resultant charge pattern is dusted with a charged powder (called toner) that is attracted to the pattern but rejected by the background. The powder pattern (the image of the original) is transferred to ordinary paper with an electrostatic charge and "fixed" on the paper by either heat or by chemicals. Finally the xerographic surface is cleaned and ready for reuse. Today, copy paper is used, which forms the photo-conductive surface and consequently becomes the final copy.

Why are most pencils yellow?

Painting pencils yellow became a sign of quality during the 1890s. The practice had originally been done as early as 1854 in Keswick, England, most likely to cover the imperfect wood used in some pencils. It was the American Koh-I-Noor Company whose successful, high-quality yellow pencil became firmly established as a sign of "pencilness" in the minds of users; any other color was assumed to indicate an inferior pencil. Yellow has thus become established as the preferred color for a writing pencil, as it has for school buses and highway signs. It makes them all highly visible, whether on a busy desk or highway.

Very early pencils were covered, square sticks of graphite (a soft black lustrous form of carbon) that left as much of a mark on the user's hand as on the paper. Later, in England and Germany, marking sticks were wrapped tightly with string to fashion pencils, and the string slowly unwound to expose more stick. In the 16th century the Germans inserted the sticks into grooves cut in narrow strips of wood.

As graphite supplies became scarce in the late 18th century, chemists sought ways to reduce the graphite content in pencils. In 1795 N.J. Conte (1755–1805) produced pencils made from ground graphite and clay that had been shaped into sticks and kiln-baked—a process still done to this very day. Today, although a pencil appears to be all one piece, it actually is formed from two pieces of wood glued together so well that the seams are not visible.

The development of the process of vulcanizing rubber in 1839 made feasible the attachment of the eraser. In 1858 Hyman L. Lipman of Philadelphia patented a pencil with an eraser glued to one end.

Who invented **Liquid Paper**®?

Betty Nesmith Graham in 1951 started using a white paint to correct her typographical errors. After her coworkers began requesting bottles of the paint, Mrs. Graham started a small company to sell the paint. Over the years her small company grew to a large industry.

AWARDS AND PRIZES

Who was the **first American woman** to win a **Nobel Prize**?

For work on glycogen conversion, Czechoslovakian-born biochemist Gerty T. Cori (1896–1957) shared the 1947 Nobel Prize for Physiology and Medicine with her husband, Carl (1896–1984), becoming the first female American winner. The first native-born American woman to be honored was Rosalyn S. Yallow (b.1921), who shared the 1977 prize in physiology and medicine with two male associates. The award was given for their work on hormones in body chemistry.

Who was the **first American-born winner** of the **Nobel Prize for Physiology or Medicine**?

The American geneticist Thomas Hunt Morgan (1866–1945) was awarded the Nobel Prize in 1933 for his study on how chromosomes function in heredity.

What were the discoveries of **John Bardeen,** the only person to win two Nobel Prizes in the same field?

In 1956 John Bardeen (b.1908) won the Nobel Prize for Physics for his research on semiconductors and discovery of the transistor effect. His second prize, awarded in 1972, was for the microscopic theory of superconductivity.

Which **women** have won the **Nobel Prize**?

As of 1991, only nine women had received the Nobel Prize. Marie Curie received the prize twice.

Chemistry

Curie, Marie S. (1911)
Joliot-Curie, Irène (1935)
Hodgkin, Dorothy Mary Crowfoot (1964)

Physics

Curie, Marie S. (1903)
Mayer, Maria Goeppert (1963)

Physiology and Medicine

Cori, Gerty (1947)
Radnitz, Theresa (1947)
Yallow, Rosalyn S. (1977)
McClintock, Barbara (1983)
Levi-Montalcini, Rita (1986)
Elion, Gertrude B. (1988)

For what accomplishments did **Marie Curie** receive a Nobel Prize?

Marie Curie (1867–1934) received two Nobel Prizes—one in physics in 1903 and one in chemistry in 1911. She and her husband Pierre (1859–1906), along with A.H. Becquerel (1852–1908), were awarded the prize in 1903 for their work on radioactivity. They separated minute amounts of two new highly radioactive chemical elements from uranium ore. The Curies named the elements "radium" and "polonium." In 1911, Marie Curie won the Nobel Prize for Chemistry for her discovery of the new elements, and for work in isolating and studying the chemical properties of radium.

Who was the **first American** to receive a **Nobel Prize**?

Albert Abraham Michelson (1852–1931), a German-born American physicist, received the 1907 Nobel Prize in Physics, becoming the first American to win the award. The Prize recognized both his design of precise optical instruments and the accurate measurements he made with them. In association with Edward Morley (1838–1923), he performed the classic Michelson-Morley experiment for light waves and made very precise determinations of the velocity of light. This work entailed the development of extremely sensitive optical equipment.

What is the **National Inventors Hall of Fame**?

The National Inventors Hall of Fame was founded in 1973 on the initiative of H. Hume Matthews, then chairman of the National Council of Patent Law Associations (now called the National Council of Intellectual Property Law Associations). The United States Patent and Trademark Office became a co-sponsor the following year. A foundation was created in 1977 to administer it. Its purpose is to honor inventors who conceived the great technological advances which contributed to the nation's welfare. Usually the recognition is for a specific patent. Thomas A. Edison (1847–1931), the first inductee, was honored for his electric lamp, United States patent number 223,898, granted January 27, 1880. Since then 93 other inventors have been inducted into the hall (as of March, 1991).

Who was the **first woman** to be named to the **National Inventors Hall of Fame?**

Gertrude Belle Elion (b.1918) was named to the National Inventors Hall of Fame in 1991 for her research at the Burroughs Wellcome Company, which led to the development of drugs to combat leukemia, septic shock, and tissue rejection in patients undergoing kidney transplants.

Which astronauts have won the **Congressional Space Medal of Honor?**

The Congressional Space Medal of Honor recognizes astronauts who have made outstanding contributions to their field and to the welfare of the United States. The President of the United States presents the award to the recipient in the name of the Congress.

Recipient	Award Date
Neil A. Armstrong	1 Oct 1978
Frank Borman	1 Oct 1978
Charles P. Conrad	1 Oct 1978
John H. Glenn, Jr.	1 Oct 1978
Virgil "Gus" Grissom	1 Oct 1978
Alan B. Shepard, Jr.	1 Oct 1978
John W. Young	19 May 1981

What is the **Guzman prize?**

It is a prize offered in France beginning in 1901 to the first person to make contact with beings from another planet. Mars is excluded as being "too easy." The prize is as yet unclaimed.

HOAXES, ENIGMAS, AND ACCIDENTS

What was the **world's greatest scientific fraud?**

Probably the most famous fraud was the Piltdown man, a phony human fossil "discovered" in a gravel formation in southern England in 1912. An unknown person combined an ape jaw with skull fragments of a modern man so skillfully that scientists argued over Piltdown man for about 40 years. The forger filed the teeth in the ape jaw to give them the appearance of a human wear pattern. Chemical investigations in the 1950s showed the cranium and jaw were of different ages.

What is the **Bermuda Triangle**?

The Bermuda Triangle is an area of the Atlantic Ocean with the corners of the triangle being at the southern Virginia coast, the Bermuda Islands, and the Florida Keys, covering about 14,000 square miles (36,260 square kilometers). On a number of occasions ships and airplanes have vanished in this area, often without a trace.

Explanations for these disappearances have ranged from the plausible (extreme air turbulence, powerful ocean currents, waterspouts, magnetic anomalies, electromagnetic storms) to the absurd (kidnappings by alien flying saucers, gateways to other dimensions). However, scientific studies have revealed no significant peculiarities about the area. Many of the so-called incidents within the Bermuda Triangle can be attributed to other causes, including bad weather, malfunctioning instruments, and faulty navigation. It is also not the only part of the ocean where ships and planes have disappeared.

Does dowsing have any scientific basis?

Dowsing is searching for water, metal ores, or other materials hidden beneath the Earth's surface by using a divining rod or dowsing rod. These rods are usually forked or Y-shaped and can be made of wood or metal. The rods are carried by their tips and are believed to bend downward when they pass over hidden materials. Although studies have failed to explain scientifically the basis for dowsing, the practice is still popular throughout the world. Some civil engineering companies use dowsers to locate existing pipelines or cables when making site surveys, because often the companies find the results more reliable than those from standard detectors.

Which **industrial accidents** have become major disasters?

Date	Place	Event	Result
9/21/21	Oppau, Germany	Explosion at a nitrate manufacturing plant destroyed plant and nearby village	561 deaths; more than 1,500 persons injured

Date	Place	Event	Result
4/16/47	Texas City, Texas	Explosion in freighter being loaded with ammonium nitrate	561 deaths; much of city destroyed
7/28/48	Ludwigshafen, Federal Democratic Republic of Germany	Vapor explosion from dimethyl ether	209 deaths
7/10/76	Seveso, Italy	Chemical reactor explosion releasing 2,3,7,8-TCDD	100,000 animals killed; 760 people evacuated; 4,450 acres contaminated
2/25/84	Cubatao, Sao Paulo, Brazil	Explosion from a gasoline leak in a pipeline burned a nearby shanty town	Greater than 500
11/19/84	San Juan, Ixtaheupec, Mexico City, Mexico	5 million liters of liquified butane exploded at a storage facility	More than 400 deaths; 7,231 persons injured; 700,000 evacuated
12/3/84	Bhopal, India	Release of methyl isocyanate from pesticide plant	More than 2,000 deaths; 100,000 injured

What is **pyramid power**?

An interesting belief about the strange power of pyramids was popular during the 1970s. The power of the pyramid is based on the belief that a pyramid shaped like that of the Great Pyramid of Giza is able to focus an unspecified energy in a mystical way to bring about a beneficial result. Some of the claims of its power includes keeping food fresh, making meat tender, improving the taste of wine and tobacco, purifying water, curing headaches, shortening healing time, making plants grow better, and keeping razor blades sharp. This was done by placing the item under study inside a hollow pyramid. A physicist, Dr. G. Pat Flanagan, describes experiments he conducted in his book *Pyramid Power,* published in 1973. Later researchers have debunked this theory.

What is a **jackalope**?

This imaginary animal is supposed to be a cross between a jack rabbit and a deer. It probably was created by a taxidermist.

Further Reading

Books

Abell, George O. *Realm of the Universe*. New York: Holt, Rinehart and Winston, Inc., 1976.

Academic American Encyclopedia. Danbury, CT: Grolier, 1991. 21 vols.

Ackerknecht, Erwin H. *A Short Story of Medicine*. New York: Ronald Press, 1955.

Acronyms Initialisms & Abbreviations Dictionary 1993. 17th ed. Detroit, MI: Gale Research, 1992.

Adams, Ramon F. *The Language of the Railroader*. Norman, OK: University of Oklahoma Press, 1977.

Adler, Bill. *Whole Earth Quiz Book: How Well Do You Know the Planet*. New York: Quill, 1991.

Agent Orange and Its Associated Dioxin: Assessment of a Controversy. New York: Elsevier, 1988.

Agress, Clarence M. *Energetics*. New York: Grosset & Dunlap, 1978.

Ahrens, C. Donald. *Meteorology Today*. 2nd ed. St. Paul, MN: West Publishing, 1985.

Ainsworth, G.C. *Ainsworth's & Bisby's Dictionary of the Fungi*. 5th ed. London: Commonwealth Agricultural Bureau, 1961.

Ali, Sheikh R. *The Peace and Nuclear War Dictionary*. Santa Barbara, CA.: ABC-CLIO, 1989.

Allaby, Michael. *Dictionary of the Environment*, 3rd ed. New York: New York University Press, 1989.

The Almanac of Science and Technology. San Diego, CA: Harcourt Brace Jovanovich, 1990.

American Academy of Dermatology. *Poison Ivy* (Pamphlet). Washington, DC: American Academy of Dermatology, 1990.

The American Geological Institute. *Dictionary of Geological Terms*. Rev. ed. Garden City, NY: Anchor Press, 1976.

The American Medical Association Encyclopedia of Medicine. New York: Random House, 1989.

American Medical Association Family Medical Guide. Rev. ed. New York: Random House, 1987.

Anderson, Norman D., and Walter R. Brown. *Ferris Wheels*. New York: Pantheon Books, 1983.

Angelo, Joseph A. *The Extraterrestrial Encyclopedia*. Rev. & updated ed. New York: Facts On File, 1991.

Annual Energy Review 1990. Washington, DC: U.S. Department of Energy, Energy Information Administration, 1990.

Arem, Joel E. *Color Encyclopedia of Gemstones*. 2nd ed. New York: Van Nostrand Reinhold, 1987.

Argenzio, Victor. *Diamonds Eternal*. New York: David McKay, 1974.

Armstrong, Joseph E. *Science in Biology*. Prospect Heights, IL: Waveland Press, 1990.

The ARRL Handbook for Radio Amateurs. 68th ed. Newington, CT: American Radio Relay League, 1991.

Ashby, W. Ross. *Introduction to Cybernetics*. New York: John Wiley & Sons, 1958.

Ashworth, William. *The Encyclopedia of Environmental Studies*. New York: Facts On File, 1991.

Asimov, Isaac. *Asimov On Numbers*. New York: Doubleday, 1977.

Asimov, Isaac. *Asimov's Biographical Encyclopedia of Science and Technology*. 2nd rev. ed. Garden City, NY: Doubleday & Company, Inc., 1982.

Asimov, Isaac. *Asimov's Chronology of Science and Discovery*. New York: Harper and Row, 1989.

Asimov, Isaac. *The Human Body*. New rev. ed. New York: A Mentor Book, 1992.

Asimov, Isaac. *Isaac Asimov's Guide to Earth and Space*. New York: Random House, 1991.

Asimov, Isaac. *Understanding Physics*. New York: Dorset Press, 1988. 3 vols. in one.

Astronauts and Cosmonauts Biographical and Statistical Data Report to the Committee on Science, Space, and Technology U.S. House of Representatives, 1989. Washington, DC: Committee on Science, Space and Technology, 1989.

Automotive Encyclopedia. Rev. ed. South Holland, IL: Goodheart-Willcox Company, 1989.

Babbitt, Harold E. *Sewage and Sewage Treatment*. 8th ed. New York: Wiley, 1958.

Bagenal, Philip, and Jonathan Meades. *The Illustrated Atlas of the World's Great Buildings*. London: Salamander Books, Ltd., 1980.

Bair, Frank E. *The Weather Almanac*. 6th ed. Detroit, MI: Gale Research Inc., 1992.

Baker, David. *The History of Manned Space Flight*. New York: Crown Publishers, Inc., 1982.

Baker, Susan P., et al. *The Injury Fact Book*. 2nd ed. New York: Oxford University Press, 1992.

Balfour, Henry H. *Herpes Diseases and Your Health*. Minneapolis, MN: University of Minnesota Press, 1984.

Bali, Mrinal. *Space Exploration: A Reference Handbook*. Santa Barbara, CA: ABC-CLIO, 1990.

Ballast, David Kent. *Architects' Handbook of Formulas, Tables and Mathematical Calculations*. New York: Prentice-Hall, 1988.

Bamberger, Richard. *Physics Through Experiment*. New York: Sterling, 1969.

Barnard, Christiaan. *The Body Machine*. New York: Crown, 1981.

Barnhart, Robert K. *The American Heritage Dictionary of Science*. Boston, MA: Houghton, Mifflin, 1986.

Barry, John. *Technobabble*. Cambridge, MA: MIT Press, 1991.

Bartholomew, Mel. *Square Foot Gardening*. Emmaus, PA: Rodale Press, 1981.

Bates, Robert, and Julia A. Jackson. *Glossary of Geology*. 3rd ed. Alexandria, VA: American Geological Institute, 1987.

Baylin, Frank, and Brent Gale. *Home Satellite TV Installation and Troubleshooting Manual*. 1986 ed. Boulder, CO: Baylin/Gale Productions, 1985.

Beatty, J. Kelly, Brian O'Leary, and Andrew Chaikin. *The New Solar System*. Cambridge, MA: Sky Publishing Corp., 1981.

Beck, James H. *Rail Talk*. Gretna, NE: James Publications, 1978.

Beeching, W.A. *Century of the Typewriter*. New York: St. Martin's Press, 1974.

Berg, Richard E. *The Physics of Sound*. Englewood Cliffs, NJ: Prentice-Hall, 1982.

Berliner, Barbara. *The Book of Answers*. Englewood Cliffs, NJ: Prentice-Hall, 1990.

Bernard, Josef. *The Cellular Connection*. Mendocino, CA: Quantum Publishing, 1987.

Berry, James. *Exploring Crystals*. New York: Crowell-Collier Press, 1969.

Best, Charles H. *Best and Taylor's Physiological Basis of Medical Practice*. Baltimore, MD: Williams & Wilkins, 1985.

Beyer, Don E. *The Manhattan Project*. New York: Watts, 1991.

The Biographical Dictionary of Scientists: Astronomers. New York: Peter Bedrick Books, 1984.

Biographical Dictionary of Scientists: Biologists. New York: Peter Bedrick Books, 1984.

The Biographical Dictionary of Scientists: Chemists. New York: Peter Bedrick Books, 1983.

The Biographical Dictionary of Scientists: Mathematicians. New York: Peter Bedrick Books, 1986.

The Biographical Dictionary of Scientists: Physicists. New York: Peter Bedrick Books, 1984.

Biographical Encyclopedia of Scientists. New York: Facts On File, 1981. 2 vols.

The Birds Around Us. San Ramon, CA: Ortho, 1986.

Bishop, Peter. *Fifth Generation Computers*. Hempstead, Eng.: Ellis Horwood, Ltd., 1986.

Blackwell, Will H. *Poisonous and Medicinal Plants*. Englewood Cliffs, NJ: Prentice-Hall, 1990.

Blair, Ian. *Taming the Atom*. Bristol, Eng.: Adam Hilger, 1983.

Block, E.B. *Fingerprinting: Magic Weapon Against Crime*. New York: McKay, 1969.

Blocksma, Mary. *Reading the Numbers*. New York: Penguin, 1989.

The BOCA National Building Code 1990. 11th ed. Country Club Hills, IL: Building Officials and Code Administrators International, 1989.

Bogner, Bruce F. *Vehicular Traffic Radar Handbook for Attorneys*. Mount Holly, NJ: The Brehn Corporation, 1979.

Bohren, Craig F. *Clouds in a Glass of Beer*. New York: Wiley, 1987.

Bolton, W.C. *Physics Experiments and Projects*. Elmsford, NY: Pergamon, 1968.

Bomberger, Audrey S., and Betty A. Dannenfelser. *Radiation and Health*. Gaithersburg, MD: Aspen Pubs., Inc., 1984.

Bonnet, Robert L. *Botany: 49 Science Fair Projects*. Blue Ridge Summit, PA: Tab Books, 1989.

Booth, Nicholas. *The Concise Illustrated Book of Planets and Stars*. New York: Gallery Books, 1990.

Bowler, Peter J. *Evolution, the History of an Idea*. Berkeley, CA: University of California Press, 1984.

Boyer, Rick. *Places Rated Almanac*. Englewood Cliffs, NJ: Prentice-Hall, 1989.

Bradbury, Savile. *The Evolution of the Microscope*. Elmsford, NY: Pergamon Press, 1967.

Bradford, Gershom. *A Glossary of Sea Terms*. New York: Dodd, Mead, 1942.

Brady, George S. *Materials Handbook*. 12th ed. New York: McGraw-Hill, 1986.

Brandreth, Gyles. *Your Vital Statistics*. New York: Citadel, 1986.

Branson, Gary D. *The Complete Guide to Recycling at Home*. Whitehall, VA: Betterway Publications, 1991.

Braun, Wernher von, and Frederick I. Ordway III. *Space Travel: A History*. New York: Harper & Row, 1985.

Brennan, Richard P. *Dictionary of Scientific Literacy*. New York: John Wiley & Sons, Inc., 1992.

Brennan, Richard P. *Levitating Trains and Kamikaze Genes*. New York: Harper Perennial, 1990.

Britannica Book of the Year 1991. Chicago, IL: Encyclopaedia Britannica, 1991.

Britten, Frederick J. *Britten's Old Clocks and Watches and Their Makers*. 8th ed. New York: Dutton, 1973.

Broadcasting Yearbook 1991. New York: Broadcasting Publications, Inc., 1991.

Brody, Jane E. *Jane Brody's Good Food Book*. New York: W.W. Norton, 1985.

Brown, Theodore L. *Chemistry*. 4th ed. Englewood Cliffs, NJ: Prentice-Hall, 1988.

Brown, Victor J. *Engineering Terminology*. Chicago, IL: Gillette, 1938.

Buchman, Dian Dincin. *Dian Dincin Buchman's Herbal Medicine*. New York: Gramercy, 1980.

Burnam, Tom. *The Dictionary of Misinformation*. New York: Perennial Library, 1986.

Burton, J.L. *Essentials of Dermatology*. 3rd ed. New York: Churchill Livingstone, 1990.

Burton, Maurice. *Encyclopaedia of Animals in Colour*. London: Octopus Books, 1972.

Burton, Maurice, and Robert Burton. *Encyclopedia of Insects and Arachnids*. New York: Crescent Books, 1975.

Bynum, W.F., et al. *Dictionary of the History of Science*. Princeton, NJ: Princeton University Press, 1985.

Byrne, Austin T. *A Treatise on Highway Construction*. New York: John Wiley & Sons, 1896.

Cairis, Nicholas. *Cruise Ships of the World*. Boston, MA: Pegasus, 1988.

Calasibetta, Charlotte M. *Fairchild's Dictionary of Fashion*. New York: Fairchild Publications, 1988.

Calder, Nigel. *The Comet Is Coming*. New York: Penguin Books, 1982.

Callahan Philip S. *Bird Behavior*. New York: Four Winds Press, 1975.

The Cambridge Encyclopedia of Ornithology. New York: Cambridge University Press, 1991.

The Cambridge Encyclopedia of Space. New York: Cambridge University Press, 1990.

Campbell, Neil A. *Biology*. 2nd ed. Redwood City, CA: Benjamin/Cummings Publishing Co., Inc., 1990.

Can Elephants Swim? New York: Time-Life Books, 1969.

Carlson, Neil R. *Foundations of Physiological Psychology*. Needham Heights, MA: Allyn Bacon, Inc., 1988.

Carroll, Anstice, and Embree De Persiis Vona. *The Health Food Dictionary with Recipes*. Englewood Cliffs, NJ: Prentice-Hall, Inc., 1973.

Carter, E.F. *Dictionary of Inventions and Discoveries*. New York: Crane Russak, 1974.

Cartnell, Robert. *The Incredible Scream Machine: A History of the Roller Coaster*. Fairview Park, OH: Amusement Park Books, 1987.

Carwell, Hattie. *Blacks in Science: Astrophysicist to Zoologist*. Oakland, CA: Exposition Press, 1977.

Cary, James. *Tanks and Armor in Modern Warfare*. New York: Franklin Watts, 1966.

Cassel, Don. *Understanding Computers*. Englewood Cliffs, NJ: Prentice-Hall, 1990.

Catalog of American Car ID Numbers 1970-79. Sidney, OH: Amos Press, Inc., 1991.

Cazeau, Charles J. *Science Trivia*. New York: Berkley Books, 1986.

Chambers Science and Technology Dictionary. Cambridge, Eng. & Edinburgh, Scotland: W&R Chambers, Ltd., and Cambridge University Press, 1988.

Charney, Leonard. *Build A Yurt*. New York: Macmillan, 1974.

Childs, W.H.J. *Physical Constants*. 9th ed. New York: Chapman and Hall, 1972.

Chinn, George M. *The Machine Gun*. Washington, DC: U.S. Department of the Navy, 1951.

Choukas-Bradley, Melanie, and Polly Alexander. *City of Trees*. Rev. ed. Baltimore, MD: Johns Hopkins University Press, 1987.

Churchill, James E. *The Backyard Building Book*. Harrisburg, PA: Stackpole Books, 1976.

Churchman, Lee W. *Survey of Electronics*. San Francisco, CA: Rinehart Press, 1971.

Cipolla, Carlo M., and Derek Birdsall. *The Technology of Man*. New York: Holt, Rinehart and Winston, 1980.

Cities of the United States: The West. Detroit, MI: Gale Research, 1989.

Clearing the Air: Perspectives on Environmental Tobacco Smoke. Lexington, MA: Lexington Books, 1988.

Coated Abrasives: Modern Tool of Industry. New York: McGraw-Hill, 1958.

Cody, John. *Visualizing Muscles.* Lawrence, KS: University Press of Kansas, 1990.

Cohen, I. Bernard. *Revolution in Science.* Cambridge, MA: Belknap Press, 1985.

Collin, P.H. *Dictionary of Ecology and the Environment.* Teddington, Eng.: Peter Collin Publishing, 1988.

Columbia University College of Physicians and Surgeons. *Complete Home Medical Guide.* New York: Crown, 1985.

Comprehensive Textbook of Psychiatry. 5th ed. Baltimore, MD: Williams & Wilkins, 1989. 2 vols.

Compton, William David. *Where No Man Has Gone Before.* Washington, DC: Superintendent of Documents, 1980.

Conant, Roger. *A Field Guide to Reptiles and Amphibians: Eastern and Central North America.* 3rd updated ed. Boston, MA: Houghton Mifflin Company, 1991.

Concise Chemical and Technical Dictionary. 4th enlarged ed. New York: Chemical Publ., Co., Inc., 1986.

Cone, Robert J. *How the New Technology Works.* Phoenix, AZ: Oryx Press, 1991.

Congram, Marjorie. *Horsehair: A Textile Resource.* Martinsville, NJ: Dockwra Press, 1987.

Considine, Douglas M. *Energy Technology Handbook.* New York: McGraw-Hill, 1977.

Corliss, William R. *Tornadoes, Dark Days, Anomalous Precipitation, with Related Weather Phenomena.* Glen Arm, MD: Sourcebook Project, 1983.

Cornell, Felix. *American Merchant Seaman's Manual.* Centreville, MD: Cornell, Maritime, 1964.

Cornell, James. *The Great International Disaster Book.* 3rd ed. New York: Charles Scribner's, 1982.

Cortada, James W. *Historical Dictionary of Data Processing: Technology.* Westport, CT: Greenwood Press, 1987.

Costello, David F. *The World of the Porcupine.* Philadelphia, PA: Lippincott, 1966.

Council on Environmental Quality. *Environmental Quality.* Washington, DC: Council on Environmental Quality, 1992.

Cox, James M. *Mark Twain: The Fate of Humor.* Princeton, NJ: Princeton University Press, 1966.

CRC Handbook of Physics and Chemistry. 72nd ed. St. Louis, MO: CRC Press, 1991.

Crockett, James U. *Crockett's Indoor Garden.* Boston, MA: Little, Brown, 1978.

Crocodiles and Alligators. New York: Facts On File, 1989.

Crowson, Phillip. *Minerals Handbook, 1990-1991.* New York: Stockton Press, 1990.

Cruickshank, Allan D. *1001 Questions Answered About Birds.* New York: Dodd, Mead, 1958.

Cuff, David J. *The United States Energy Atlas.* 2nd ed. New York: Macmillan, 1986.

Cunningham, William P., and Barbara Woodhouse Saigo. *Environmental Science: A Global Concern.* Dubuque, IA: Wm. C. Brown Publishers, 1990.

Current Medical Diagnosis and Treatment 1991. 30th ed. Norwalk, CT: Appleton & Lange, 1991.

Curtis, Anthony R. *Space Almanac.* Woodsboro, MD: ARCsoft, 1990.

Curtis, Helena, and N. Sue Barnes. *Invitation to Biology.* 4th ed. New York: Worth Publ., Inc., 1985.

Daintith, John. *The Facts on File Dictionary of Physics.* New York: Facts On File, 1988.

Dalley, Robert L. *Are You Burning Money?* New York: Reston, 1982.

Darwin, Charles. *On the Origin of Species.* Cambridge, MA: Harvard University Press, 1964.

Davie, Michael. *Titanic: The Death and Life of a Legend.* New York: Alfred Knopf, 1987.

Davis, G.J. *Automotive Reference.* Boise, ID: Whitehorse, 1987.

Day, David. *The Doomsday Book of Animals*. New York: Viking, 1983.

Day, John, and C. Eng. *The Bosch Book of the Motor Car*. New York: St. Martin's Press, 1976.

Dean, John A. *Lange's Handbook of Chemistry*. 13th ed. New York: McGraw Hill, 1985.

De Bono, Edward. *Eureka!* New York: Holt, Rinehart and Winston, 1974.

Deming, Richard. *Metric Power*. Nashville, TN: Nelson, 1974.

De Voney, Chris. *MS-DOS User's Guide*. 2nd ed. Indianapolis, IN: Que Corporation, 1987.

Diagram Group. *Comparisons*. New York: St. Martin's Press, 1980.

Diamond, Freda. *The Story of Glass*. San Diego, CA: Harcourt, Brace and Co., 1953.

Dickerson, Richard Earl. *Chemical Principles*. 3rd ed. Menlo Park, CA: The Benjamin/Cummings Publ., Co., Inc., 1979.

Dickinson, Terence, and Alan Dyer. *The Backyard Astronomer's Guide*. Ontario, Can.: Camden House, Publishing, 1991.

Dictionary of American Medical Biography. Westport, CT: Greenwood, 1984. 2 vols.

Dictionary of Scientific Biography. New York: Charles Scribner's Sons, 1973.

Dictionary of Visual Science. 4th ed. Radnor, PA: Chilton Trade Book, 1989.

Diseases and Disorders Handbook. Springhouse PA: Springhouse Corporation, 1990.

Donovan, Richard T. *World Guide to Covered Bridges*. Rev. ed. Worcester, MA: The National Society for the Preservation of Covered Bridges, 1980.

Dougans, Inge. *Reflexology*. New York: Element, 1991.

Douglas, R.W., and Susan Frank. *A History of Glassmaking*. London: G.T. Foulis & Co., 1972.

Dowling, Harry F. *Fighting Infection*. Cambridge, MA: Harvard University Press, 1977.

Downing, Douglas, and Michael Covington. *Dictionary of Computer Terms*. 2nd ed. Hauppauge, NY: Barron, 1989.

Downs, Robert B. *Landmarks in Science: Hippocrates to Carson*. Littleton, CO: Libraries Unlimited, Inc., 1982.

Downs, Robert B. *Scientific Enigmas*. Littleton, CO: Libraries Unlimited, Inc., 1987.

Dowson, Gordon. *Powder Metallurgy: The Process and Its Products*. London: Adam Hilger, 1990.

Drake, George R. *Weatherizing Your Home*. New York: Reston Publishing Co., 1978.

Dreisbach, Robert H. *Handbook of Poisoning*. 12th ed. Norwalk, CT: Appleton & Lange, 1987.

Drimmer, Frederick. *The Elephant Man*. New York: Putnam, 1985.

Dublin, Louis I. *Factbook on Man from Birth to Death*. 2nd ed. New York: Macmillan, 1965.

Duensing, Edward. *Talking to Fireflies, Shrinking the Moon*. New York: Penguin, 1990.

Duplaix, Nicole, and Noel Simon. *World Guide to Mammals*. New York: Crown Publishers, 1976.

DuVall, Nell. *Domestic Technology*. Boston, MA: G.K. Hall, 1988.

Duxbury, Alyn C., and Alison Duxbury. *An Introduction to the World's Oceans*. Reading, MA: Addison-Wesley Publishing Company, Inc., 1984.

Dyson, James L. *The World of Ice*. New York: Knopf, 1962.

Eating to Lower Your High Blood Cholesterol (Pamphlet). Washington, DC: U.S. Department of Health and Human Services, 1989.

Edelson, Edward. *Sports Medicine*. New York: Chelsea House, 1988.

Eden, Maxwell. *Kiteworks*. New York: Sterling, 1989.

Edmond's Car Savvy. Alhambra, CA: Edmund Publications, 1991.

Edmunds, Robert A. *The Prentice-Hall Encyclopedia of Information Technology*. Englewood Cliffs, NJ: Prentice-Hall, Inc., 1989.

Ellis, John. *The Social History of the Machine Gun*. London: Croom Helm, Ltd., 1975.

Ellis, Keith. *Thomas Telford*. Duluth, MN: Priory Press, 1974.

Emiliani, Cesare. *The Scientific Companion*. New York: Wiley Science Editions, 1988.

Encyclopaedic Dictionary of Physical Geography. London: Blackwell, 1985.

Encyclopedia Americana. Danbury, CT: Grolier, 1990. 30 vols.

Encyclopedia of Associations. 27th ed. Detroit, MI: Gale Research, 1993.

Encyclopedia of Aviation. New York: Scribners, 1977.

Encyclopedia of Chemical Technology. 3rd ed. New York: Wiley, 1978. 24 vols.

Encyclopedia of Human Evolution and Prehistory. New York: Garland Publ., 1988.

The Encyclopedia of Insects. New York: Facts On File, 1986.

Encyclopedia of Physical Science and Technology. San Diego, CA: Academic Press, 1987. 15 vols.

Encyclopedia of Twentieth Century Warfare. New York: Orion, 1989.

Encyclopedic Dictionary of Science. New York: Facts On File, 1988.

Engineering and the Advancement of Human Welfare. Washington, DC: National Academy Press, 1989.

English Translations of German Standards Catalog, 1991. Braintree, MA: Beuth Verlag Gmbh., 1991.

Erickson, Joan Beth. *Flower Garden Plans*. San Ramon, CA: Ortho Books, 1991.

Everett, Thomas H. *The New York Botanical Garden Illustrated Encyclopedia of Horticulture*. Hamden, CT: Garland, 1981.

The Facts On File Dictionary of Chemistry. Rev. and enl. ed. New York: Facts On File, Inc., 1988.

Facts On File Dictionary of Physics. New York: Facts On File, Inc., 1988.

Farb, Peter. *The Insects*. Alexandria, VA: Time-Life Books, 1977.

Farber, Edward. *Nobel Prize Winners in Chemistry 1901-1961*. Rev. ed. New York: Ablard-Schuman, 1963.

Fejer, Eva, and Cecilia Fitzsimons. *An Instant Guide to Rocks and Minerals*. Stamford, CT: Longmeadow Press, 1988.

Feldman, David. *Do Penguins Have Knees?* New York: Harper Perennial, 1991.

Feldman, David. *Why Do Clocks Run Clockwise? and Other Imponderables*. New York: Perennial Library, 1988.

Feldman, David. *Why Do Dogs Have Wet Noses? and Other Imponderables of Everyday Life*. New York: Harper Perennial, 1991.

Feltwell, John. *The Natural History of Butterflies*. New York: Facts On File, 1986.

Fenton, Carroll L. *The Fossil Book*. New York: Doubleday, 1989.

Field, Frank. *Doctor Frank Field's Weather Book*. New York: Putnam, 1981.

Field, Gary C. *Color and Its Reproduction*. Pittsburgh, PA: Graphic Arts Technical Foundation, 1988.

Field, Leslie. *The Queen's Jewels*. New York: Harry N. Abrams, 1987.

50 Simple Things You Can Do to Save the Earth. Berkeley, CA: Earth Works Press, 1989.

Fisher, Arthur. *The Healthy Heart*. New York: Time-Life Books, 1981.

Fisher, David J. *Rules of Thumb for Engineers and Scientists*. Houston, TX: Gulf, 1991.

Flaste, Richard. *The New York Times Book of Science Literacy*. New York: Times Books, Random, 1991.

Flatow, Ira. *Rainbows, Curve Balls, and Other Wonders of the Natural World Explained*. New York: Harper & Row Publishers, 1988.

Fletcher, Edward. *Pebble Collecting and Polishing*. New York: Sterling, 1973.

Forrester, Frank H. *1001 Questions Answered About the Weather*. New York: Grosset & Dunlap, 1957.

Frankel, Edward. *Poison Ivy, Poison Oak, Poison Sumac, and Their Relatives*. Pacific Grove, CA: Boxwood Press, 1991.

Freedman, Alan. *The Computer Glossary*. 5th ed. New York: AMACOM, 1991.

Freiberger, Paul, and Michale Swaine. *Fire in the Valley: The Making of the Personal Computer*. Berkeley, CA: Osborne/McGraw-Hill, 1984.

Frew, Timothy. *Salmon*. New York: Mallard Press, 1991.

Fritts, Harold C. *Tree Rings and Climate*. New York: Academic Press, 1976.

Frost, Harwood. *The Art of Roadmaking*. New York: Engineering News Publishing Company, 1910.

Fruits and Vegetables: 1001 Gardening Questions Answered. Pownal, VT: Storey Communications, Inc., 1990.

Funk & Wagnalls New Standard Dictionary of the English Language. New York: Funk & Wagnalls, 1959.

Funk, Charles E. *Horse Feathers and Other Curious Words*. New York: Harper, 1958.

Galperin, Anne. *Gynecological Disorders*. New York: Chelsea House, 1991.

Gardiner, Mary S. *The Biology of Invertebrates*. New York: McGraw-Hill, 1972.

Gascoigne, Robert M. *A Chronology of the History of Science 1450-1900*. New York: Garland Publishing, Inc., 1987.

Gatland, Kenneth. *The Illustrated Encyclopedia of Space Technology*. New York: Orion Books, 1989.

Gay, Kathlyn. *The Greenhouse Effect*. New York: Franklin Watts, 1986.

Gay, Kathlyn. *Ozone*. New York: Franklin Watts, 1989.

Gene Hughes' Police Call Radio Guide. Los Angeles, CA: Hollins Radio Data, 1990.

George, L. David, and Jennifer J. George. *Marine Life*. New York: Wiley Interscience, 1979.

Gerken, Louis C. *Airships: History and Technology*. Chula Vista, CA: American Scientific Corporation, 1990.

Giedion, Siegfried. *Mechanization Takes Command*. New York: Oxford University Press, 1948.

Giscard d'Estaing, Valerie-Anne. *The World Almanac Book of Inventions*. New York: World Almanac Publications, 1985.

Giscard d'Estaing, Valerie-Anne. *The Second World Almanac Book of Inventions*. New York: World Almanac, 1986.

Godish, Thad. *Indoor Air Pollution Control*. Chelsea, MI: Lewis Publishers, 1989.

Gong, Victor. *AIDS: Facts and Issues*. New Brunswich, NJ: Rutgers University Press, 1980.

Goulty, George A. *A Dictionary of Landscape*. Brookfield, VT: Gower Publishing Co., 1991.

Graedon, Joe, and Dr. Teresa Graedon. *Graedons' Best Medicine*. New York: Bantam Book, 1991.

Graf, Rudolf F., and George J. Whalen. *The Reston Encyclopedia of Biomedical Engineering Terms*. New York: Reston Publishing, 1977.

Graham, John. *Facts on File Dictionary of Telecommunications*. Rev. ed. New York: Facts On File, 1991.

Gray, Henry. *Anatomy of the Human Body*. 28th ed. Malvern, PA: Lea & Febiger, 1966.

Gray, Peter. *The Encyclopedia of the Biological Sciences*. New York: Van Nostrand Reinhold, 1970.

Great Britain Meteorological Office. *Meteorological Glossary*. New York: Chemical Publishing, 1972.

Great Disasters. New York: Reader's Digest Association, 1989.

Great Engineers and Pioneers in Technology. New York: St. Martin's Press, 1981.

The Great Scientists. Danbury, CT: Grolier, 1989.

Green, James Harry. *The Dow Jones-Irwin Handbook of Telecommunications*. Homewood, IL: Dow Jones-Irwin, 1986.

Greenfield, Ellen J. *House Dangerous*. New York: Vintage Books, 1987.

Grimm, William C. *The Trees of Pennsylvania*. Harrisburg, PA: Stackpole, 1950.

Groves, Don. *The Ocean Book*. New York: Wiley, 1989.

Grzimek's Animal Life Encyclopedia. New York: Van Nostrand Reinhold Company, 1973. 13 vols.

Grzimek's Encyclopedia of Mammals. 2nd ed. New York: McGraw-Hill, 1990. 5 vols.

Guedes, Pedro. *The Macmillan Encyclopedia of Architecture and Technological Change*. London: Macmillan Press, 1979.

Guiley, Rosemary E. *Moonscapes*. Englewood Cliffs, NJ: Prentice-Hall, 1991.

The Guinness Book of Answers. 8th ed. Enfield, Eng.: Guinness Publishing, 1991.

Guinness Book of Records 1992. New York: Bantam Book, 1992.

Gurney, Gene. *Space Shuttle Log*. Blue Ridge Summit, PA: TAB, 1988.

Guyton, Arthur C. *Basic Human Physiology*. Philadelphia, PA: Saunders, 1977.

Guyton, Arthur C. *Textbook of Medical Physiology*. 8th ed. Philadelphia, PA: Saunders, 1991.

Haber, Louis. *Odyssey Black Pioneers of Science and Invention*. San Diego, CA: An Odyssey Book, Harcourt Brace Jovanovich, Publishers, 1970.

Hackh, Ingo W.D. *Grant & Hackh's Chemical Dictionary*. 5th ed. New York: McGraw-Hill, 1987.

Haggard, Howard W. *The Lame, the Halt, and the Blind*. New York: Harper, 1932.

Hale, Mason E., Jr. *The Biology of Lichens*. 2nd ed. London: Edward Arnold, Ltd., 1974.

Halliday, William R. *Depths of the Earth*. New York: Harper, 1976.

Hamilton, William R. *The Henry Holt Guide to Minerals, Rocks and Fossils*. New York: Henry Holt, 1989.

Hampel, Clifford A. *Glossary of Chemical Terms*. 2nd ed. New York: Van Nostrand Reinhold Co., Inc., 1982.

Hand, A.J. *Home Energy How-To*. New York: Harper & Row, 1977.

Handbook of Glass Manufacture. 3rd ed. New York: Ashlee Publishing, 1984.

Hapgood, Charles H. *Maps of the Ancient Sea Kings*. Radnor, PA: Chilton, 1966.

Harding, Anthony. *The Guinness Book of Car Facts and Feats*. London: Guinness Superlatives, 1980.

Harding, Anthony. *The Guinness Book of the Car*. London: Guinness Superlatives, 1987.

Harris, Harry. *Good Old-Fashioned Yankee Ingenuity*. Chelsea, MI: Scarborough House, 1990.

Harte, John. *Toxics A to Z*. Berkeley, CA: University of California Press, 1991.

Hartmann, William K., and Ron Miller. *Cycles of Fire*. New York: Workman Publishing, 1987.

Hawkes, Nigel. *Structures*. New York: Macmillan, 1990.

Hawley's Condensed Chemical Dictionary. 11th ed. New York: Van Nostrand Reinhold, 1987.

Haygreen, John G. *Forest Products and Wood Science*. 2nd ed. Ames, IA: Iowa State University Press, 1989.

Hazen, Robert M., and James Trefil. *Science Matters: Achieving Scientific Literacy*. New York: Anchor Books, 1991.

Headstrom, Richard. *Spiders of the United States*. Stamford, CT: A.S. Barnes, 1973.

Hechtlinger, A. *Modern Science Dictionary*. 2nd ed. Palisade, NJ: Franklin Publ. Co., Inc., 1975.

Hegstad, Lorrie N. *Essential Drug Dosage Calculations*. Bowie, MD: R.J. Brady, 1983.

Heintzelman, Donald S. *A Guide to Eastern Hawk Watching*. University Park, PA: Pennsylvania State University Press, 1976.

Heloise Hints for a Healthy Planet. New York: Perigee, 1990.

Henrickson, Charles H. *Chemistry for the Health Professions*. New York: Van Nostrand, 1980.

Herbert, Don. *Mr. Wizard's Experiments for Young Scientists*. New York: Doubleday, 1959.

Hillman, Harold. *Kitchen Science*. Rev. ed. Boston, MA: Houghton Mifflin Company, 1989.

Hiscox, Gardner D. *Henley's Twentieth Century Book of Formulas, Processes, and Trade Secrets*. New York: NY Books, Inc., 1963.

Hofstadter, Douglas R. *Gödel, Escher, Bach: An Eternal Golden Braid*. New York: Vintage Books, 1979.

Hogg, Ian V. *The Illustrated Encyclopedia of Artillery*. London: Stanley Paul & Co., 1987.

Hooper, Meredith. *Everyday Inventions*. London: Angus & Robertson, 1972.

Hooper, Meredith. *More Everyday Inventions*. London: Angus and Robertson, 1974.

Hopkins, Jeanne. *Glossary of Astronomy and Astrophysics*. Chicago, IL: University of Chicago Press, 1976.

Hopkins, Nigel J., John W. Mayne, and John R. Hudson. *The Numbers You Need*. Detroit, MI: Gale Research, 1992.

Horton, Edward. *The Illustrated History of the Submarine*. London: Sidgwick & Jackson, 1974.

How in the World. Pleasantville, NY: The Reader's Digest Association, 1990.

How Things Work in Your Home. New York: Holt, Rinehart, and Winston, 1985.

How Things Work: Structures. New York: Time-Life Books, 1991.

How to Do Just About Anything. Pleasantville, NY: Reader's Digest Association, 1986.

Howard, A.V. *Chamber's Dictionary of Scientists*. London: W. & R. Chambers, Ltd., 1955.

Howes, F.N. *A Dictionary of Useful and Everyday Plants and Their Common Names*. Cambridge, Eng.: Cambridge University Press, 1974.

Hudgeons, Marc. *The Official Investors Guide; Buying, Selling Gold, Silver, Diamonds*. Orlando, FL: House of Collectibles, 1981.

Hunnicutt, R.P. *Sherman: A History of the American Medium Tank*. San Rafeal, CA: Taurus Enterprises, 1978.

Hunt, V. Daniel. *The Gasohol Handbook*. New York: Industrial Press, 1981.

Hunter, Linda Mason. *The Healthy Home*. New York: Pocket Books, 1989.

Huxley, Thomas H. *The Crayfish*. Cambridge, MA: MIT Press, 1974.

Hyne, Norman J. *Dictionary of Petroleum Exploration, Drilling & Production*. Tulsa, OK: PennWell, 1991.

Illingworth, Valerie. *The Facts On File Dictionary of Astronomy*. New York: Facts On File, 1979.

Illustrated Dictionary of Botany. Chestnut Ridge, NY: Triune Books, 1979.

The Illustrated Encyclopedia of Wildlife. Lakeville, CT: Grey Castle Press, 1991. 15 vols.

The Illustrated Science and Invention Encyclopedia. International ed. Westport, CT: H.S. Stuttman Publishers, 1983, 23 vols.

Inglis, Andrew F. *Behind the Tube: A History of Broadcasting Technology and Business*. London: Focal Press, 1990.

Inventive Genius. New York: Time-Life Books, 1991.

The Inventive Yankee. Camden, ME: Yankee Books, 1989.

Involuntary Smoking. Washington, DC: U.S. Public Health Service, 1979.

Iver, David F. *Dictionary of Astronomy, Space, and Atmospheric Phenomena*. New York: Van Nostrand Reinhold Co., 1979.

Jackson, Donald C. *Great American Bridges and Dams*. Washington, DC: The Preservation Press, 1988.

Jane's Encyclopedia of Aviation. New York: Portland House, 1989.

Jayne, Kate Lindley, and Claudette Suzanne Mautor. *Living with Potpourri*. New York: Peter Pauper, 1988.

Jelinek, Jan. *The Pictorial Encyclopedia of the Evolution of Man*. London: Hamlyn, 1975.

Jerram, Mike. *The World's Classic Aircraft*. London: Frederick Muller, Ltd., 1981.

Jerrard, H.G. *A Dictionary of Scientific Units*. 4th ed. New York: Chapman and Hall, 1980.

Jespersen, James. *RAMS, ROMS and Robots*. New York: Atheneum, 1984.

Johnson, Leland G. *Biology*. Dubuque, IA: Wm. C. Brown, 1983.

Jones, Julia, and Barbara Deer. *Royal Pleasures and Pastimes*. Devon, Eng.: David and Charles, 1990.

Kane, Joseph N. *Famous First Facts*. 4th ed. New York: Wilson, 1981.

Kaplan, Eugene H. *Field Guide to Coral Reefs*. Boston, MA: Houghton Mifflin Co., 1982.

Karlen, Arno. *Napoleon's Glands*. Boston, MA: Little, Brown, 1984.

Kaufman, Wallace. *The Beaches Are Moving*. New York: Anchor Press, 1979.

Keeler, Harriet L. *Our Early Wild Flowers*. New York: Charles Scribner's Sons, 1916.

Kemp, Peter. *Encyclopedia of Ships and Sailing*. Dobbs Ferry, NY: Stanford Maritime, 1989.

Kendig, Frank, and Richard Hutton. *Life-Spans*. New York: Holt, Rinehart and Winston, 1979.

Kennedy, Kenneth A.R. *Neanderthal Man*. Minneapolis, MN: Burgess, 1975.

Kerrod, Robin. *The Concise Dictionary of Science*. New York: Arco Publ. Inc., 1985.

Klinowska, Margaret. *Dolphins, Porpoises and Whales of the World*. Gland, Switz.: IUCN, 1991.

Krantz, Les. *The Best and Worst of Everything*. New York: Prentice Hall General Reference, 1991.

Kress, Stephen W. *The Audubon Society Guide to Attracting Birds*. New York: Charles Scribner's Sons, 1985.

Kroschivitz, Jacqueline I. *Chemistry*. New York: McGraw-Hill, 1990.

Labatut, Jean, and J.L. Wheaton. *Highways in Our National Life*. Princeton, NJ: Princeton University Press, 1950.

Lambert, David. *Field Guide to Early Man*. New York: Facts On File, 1987.

Lane, Ferdinand C. *Earth's Grandest Rivers*. New York: Doubleday, 1949.

Lawrence, Eleanor. *Henderson's Dictionary of Biological Terms*. 10th ed. New York: John Wiley and Sons, 1989.

Lean, Geoffrey, Don Hinrichsen, and Adam Markham. *WWF Atlas of the Environment*. Boston, MA: Willard Grant Press, 1991.

LeBlanc, Raymond. *Gold-Leaf Techniques*. Cincinnati, OH: ST Publications, 1986.

Lee, Sally. *Predicting Violent Storms*. New York: Franklin Watts, 1989.

Leet, L. Don, and Sheldon Judson. *Physical Geology*. 4th ed. Englewood Cliffs, NJ: Prentice-Hall, 1971.

Leggett, Jeremy. *Global Warming*. Oxford, Eng.: Oxford University Press, 1990.

Leopold, Luna B. *Water*. New York: Time, 1966.

Levy, Richard C. *The Inventor's Desktop Companion*. Detroit, MI: Visible Ink Press, 1991.

Lincoln, John W. *Driving Without Gas*. Pownal, VT: Garden Way, 1980.

Lincoln, R.J. *A Dictionary of Ecology, Evolution, and Systematics*. New York: Cambridge University Press, 1982.

Living Invertebrates. Palo Alto, CA: Blackwell Scientific Publications, 1987.

Lloyd, Elizabeth J. *Enchanted Circles*. New York: Simon and Schuster, 1991.

Logan, Carolynn. *Logan's Medical and Scientific Abbreviations*. Philadelphia, PA: Lippincott, 1987.

Longley, Dennis, and Michael Shain. *Van Nostrand Reinhold Dictionary of Information Technology*. 3rd. ed. New York: Van Nostrand Reinhold, 1989.

Loomer, Alice. *Famous Flaws*. New York: Macmillan, 1976.

Mabberley, D.J. *The Plant Book*. New York: Cambridge University Press, 1987.

Macauley, David. *The Way Things Work*. Boston, MA: Houghton Mifflin Co., 1988.

MacEachern, Diane. *Save Our Planet*. New York: Dell, 1990.

Maclean, Norman. *Dictionary of Genetics and Cell Biology*. New York: New York University Press, 1987.

Macmillan Illustrated Animal Encyclopedia. New York: MacMillan, 1984.

Maerz, A. *A Dictionary of Color*. 2nd ed. New York: McGraw-Hill, 1950.

Magill, Frank N. *Great Events from History II: Science and Technology Series*. Englewood Cliffs, NJ: Salem Press, 1991. 5 vols.

Magill, Frank N. *Magill's Survey of Science: Earth Science Series*. Englewood Cliffs, NJ: Salem Press, 1990.

Magill, Frank N. *Magill's Survey of Science: Life Science Series*. Englewood Cliffs, NJ: Salem Press, 1991. 6 vols.

Magill, Frank N. *Magill's Survey of Science. Physical Science Series*. Englewood Cliffs, NJ: Salem Press, 1992. 6 vols.

Magill, Frank N. *Magill's Survey of Science: Space Exploration Series*. Englewood Cliffs, NJ: Salem Press, 1989.

Magill, Frank N. *Nobel Prize Winners, Physiology or Medicine*. Englewood Cliffs, NJ: Salem Press, 1991.

Maginley, C.J. *Models of America's Past*. San Diego, CA: Harcourt, Brace & World, 1969.

Magner, Lois N. *A History of the Life Sciences*. New York: Marcel Dekker, 1979.

The Making, Shaping and Treating of Steel. 10th ed. Pittsburgh, PA: United States Steel, 1985.

Managing the Future of America's Forests (Pamphlet). Washington, DC: American Forest Council, 1989.

Manchester, William R. *The Arms of Krupp, 1587-1968*. Boston, MA: Little Brown, 1968.

Manko, Howard H. *Solders & Soldering*. 2nd ed. New York: McGraw-Hill, 1979.

Mansfield, George Rogers. *Origin of the Brown Mountain Light in North Carolina*. Washington, DC: U.S. Geological Survey, 1971.

The Map Catalog. New York: Vintage, 1986.

Margo. *Growing New Hair*. Brookline, MA: Autumn Press, 1980.

Margulis, Lynn. *Five Kingdoms*. New York: W.H. Freeman and Company, 1988.

Mark's Standard Handbook for Mechanical Engineers. 9th ed. New York: McGraw-Hill, 1987.

Marshall Cavendish Illustrated Encyclopedia of Family Health. London: Marshall Cavendish, 1984. 24 vols.

Marshall Cavendish International Wildlife Encyclopedia. London: Marshall Cavendish, 1989. 24 vols.

Marshall, John. *The Guinness Railway Book*. London: Guinness Books, 1989.

Marshall, John. *Rail: The Records*. London: Guinness Books, 1985.

Matthews, L. Harrison. *The Life of Mammals*. New York: Universe Books, 1971.

Matthews, Rupert O. *The Atlas of Natural Wonders*. New York: Facts On File, 1988.

May, John. *The Greenpeace Book of Antarctica: A New View of the Seventh Continent*. 1st ed. New York: Doubleday, 1989.

May, John. *The Greenpeace Book of the Nuclear Age*. New York: Pantheon, 1989.

Mayhew, Susan, and Anne Penny. *The Concise Oxford Dictionary of Geography*. New York: Oxford University Press, 1992.

Mayo Clinic Family Health Book. New York: Morrow, 1990.

McAleer, Neil. *The Body Almanac*. New York: Doubleday, 1985.

McAleer, Neil. *The OMNI Space Almanac*. New York: World Almanac, 1987.

McCarthy, Eugene J., et al. *The Second Opinion Handbook*. New York: Nick Lyons Books, 1987.

McElroy, Thomas P. *The New Handbook of Attracting Birds*. New York: Knopf, 1960.

McEwan, W.A., and A.H. Lewis. *Encyclopedia of Nautical Knowledge*. Centreville, MD: Cornell Maritime Press, 1953.

McGraw-Hill Dictionary of Physics and Mathematics. New York: McGraw-Hill Book Co., 1978.

The McGraw-Hill Dictionary of Scientific and Technical Terms. 4th ed. New York: McGraw-Hill, 1989.

McGraw-Hill Encyclopedia of Science and Technology. 7th ed. New York: McGraw Hill, Inc., 1992. 20 vols.

McKinnell, Robert G. *Cloning of Frogs, Mice, and Other Animals*. Rev. ed. Minneapolis, MN: University of Minnesota Press, 1985.

McNeil, Ian. *An Encyclopedia of the History of Technology*. London: Routledge, 1990.

Means Illustrated Construction Dictionary. Kingston, MA: R.S. Means, 1985.

Medawar, P.B., and J.S. Medawar. *Aristotle to Zoos*. Cambridge, MA: Harvard University Press, 1983.

Medicine. New York: Time-Life Books, 1991.

Melaragno, Michele. *An Introduction to Shell Structures*. New York: Van Nostrand Reinhold, 1991.

Melonakos, K. *Saunders Pocket Reference for Nurses*. Philadelphia, PA: Saunders, 1990.

Memmler, Ruth L. *Structure and Function of the Human Body*. Philadelphia, PA: Lippincott, 1987.

Menzel, Donald H., and Jay M. Pasachoff. *A Field Guide to Stars and Planets*. 2nd ed. Boston, MA: Houghton Mifflin Co., 1990.

Merck Manual of Diagnosis and Therapy. 15th ed. West Point, PA: Merck, Shark & Dohme, 1987.

Michard, Jean-Guy. *The Reign of the Dinosaurs*. New York: Harry N. Abrams, Incorporated, 1992.

Mierhof, Annette. *The Dried Flower Book*. New York: Dutton, 1981.

Milestones of Aviation. Washington, DC: Smithsonian Institution, 1989.

Miller, E. Willard. *Environmental Hazards: Toxic Waste and Hazardous Material*. Santa Barbara, CA: ABC-CLIO, 1991.

Miller, Ron, and William K. Hartmann. *The Grand Tour: A Traveler's Guide to the Solar System*. New York: Workman Publishing, 1981.

Millichap, J. Gordon. *Dyslexia as the Neurologist and Educator Read It*. Springfield, IL: Thomas, 1986.

Minerals Yearbook, 1989. Washington, DC: U.S. Bureau of Mines, 1991. 3 vols.

Mondey, David. *The Guinness Book of Aircraft*. London: Guinness, 1988.

Moore, John E. *Submarine Warfare*. Bethesda, MD: Alder & Alder, 1987.

Moore, Patrick, et al. *The Atlas of the Solar System*. New York: Crescent Books with the Royal Astronomical Society, 1990.

Moore, Patrick. *International Encyclopedia of Astronomy*. New York: Orion Books, 1987.

Moore-Landecker, Elizabeth. *Fundamentals of the Fungi*. 3rd ed. Englewood Cliffs, NJ: Prentice Hall, 1990.

Morgan, George W. *Geodesic and Geodetic Domes and Space Structures*. Madison, WI: Sci-Tech Publications, 1985.

Morgans, W.M. *Outlines of Paint Technology*. New York: Halsted Press, 1990.

Morlan, Michael. *Kitty Hawk to NASA*. Shawnee Mission, KS: Bon a Tirer, 1991.

Mosby's Medical, Nursing and Allied Health Dictionary. 3rd ed. St. Louis, MO: Mosby, 1990.

The Motor Gasoline Industry. Washington, DC: U.S. Department of Energy. Energy Information Administration, 1991.

Murmurs of Earth: The Voyager Interstellar Record. New York: Random House, 1978.

MVMA Motor Vehicle Facts and Figures '91. Detroit, MI: Motor Vehicle Manufacturers Association of the United States, 1991.

Naar, Jon. *Design for a Livable Planet*. New York: Harper & Row, Publishers, 1990.

Nassau, Kurt. *Gems Made by Man*. Radnor, PA: Chilton, 1980.

The National Inventors Hall of Fame. Washington, DC: U.S. Patent and Trademark Office, 1990.

National Safety Council Accident Facts, 1989. Washington, DC: National Safety Council, 1989.

Nature on the Rampage. Washington, DC: National Geographic Society, 1986.

Nayler, Joseph L. *Aviation: Its Technical Development*. Chester Springs, PA: Dufour Editions, 1965.

Nebel, Bernard J. *Environmental Science*. Englewood Cliffs, NJ: Prentice-Hall, 1990.

The New American State Papers, Science and Technology, vol. 4: Patents. Wilmington, DE: Scholarly Resources, 1973.

The New Book of Popular Science. Danbury, CT: Grolier, 1988. 6 vols.

A New Dictionary of Physics. Bristol, Eng.: Longman Group, 1975.

New Encyclopaedia Britannica. 15th ed. Chicago, IL: Encyclopaedia Britannica, 1990. 29 vols.

The New Good Housekeeping Family Health and Medical Guide. New York: Hearst Books, 1989.

New Illustrated Science and Invention Encyclopedia. Westport, CT: Stuttman, 1988. 23 vols.

New York Public Library Desk Reference. New York: Webster's New World, 1989.

New York Times Book of Indoor and Outdoor Gardening Questions. New York: Quadrangle, 1975.

Newmark, Joseph. *Mathematics as a Second Language*. 4th ed. Redding, MA: Addison-Wesley, 1987.

Newton, Michael. *Armed and Dangerous*. Cincinnati, OH: Writer's Digest Books, 1990.

The Next Step: 50 More Things You Can Do To Save the Earth. Berkeley, CA: Earth Works Group, 1991.

Nichols, Herbert L. *Moving the Earth*. 3rd ed. Greenwich, CT: North Castle Books, 1976.

Nickon, Alex, and Ernest F. Silversmith. *Organic Chemistry: The Name Game*. New York: Pergamon Press, 1987.

Niebel, Benjamin W. *Motion and Time Study*. 7th ed. Homewood, IL: Irwin, 1982.

The 1992 Information Please Environmental Almanac. Boston, MA: Houghton Mifflin Company, 1992.

Nobile, Philip. *Complete Ecology Fact Book*. New York: Doubleday, 1972.

Nock, O.S. *Encyclopedia of Railways*. London: Octopus Books, 1977.

Norman, Bruce. *Secret Warfare: The Battle of Codes and Ciphers*. Reston, VA: Acropolis Books, 1973.

Norman, David. *Dinosaur!* New York: Prentice-Hall, 1991.

Nowak, Robert M. *Walker's Mammals of the World*. 5th ed. Baltimore, MD: The Johns Hopkins University Press, 1991. 2 vols.

The Nuclear Waste Primer. New York: Nick Lyons Books, 1985.

Nunn, Richard V. *Saving Home Energy*. Rev. ed. Birmingham, AL: Oxmoor House, 1978.

O'Brien, Robert. *Machines*. New York: Time Inc., 1964.

The Odds on Virtually Everything. New York: G.P. Putnam's Sons, 1980.

Odum, Eugene. *Fundamentals of Ecology*. 3rd ed. Philadelphia, PA: Saunders, 1971.

Oglesby, Clarkson H., and R. Gary Hicks. *Highway Engineering*. New York: John Wiley & Sons, 1982.

Ojakargas, Richard. *Schaum's Outline of Theory and Problems of Introductory Geology*. New York: McGraw-Hill, 1991.

Ortho's Complete Guide to Successful Gardening. San Ramon, CA: Ortho Books, 1983.

Otto, James H. *Modern Biology*. Fort Worth, TX: Holt, Rinehart and Winston, 1981.

The Oxford Dictionary for Scientific Writers and Editors. Oxford, Eng.: Oxford University Press, 1991.

Oxford Illustrated Encyclopedia of Invention and Technology. Oxford, Eng.: Oxford University Press, 1992.

Pais, Abraham. *'Subtle Is the Lord —'* New York: Oxford University Press, 1982.

Palmer, Ephram Laurence. *Fieldbook of Natural History*. 2nd ed. New York: McGraw-Hill, 1974.

Panati, Charles. *Panati's Browser's Book of Beginnings*. Boston, MA: Houghton Mifflin, 1984.

Panati, Charles. *Panati's Extraordinary Origins of Everyday Things*. New York: Perennial Library, 1987.

Parker, Sybil P. *McGraw-Hill Concise Encyclopedia of Science and Technology*. 2nd ed. New York: McGraw-Hill, 1989.

Parker, Sybil P. *Synopsis & Classification of Living Organisms*. New York: McGraw-Hill, 1982.

Parkinson, Clair L. *Breakthroughs: A Chronicle of Great Achievements in Science and Mathematics*. Boston, MA: G.K. Hall, 1985.

Partington, J.R. *A Short History of Chemistry*. 3rd ed. New York: Dover, 1989.

Pasachoff, Jay M. *Contemporary Astronomy*. Philadelphia, PA: W.B. Saunders, Co., 1977.

Passarin, d'Entreves P. *The Secret Life of Insects*. New York: Chartwell, 1976.

Passport to World Band Radio. Penn's Park, PA: International Broadcasting Services, 1992.

Pawley, Martin. *Building for Tomorrow: Putting Waste to Work*. San Francisco, CA: Sierra Club Books, 1982.

Pawley, Martin. *Garbage Housing*. New York: Halsted, 1975.

Pearce, B.G. *Health Hazards of VDT's*. New York: John Wiley & Sons, 1984.

Pearl, Richard M. *1001 Questions Answered About the Mineral Kingdom*. New York: Dodd, Mead, 1959.

Penny, Malcolm. *Rhinos: Endangered Species*. New York: Facts On File, 1988.

Pennycook, Bob. *Building with Glass Blocks*. New York: Doubleday, 1987.

The Pentagon: A National Institution. Berlin, MD: D'OR Press, 1986.

Peters, George H. *The Plimsoll Line*. London: Barry Rose, Ltd., 1975.

Petroski, Henry. *The Pencil: A History of Design and Circumstance*. New York: Knopf, 1990.

Pettingill, O.E. *Born to Run*. New York: Arco, 1973.

Pfadt, Robert E. *Fundamentals of Applied Entomology*. 3rd ed. New York: Macmillan, 1978.

Pfeffer, Pierre. *Predators and Predation*. New York: Facts On File, 1989.

Phipps, William E. *Cremation Concerns*. Springfield, IL: CC Thomas, 1989.

Pickering, James S. *1001 Questions Answered About Astronomy*. New York: Dodd, Mead, 1958.

Pinkney, Cathey. *The Patient's Guide to Medical Tests*. 3rd ed. New York: Facts On File, 1986.

Planetary and Lunar Coordinates for the Years 1984-2000. London: H.M. Stationery Office, 1983.

Plants: Their Biology and Importance. New York: Harper & Row, Publishers, 1989.

Platt, Rutherford. *1001 Questions Answered About Trees*. New York: Dodd, Mead, 1959.

Plumridge, John H. *Hospital Ships and Ambulance Trains*. London: Seeley, 1975.

Plunkett, Edward R. *Folk Name & Trade Diseases*. Stamford, CT: Barrett Book Co., 1978.

Pogue, William R. *How Do You Go to the Bathroom in Space?* New York: Tom Doherty Associates, 1985.

Poirier, René. *The Fifteen Wonders of the World*. New York: Random House, 1961.

Popular Encyclopedia of Plants. New York: Cambridge University Press, 1982.

Power, Rex. *How to Beat Police Radar*. New York: Arco Publishing, 1977.

Practical Botany. Reston, VA: Reston Publishing Co., Inc., 1983.

Press, Frank. *Earth*. 2nd ed. San Francisco, CA: W.H. Freeman, 1978.

Prevention's Giant Book of Health Facts. Emmaus, PA: Rodale Press, 1991.

The Public Health Consequences of Disasters, 1989. Atlanta, GA: Centers for Disease Control, 1989.

Pugh, Anthony. *Polyhedra*. Berkeley, CA: University of California Press, 1976.

Putnam, Robert E. *Builder's Comprehensive Dictionary*. 2nd ed. Carlsbad, CA: Craftsman Book Company, 1989.

Rackensky, Stanley. *Getting Pests to Bug Off*. New York: Crown, 1978.

Raymond, Eric S. *The New Hacker's Dictionary*. Cambridge, MA: The MIT Press, 1991.

Reader's Digest Consumer Advisor. Pleasantville, NY: Reader's Digest Association, 1989.

Reader's Digest Fix-It-Yourself Manual. Pleasantville, NY: Reader's Digest Association, 1978.

Reader's Digest Practical Problem Solver. Pleasantville, NY: Reader's Digest Association, 1991.

Renmore, C.D. *Silicon Chips and You*. New York: Beaufort Books, 1980.

Retallack, Dorothy L. *The Sound of Music and Plants*. Marina del Ray, CA: DeVorss, 1973.

Rheingold, Howard. *Tools for Thought*. New York: Simon & Schuster, 1985.

Rhodes, Frank H.T. *Geology*. New York: Golden Press, 1972.

Rhodes, Richard. *The Making of the Atomic Bomb*. New York: Simon and Schuster, 1986.

Ricciuti, Edward R. *The Devil's Garden: Facts and Folklore of Perilous Plants*. New York: Walker, 1978.

Richardson, Robert O. *The Weird and Wondrous World of Patents*. New York: Sterling, 1990.

Rickard, Teresa. *Barnes & Noble Thesaurus of Physics*. New York: Harper & Row, 1984.

Ringler, Carol Ann. *Are You at Risk?* New York: Facts On File, 1991.

Roberts, Kenneth L. *The Seventh Sense*. New York: Doubleday, 1953.

Roberts, Royston M. *Serendipity: Accidental Discoveries in Science*. New York: John Wiley & Sons, 1989.

Rochester, Jack B. *The Naked Computer*. New York: Morrow, 1983.

Rodale's Illustrated Encyclopedia of Gardening and Landscaping Techniques. Emmaus, PA: Rodale Press, 1990.

Room, Adrian. *Dictionary of Astronomical Names*. New York: Routledge, 1988.

Root, Waverly L. *Food*. New York: Simon and Schuster, 1980.

Rosenberg, Jerry M. *Dictionary of Computers, Data Processing, and Telecommunications*. New York: John Wiley & Sons, 1984.

Ross, Frank Xavier. *The Metric System—Measures for All Mankind*. New York: Phillips, 1974.

Roth, Charles E. *The Plant Observer's Guidebook*. Englewood Cliffs, NJ: Prentice-Hall, Inc., 1984.

Rothenberg, Mikel A. *Dictionary of Medical Terms for the Non-Medical Person*. 2nd ed. Hauppauge, NY: Barron, 1989.

Rovin, Jeff. *Laws of Order*. New York: Ballantine, 1992.

Rush to Burn. Washington, DC: Island Press, 1989.

Sagan, Carl. *Broca's Brain*. New York: Random House, 1979.

Sagan, Carl. *Cosmos*. New York: Random House, 1980.

Sagan, Carl, and Ann Druyan. *Comet*. New York: Random House, 1985.

Sammons, Vivian O. *Blacks in Science and Medicine*. New York: Hemisphere Publishing, 1990.

Sanders, Dennis. *The First of Everything*. New York: Delacorte, 1981.

Sanders, Ti. *Weather*. South Bend, IN: Icarus Press, 1985.

Savitskii, E.M. *Handbook of Precious Metals*. New York: Hemisphere Publishing Corp., 1989.

Sawyer, L.A., and W.H. Mitchell. *The Liberty Ships*. Cambridge, MD: Cornell Maritime Press, 1970.

Schaefer, Vincent J., and John A. Day. *A Field Guide to the Atmosphere*. Boston, MA: Houghton Mifflin Company, 1981.

Schneck, Marcus. *Butterflies*. Emmaus, PA: Rodale Press, 1990.

Schneck, Marcus. *Elephants*. Stamford, CT: Longmeadow Press, 1992.

Schneider, Herman. *The Harper Dictionary of Science in Everyday Language*. New York: Harper & Row, 1988.

Schodek, Danie L. *Landmarks in American Civil Engineering*. Cambridge, MA: MIT Press, 1987.

Schorger, A.W. *The Passenger Pigeon*. Madison, WI: University of Wisconsin Press, 1955.

Schremp, Gerry. *Kitchen Culture*. New York: Pharos, 1991.

Schweighauser, Charles A. *Astronomy from A to Z*. Springfield, IL: Illinois Issues, 1991.

Schweitzer, Glenn E. *Borrowed Earth, Borrowed Time*. New York: Plenum, 1991.

Science and Technology Illustrated. Chicago, IL: Encyclopaedia Britannica, Inc., 1984, 28 vols.

Scientific Quotations: The Harvest of a Quiet Eye. New York: Crane, Russak, 1977.

Scott, John S. *Dictionary of Civil Engineering*. New York: Halsted Press, 1981.

Search for Immortality. New York: Time Life Books, 1992.

Sedenko, Jerry. *The Butterfly Garden*. New York: Villard Books, 1991.

Self, Charles. *Wood Heating Handbook*. Blue Ridge Summit, PA: TAB, 1977.

Selkurt, Ewald E. *Physiology*. 5th ed. Boston, MA: Little, Brown, 1984.

Shacket, Sheldon R. *The Complete Book of Electric Vehicles*. Northbrook, IL: Domus Books, 1979.

Shafritz, Jay M. *The Facts on File Dictionary of Military Science*. New York: Facts On File, 1989.

Sherman, Irwin W. *Biology: A Human Approach*. 4th ed. New York: Oxford University Press, 1989.

Sherwood, Gerald E., and Robert C. Stroh. *Wood-Frame House Construction*. Rev. ed. Washington, DC: U.S. Department of Agriculture. Forest Service, 1989.

Shipley, Robert M. *Dictionary of Gems and Gemology*. 6th ed. Santa Monica, CA: Gemological Institute of America, 1974.

Shipman, James T., and Jerry D. Wilson. *An Introduction to Physical Science*. 6th ed. Lexington, MA: D.C. Heath and Company, 1990.

Shore, John. *The Sachertorte Algorithm*. New York: Viking Penguin, Inc., 1985.

Shores, Christopher F. *Fighter Aces*. London: Hamlyn, 1975.

Siegman, Gita. *Awards, Honors & Prizes*. 9th ed. (1991-92). Detroit, MI: Gale Research Inc., 1991.

Sikorsky, Robert. *How To Get More Miles Per Gallon in the 1990's*. Blue Ridge Summit, PA: TAB, 1991.

Silverman, Sharon H. *Going Underground*. Philadelphia, PA: Camino Books, 1991.

Simon, Gilbert. *The Parent's Pediatric Companion*. New York: Morrow, 1985.

Sinclair, Ian R. *The HarperCollins Dictionary of Computer Terms*. New York: HarperPerennial, 1991.

Singleton, Paul, and Diana Sainsbury. *Dictionary of Microbiology*. New York: John Wiley and Sons, 1978.

Sinnes, A. Cort. *All About Perennials*. San Ramon, CA: Ortho Books, 1981.

Skinner, Brian J. *The Dynamic Earth*. New York: John Wiley & Sons, 1989.

Smallwood, Charles A., et al. *The Cable Car Book*. Berkeley, CA: Celestial Arts, 1980.

Smith, Anthony. *The Body*. New York: Viking Press, 1986.

Smith, Marcia. *Space Activities of the United States and Other Launching Countries/Organizations: 1957-1991*. Washington, DC: Library of Congress, Science Policy Research Division, 1992.

Smith, Michael D. *All About Bulbs*. San Ramon, CA: Ortho Books, 1986.

Smith, Richard Furnald. *Chemistry for the Million*. New York: Charles Scribner's Sons, 1972.

Solomon, Eldra Pearl, and Gloria A. Phillips. *Understanding Human Anatomy and Physiology*. Philadelphia, PA: W.B. Saunders Company, 1987.

Space Flight: The First 30 Years. Washington, DC: NASA, 1991.

Spangenburg, Ray, and Diane Moser. *Space People From A-Z*. New York: Facts On File, 1990.

Spar, Jerome. *The Way of the Weather*. Mankato, MN: Creative Educational Society, 1967.

Stacey, Tom. *The Hindenberg*. San Diego, CA: Lucent, 1990.

Stamper, Eugene. *Handbook of Air Conditioning, Heating and Ventilating*. New York: Industrial Press, 1979.

Standard Handbook for Civil Engineers. 3rd ed. New York: McGraw-Hill, 1983.

Starr, Cecie, and Ralph Taggart. *Biology*. 6th ed. Belmont, CA: Wadsworth Publ. Co., 1992.

Stedman's Medical Dictionary. 25th ed. Baltimore, MD: Williams & Wilkins, 1990.

Stein, Edwin I. *Arithmetic for College Students*. Rev. ed. Needham Heights, MA: Allyn and Bacon, 1961.

Stephens, John H. *The Guinness Book of Structures*. London: Guinness Superlatives, Ltd., 1976.

Stephenson, D.J. *Newnes Guide to Satellite TV*. London: Newnes, 1991.

Stevenson, L. Harold. *The Facts On File Dictionary of Environmental Science*. New York: Facts On File, 1991.

Stiegeler, Stella F. *A Dictionary of Earth Sciences*. Cavaye Place, London: Pan Books, Ltd., 1978.

Stilwell, E. Joseph, et al. *Packaging for the Environment*. New York: AMACOM, 1991.

Stimpson, George. *Information Roundup*. New York: Harper, 1948.

Stokes, Donald. *The Bluebird Book*. Boston, MA: Little, Brown, 1991.

Stories Behind Everyday Things. Pleasantville, NY: Reader's Digest, 1980.

Sutton, Caroline, and Duncan M. Anderson. *How Do They Do That?* New York: Quill, 1982.

Swank, James M. *History of the Manufacture of Iron in All Ages*. New York: Burt Franklin, 1965.

Swartz, Delbert. *Collegiate Dictionary of Botany*. Ridgefield, CT: Ronald Press, 1971.

Taber, Robert W. *1001 Questions Answered About the Oceans and Oceanography*. New York: Dodd, Mead, 1972.

Taber's Cyclopedic Medical Dictionary. 16th ed. Philadelphia, PA: Davis, 1989.

Taylor, David. *You & Your Cat*. New York: Alfred Knopf, 1988.

Taylor, Michael J.H., and John W.R. Taylor. *Encyclopedia of Aircraft*. London: Weidenfeld & Nicolson, 1978.

Taylor, Walter H. *Concrete Technology and Practice*. New York: McGraw-Hill, 1977.

Temple, Robert K.G. *The Genius of China*. New York: Simon and Schuster, 1986.

Terres, John K. *The Audubon Society Encyclopedia of North American Birds*. New York: Wings Book, 1991.

Thermal and Sound Control. Valley Forge, PA: Certainteed Corporation, 1991.

Thomas, Robert B. *The Old Farmer's Almanac 1988*. Dublin, NH: Yankee Publishing, 1989.

Thrush, Paul W. *A Dictionary of Mining, Mineral, and Related Terms*. Washington, DC: U.S. Bureau of Mines, 1968.

Thygerson, Alton L. *First Aid Essentials*. Boston, MA: Jones and Bartlett, 1989.

Tilling, Robert I. *Eruptions of Mount St. Helens*. Rev. ed. Washington, DC: U.S. Department of the Interior, 1990.

Tilling, Robert I. *Volcanoes*. Washington, DC: U.S. Geological Survey, 1992.

The Timetable of Technology. San Diego, CA: Harvest Books, 1982.

Toothill, Elizabeth. *The Facts On File Dictionary of Biology*. New York: Facts On File Publications, 1988.

Tortora, G.J. *Principles of Anatomy and Physiology*. 4th ed. New York: Harper & Row, 1984.

Toxics in the Community. Washington, DC: U.S. Environmental Protection Agency, 1990.

Traffic Engineering Handbook. 2nd ed. Washington, DC: Institute of Traffic Engineers, 1959.

Trask, Maurice. *The Story of Cybernetics*. London: Studio Vista, 1971.

Treasures of the Tide. Vienna, VA: National Wildlife Federation, 1990.

Trefil, James. *1001 Things Everyone Should Know About Science*. New York: Doubleday, 1992.

Tufty, Barbara. *1001 Questions About Earthquakes, Avalanches, Floods and Other Natural Disasters*. New York: Dover Publications, Inc., 1978.

Tufty, Barbara. *1001 Questions Answered About Hurricanes, Tornadoes and Other Natural Air Disasters*. New York: Dover Publications, Inc., 1987.

Tunnell, James E. *Latest Intelligence*. Blue Ridge Summit, PA: TAB, 1990.

Tyning, Thomas F. *A Guide to Amphibians and Reptiles*. Boston, MA: Little, Brown & Co., 1990.

Tzimopoulos, Nicholas D., et al. *Modern Chemistry*. Fort Worth, TX: Holt, Rinehart and Winston, 1990.

Understanding Computers: Computer Languages. New York: Time-Life Books, 1986.

Understanding Computers: Illustrated Chronology and Index. New York: Time-Life Books, 1989.

U.S. Bureau of Mines. *Bulletin No. 42, 1913*. Washington, DC: U.S. Bureau of Mines, 1913.

U.S. Department of Commerce. Patent and Trademark Office. *General Information Concerning Patents*. Washington, DC: U.S. Department of Commerce, 1992.

U.S. Department of the Army. Headquarters. *Carpenter*. Washington, DC: U.S. Department of the Army. Headquarters, 1971.

U.S. Department of Health and Human Services. *Marijuana* (Pamphlet). Washington, DC: U.S. Dept. of Health and Human Services, 1984.

U.S. Department of the Interior. *The Story of the Hoover Dam*. Washington, DC: U.S. Department of the Interior, 1971.

U.S. Fish and Wildlife Service. *Backyard Bird Feeding*. Washington, DC: U.S. Fish and Wildlife Service, 1989.

U.S. Geological Survey. *Our Changing Continent* (Pamphlet). Washington, DC: U.S. Geological Survey, 1991.

The Universal Almanac 1992. Kansas City, MO: Andrews and McMeel, 1991.

Universal Healthcare Almanac. Phoenix, AZ: Silver & Cherner, Ltd., 1990.

Van Amerogen, C. *The Way Things Work Book of the Body*. New York: Simon and Schuster, 1979.

Van Andel, Tjeerd H. *New Views on an Old Planet*. New York: Cambridge University Press, 1985.

Van der Leeden, Frits. *The Water Encyclopedia*. 2nd ed. Chelsea, MI: Lewis, 1990.

Van Nostrand's Scientific Encyclopedia. 7th ed. New York: Van Nostrand Reinhold, 1989. 2 vols.

Vare, Ethlie Ann. *Mothers of Invention*. New York: Morrow, 1988.

Vergara, William C. *Science in Everyday Life*. New York: Harper & Row, Publishers, 1980.

Versatility of Trucks. Detroit, MI: Motor Vehicle Manufacturers Association, 1991.

Voelker, William. *The Natural History of Living Mammals*. Medford, NJ: Plexus, 1986.

Walkowicz, Chris. *The Complete Question and Answer Book on Dogs*. New York: Dutton, 1988.

Wallace, Irving. *The Book of Lists #2*. New York: Morrow, 1980.

Wallechinsky, David. *The Book of Lists*. New York: Morrow, 1977.

Warm House, Cool House. Yonkers, NY: Consumers Reports Books, 1991.

Weapons: An International Encyclopedia from 5000 B.C. to 2000 A.D. New York: St. Martin's Press, 1990.

Webster, John G. *Encyclopedia of Medical Devices and Instrumentation*. New York: John Wiley & Sons, 1988.

Webster's New Geographical Dictionary. Springfield, MA: Merriam-Webster, 1988.

Webster's Ninth New Collegiate Dictionary. Springfield, MA: Merriam-Webster, Inc., Publishers, 1989.

Weider, Ben, and David Hapgood. *The Murder of Napoleon*. New York: Congdon & Lattès, Inc., 1982.

Weik, Martin H. *Communications Standard Dictionary*. New York: Van Nostrand, Reinhold, 1989.

Whittick, Arnold. *Symbols, Signs and Their Meaning*. London: Leonard Hill Books Limited, 1960.

Who Was Who in American History: Science and Technology. Wilmette, IL: Marquis Who's Who, 1976.

Wilford, John N. *The Riddle of the Dinosaur*. New York: Knopf, 1986.

Williams, Gene B. *Nuclear War, Nuclear Winter*. New York: Franklin Watts, 1987.

Williams, Jack. *The Weather Book*. New York: Vintage Books, 1992.

Williams, Michael R. *A History of Computing Technology*. New York: Prentice-Hall, 1985.

Williams, T. Jeff. *Greenhouses*. San Ramon, CA: Ortho Books, 1991.

Williams, Trevor. *A Biographical Dictionary of Scientists*. London: Adam & Charles Black, 1969.

Winburne, John N. *A Dictionary of Agricultural and Allied Terminology*. East Lansing, MI: Michigan State University Press, 1962.

Winkler, Connie. *Careers in High Tech*. Englewood Cliffs, NJ: Prentice Hall Press, 1987.

Winter, Ruth. *A Consumer's Dictionary of Household, Yard and Office Chemicals*. New York: Crown Publishers, Inc., 1992.

Wise, David Burgess. *The Motor Car*. New York: Putnam, 1979.

The Wise Garden Encyclopedia. New York: HarperCollins, 1990.

Wolf, Nancy. *Plastics*. Washington, DC: Island Press, 1991.

Wolke, Robert L. *Chemistry Explained*. Englewood Cliffs, NJ: Prentice-Hall, Inc., 1980.

Wood, Gerald L. *The Guinness Book of Animal Facts and Feats*. 3rd ed. London: Guinness Superlatives, Ltd., 1982.

Woods, Geraldine. *Pollution*. New York: Franklin Watts, 1985.

The World Almanac and Book of Facts 1992. New York: World Almanac, 1991.

World and United States Aviation and Space Records. Washington, DC: National Aeronautic Association of the USA, 1991.

World Book Encyclopedia. Chicago, IL: World Book, 1990. 22 vols.

World Who's Who in Science. Chicago, IL: Marquis - Who's Who, Inc., 1968.

Wragg, David W. *A Dictionary of Aviation.* Reading, Eng.: Osprey, 1973.

Wright, R. Thomas. *Understanding Technology.* South Holland, IL: Goodheart-Willcox, 1989.

Wyatt, Allen L. *Computer Professional's Dictionary.* New York: McGraw-Hill, 1990.

Wyman, Donald. *Wyman's Gardening Encyclopedia.* 2nd ed. New York: Macmillan, 1986.

Wynbrandt, James. *The Encyclopedia of Genetic Disorders and Birth Defects.* New York: Facts On File, 1991.

Wynter, Harriet, and Anthony Turner. *Scientific Instruments.* New York: Charles Scribner's Sons, Inc., 1975.

Yearbook of Science and the Future 1991. Chicago, IL: Encyclopaedia Britannica, Inc., 1990.

Yost, Graham. *Spy Tech.* New York: Facts On File, 1985.

Zahradnik, Jiri. *A Field Guide in Color to the Animal World.* London: Octopus, 1979.

Zakrzewski, Sigmund F. *Principles of Environmental Toxicology.* Washington, DC: American Chemical Society, 1991.

Zim, Herbert S., and Paul R. Shaffer. *Rocks and Minerals.* New York: Golden Press, 1957.

Zimmerman, O.T. *Conversion Factors and Tables.* 3rd ed. Durham, NH: Industrial Research Service, 1961.

Journals and Periodicals

American Forests. Published bi-monthly by American Forestry Association, Box 2000, Washington, DC 20013.

American Health. Published monthly by Reader's Digest Association, Inc., 28 West 23rd St., New York, NY 10010.

American Scientist. Published bi-monthly by Scientific Research Society, Box 13975, 99 Alexander Dr., Research Triangle Park, NC 27709.

American Transporation Builder (now called *Transportation Builder*). Published six times a year by Transportation Builder, American Road & Transportation Builders Association, 501 School St. S.W., Washington, DC 20024.

Astronomy. Published monthly by Kalmbach Publishing Co., P.O. Box 1612, Waukesha, WI 53187.

Audubon Magazine. Published bi-monthly by National Audubon Society, 950 Third Ave., New York, NY 10022.

Automotive Industries. Published monthly by Chilton Co., Chilton Way, Radnor, PA 19089.

Aviation Week and Space Technology. Published weekly by McGraw-Hill, Inc., Aviation Week Group, 1221 Ave. of the Americas, New York, NY 10020.

Buzzworm: The Environmental Journal. Published six times a year by Buzzworm, Inc., 2305 Canyon Blvd., Ste. 206, Boulder, CO 80302.

California Geology. Published monthly by Division of Mines and Geology, 660 Bercut Dr., Sacramento, CA 95814-0131.

Chemical & Engineering News. Published weekly by American Chemical Society, 1155 16th St. N.W., Washington, DC 20036.

Compute. Published monthly by Compute Publications International, Ltd., 324 W. Wendover Ave., Ste 200, Greensboro, NC 27408.

Consumers' Research Magazine. Published monthly by Consumers' Research, Inc., 800 Maryland Ave. N.E., Washington, DC 20002.

Cornell Animal Health Newsletter. Published monthly by W.H. White Publications, 53 Park Place, New York, NY 10007.

Country Journal. Published bi-monthly by Cowles Magazines, Inc., 6405 Flank Dr., Box 8200, Harrisburg, PA 17105-8200.

Current Health. Published monthly, September through May, by General Learning Corporation, Curriculum Innovations Group, 60 Revere Dr., Northbrook, IL 60062-1563.

Discover. Published monthly by Walt Disney Magazine, Publishing Group, 500 S. Buena Vista, Burbank, CA 91521-6012.

Endangered Species Technical Bulletin Reprint. (Now called *Endangered Species Update*). Published monthly by University of Michigan, School of Natural Resources, 430 E. University, Dana Bldg., Ann Arbor, MI 48109-1115.

Environment. Published ten times a year by Heldref Publications, 4000 Albemarle St. N.W., Washington, DC 20016.

EPA Journal. Published bi-monthly by U.S. Environmental Protection Agency, Office of Public Affairs, Waterside Mall, 401 M St. S.W., Washington, DC 20460.

FDA Consumer. Published ten times a year (July-Aug. & Jan.-Feb. issues combined) by U.S. Food and Drug Administration Office of Public Affairs, 5600 Fisher Lane, Rockville, MD 20857.

Facts On File World News Digest with Index. Published weekly by Facts On File, Inc., 460 Park Ave., New York, NY 10016.

Fine Gardening. Published bi-monthly by Taunton Press, Inc., 63 S. Main St., Box 5506, Newtown, CT 06470-5506.

Fine Woodworking. Published bi-monthly by Taunton Press, Inc., 63 S. Main St., Box 5506, Newtown, CT 06470-5506.

Fire Management Notes. Published quarterly by USDA Forest Service, Box 96090, Washington, DC 20090-6090.

Flower and Garden. Published bi-monthly by KC Publishing Inc., 4251 Pennsylvania Ave., Kansas City, MO 64111-9990.

Good Housekeeping. Published monthly by Hearst Corporation, Good Housekeeping, 959 Eighth Ave., New York, NY 10019.

Harrowsmith Country Life. Published bi-monthly by Camden House Publishing, Ferry Rd., Charlotte, VT 05445.

Harvard Health Letter. Published monthly by Harvard Medical School, HMS Health Publications Group, 164 Longwood Ave., 4th Floor, Boston, MA 02115.

Harvard Medical School Health Letter (now called *Harvard Health Letter*). Published monthly by Harvard Medical School, HMS Health Publications Group, 164 Longwood Ave., 4th Floor, Boston, MA 02115.

Health. Published seven times a year by Health Magazine, 275 Madison Ave., Ste. 1314, New York, NY 10016.

Horticulture. Published monthly by Horticulture Limited Partnership, 20 Park Plaza, Ste. 1220, Boston, MA 02116-8241.

Human Behavior (ceased publication). Was published monthly by Manson Western Corp., 12031 Wilshire Blvd., Los Angeles, CA 90025.

The Journal of the American Medical Association. Published weekly by JAMA (The Journal of the American Medical Association), 535 N. Dearborn St., Chicago, IL 60610.

Life. Published monthly by The Time Inc., Magazine Company, Time & Life Bldg., Rockefeller

Center, 1271 Ave. of the Americas, New York, NY 10020.

McCall's. Published monthly by McCall's Magazine, 110 5th Ave., New York, NY 10011.

Mechanix Illustrated (now called *Home Mechanix*). Published monthly by Times Mirror Magazines, Inc., 380 Madison Ave., New York, NY 10017.

Mineral Information Service (now called *California Geology*). Published monthly by Division of Mines and Geology, 1516 Ninth St., Fourth Floor, Sacramento, CA 95814.

Motor Trend. Published monthly by Petersen Publishing Co., 8490 Sunset Blvd. Los Angeles, CA 90069.

National Geographic. Published monthly by National Geographic Society, 17th & M Sts. N.W., Washington, DC 20036.

National Geographic World. Published monthly by National Geographic Society, 17th & M Sts. N.W., Washington, DC 20036.

National Wildlife. Published bi-monthly by National Wildlife Federation, 1400 16th St. N.W., Washington, DC 20036-2266.

Natural History. Published monthly by American Museum of Natural History, Central Park W at 79th St., New York, NY 10024-5192.

Nature and Science (ceased publication). Was published bi-weekly by American Museum of Natural History, Central Park West at 79th St., New York, NY 10024-5124.

Nature Magazine (now incorporated into *Natural History*). Published monthly by American Museum of Natural History, Central Park W at 79th St., New York, NY 10024-5192.

New Scientist. Published weekly by IPC Magazines, Ltd., Holborn Group, King's Reach Tower, Stamford St., London SE9LS, England.

The New York Times. Published daily by The New York Times, 229 W. 43rd St., New York, NY 10036.

Newsweek. Published weekly by Newsweek, Inc., 444 Madison Ave., New York, NY 10022.

Nuclear News. Published monthly by American Nuclear Society, 555 N. Kensington Ave., LaGrange Park, IL 60525.

Parents. Published monthly by Gruner & Jahr, U.S.A. Publishing, 685 Third Ave., New York, NY 10017.

Pennsylvania Forests. Published quarterly by Pennsylvania Forestry Association, 56 E. Main St., Mechanicsburg, PA 17055-3851.

Pennsylvania Woodland News. Published bimonthly by Penn State University, Forest Resources Extension, 110 Ferguson Bldg., University Park, PA 16802.

Physics Today. Published monthly by American Institute of Physics, 335 E. 45th St., New York, NY 10017.

Pittsburgh Press. Published daily by Pittsburgh Press, Box 566, 34 Blvd. of the Allies, Pittsburgh, PA 15230.

The Planetary Report. Published bi-monthly by The Planetary Society, 65 North Catalina Ave., Pasedena, CA 91106.

Popular Mechanics. Published monthly by Hearst Magazines, Popular Mechanics, 224 W. 57th St., New York, NY 10019.

Popular Science. Published monthly by Times Mirror Magazines, Inc., 2 Park Ave., New York, NY 10016.

PRLC Technical Bulletin. Published six times a year by Pittsburgh Regional Library Center, 103 Yost Blvd., Pittsburgh, PA 15221.

Progressive Builder (now called *Custom Builder*). Published bi-monthly by Willows Publishing Group, Inc., 38 Laffayette St., Box 998, Yarmouth, ME 04096-0470.

Ranger Rick. Published twelve times a year by National Wildlife Federation, 1400 16th St. N.W., Washington, DC 20036-2266.

Road & Track. Published monthly by Hachette Magazines, Inc., Road & Track, 1499 Monrovia Ave., Newport Beach, CA 92663.

Science. Published weekly by American Association for the Advancement of Science, 1333 H St. N.W., Washington, DC 20005.

Science [year] (ceased publication). Was published ten times a year by William Carey, publisher, 1333 H St N.W., Washington, DC 20005.

Science Digest (now called *Breakthroughs in Health & Science*). Published monthly by Family Media, Inc., Men's and In-Home Group, 3 Park Ave., New York, NY 10016.

Science News. Published weekly by Science Service, Inc., 1719 N St. N.W., Washington, DC 20036.

Science News-Letter (now called *Science News*). Published weekly by Science Service, Inc., 1719 N St. N.W., Washington, DC 20036.

The Science Teacher. Published nine times a year by National Science Teachers Association, 1742 Connecticut Ave. N.W., Washington, DC 20009.

Scientific American. Published monthly by Scientific American, Inc., 415 Madison Ave., New York, NY 10017.

Sky and Telescope. Published monthly by Sky Publishing Corporation, Box 9111, Belmont, MA 02178.

Smithsonian. Published monthly by Smithsonian Institution, Arts & Industries Bldg., 900 Jefferson Dr., Washington, DC 20560.

Status Report. Published bi-weekly by Insurance Institute for Highway Safety, Watergate 600 Ste. 300, Washington, DC 20037.

Stone. Published irregular by Stone Press 1112-B, Ocean St., Santa Cruz, CA 95060.

Technology and Culture. Published quarterly by Society for the History of Technology, Duke University, Department of History, Durham, NC 27706.

Time. Published weekly by The Time, Inc. Magazine Company, Time & Life Building, Rockefeller Center, 1271 Ave. of the Americas, New York, NY 10020-1393.

Today's Health (incorporated into *Health*). Published monthly by Health Magazine, 275 Madison Ave., Ste. 1314, New York, NY 10016.

USA Today. Published daily by Gannett Co., Inc., P.O. Box 500, Washington, DC 20044.

Vegetarian Times. Published twelve times a year by Vegetarian Times, Inc., Box 570, Oak Park, IL 60303.

Weatherwise. Published bi-monthly by Heldref Publications, 1319 Eighteenth St. N.W., Washington, DC 20036-1802.

Index

Alcohol
    Effects *299*
    Gasohol *143*
    Poisoning *299*
Alcoholic beverages
    Effects *299*
Aldrin, Jr., Edwin E. *50, 52, 54*
Alevins (Young salmon) *245*
Algae (Diatoms) *201*
Algorithm *424*
Alimentary canal *273*
Alkali metals *12*
Alkaline-earth metals *12*
Allergies
    Dust mites *312*
    Poison ivy *312*
Alligators
    Life spans *227*
    Names for male and female *234*
    Sex determination *247*
Alpha Centauri (Star) *22, 28*
Alpha particles *8*
Alphabets
    Blind, Used by the *412*
    International phonetic *412*
    Morse code, International *413*
ALS *See* Lou Gehrig's disease
Altair (Star) *25*
Altocumulus and altostratus clouds *91*
Aluminum *117*
    Alzheimer's disease *295, 314*
    Component in human body *267*
    Health hazards *295*
    Waste, Solid *180, 183*
Aluminum cans *See* Cans
Aluminum packaging in municipal solid waste *183*
Alvarez, Luis and Walter *165*
Alvin (Research vessel) *244, 389*
Alzheimer's disease *314*
AM radio *See* Radio—AM
Amanita phalloides (Mushrooms) *302*
Amazon River
    Length *70*
    Water lilies *217*
Ambergris source *125*
America, North
    Continental divide *79*
    Diamond mine *115*
America, South
    Earliest map *87*

El Nino *159*
American Telephone and Telegraph *421*
American women in space *51, 52*
Ammann, Othmar H. *381*
Ammonia manufacture *127*
Amor asteroids *40*
Ampere, Andre Marie *19*
Amperes
    Lightning *94*
    Origin of term *19*
Amphibian planes *405, 406*
Amphibians *See* Reptiles and amphibians
Amphora (Measurement unit) *337*
AMS (American Society for Metals) geodesic dome *375*
Amundsen, Roald *74*
Amyotrophic Lateral Sclerosis (ALS), *See* Lou Gehrig's disease
Anabolic steroids, Effects of *329*
Analytical chemistry, Founders of *15*
Analytical engine *424, 430*
Androphobia *317*
Angel Falls *71*
Angel, Jimmy *67, 71*
Angstrom, Anders *6*
Angstrom (Measurement unit) *6*
Animals
    Aquatic life *See* Aquatic life
    Birds *See* Birds
    Blood types *232*
    Breath-holding capacity *257*
    Classification *199*
    Cloning *193*
    Color vision *231*
    Disease vectors *305*
    First in space *51*
    First patented *439*
    Fish *See* Fish
    Gestation period *227*
    Health benefits *292*
    Heart rate comparison *257*
    Hibernation *231*
    Hoaxes *447*
    Insects *See* Insects
    Intelligence *231*
    Kingdom category *199*
    Largest *230*
    Life spans *227*
    Mammals *See* Mammals
    Mimicry *191*
    Names for groups *237*
    Names for males and females *234*

Names for the young *235*
Rain showers *106*
Regeneration *232*
Running speed *233*
Smallest *230*
Snoring *234*
Use in carbon monoxide detection *140*
Zoonosis *305*
Anorexia *313*
Antarctica
    Discoverers *74*
    Ice thickness *73*
Anteaters, Reproduction of *258*
Antelopes
    Names for the young *235*
    Speed *233*
Anthophobia *317*
Antibiotics *327, 328*
Antibodies *268, 269, 327, 328*
    Diseases *203*
    See also Monoclonal antibodies
Antigens *329*
    Antigen-antibody reaction *268, 269*
Antimony (Stibnite) *115*
Antimony glance *115*
Antiquarks *9*
Ants
    Comparison to termites *240*
    Life span *227*
    Names for groups *237*
    Names for male and female *234*
    Names for the young *235*
Anus *273*
Anvil bone of ear *290*
Aphelion, Earth *35*
Apollo missions *50, 52, 56*
Appendix, Purpose of *281*
Appetite, Loss of *313*
Apple maggots *223*
Apple trees planted by Johnny Appleseed *226*
Appleton Layer *59*
April, Derivation of name *352*
Aqua regia *127*
Aquatic life
    Corals *244*
    Eel *246*
    Krill *244*
    Marine mammals *See* Marine mammals
    Tube worms, Giant *244*
    Whelks *245*

481

INDEX

J

K

L

497

Phosphorous as component in human body *267*

Photochemical air pollution *See* Smog

Photodegradable plastic *135*

Photography (Holography) *363*

Photons *362*

Photosynthesis *197, 222*

Phototropism *206*

Photovoltaic energy production, U.S. *152*

Phylae of living things *200*

Phyllophobia *317*

Physical effects of spaceflight *48*

Physical science, Founders of *19*

Physicians *319*
    Ancient *320*
    Female *319*
    First black surgeon to perform heart surgery *332*
    Hippocratic Oath *318*
    Illnesses caused by physicians *314*
    Number in U.S. *321*
    Ophthalmologists *321*

Physics, Atomic *See* Atomic physics

Physics, Theoretical, Founders of *16*

Physiology, Founder of *273*

Piazzi, Giuseppe *40*

Pica *316*

Piccard, Jacques *66*

Picture elements *433*

PID *See* Pelvic inflammatory disease

Pigeons
    Homing *251*
    Names for male and female *234*
    Names for the young *235*

Pigments
    Carbon black *133*
    Titanium dioxide *133*

Pigs
    Blood types *232*
    Diseases transmitted to humans *305*
    Gestation period *254*
    Intelligence *231*

Pilkington, Alistair *130*

Piltdown man *445*

Pine trees
    Lifespans *207*
    Products *125*
    Tallest *208*

Pint (Measurement unit) *341*

Pioneer Dollars *113*

Pioneer 3 *60*

Pioneer 10 *36*

Pioneer 11 *36*

Pipe and cigar smoking, Risks of *299*

Piri Re'is map *87*

Pisa, Leaning tower of *374*

Pisano, Bonanno *374*

Pistils *213*

Pitchblende *114*

Pitcher plants (Carnivorous) *214*

Pittsburgh, Bridges in *213*

Pituitary glands, Functions of *282*

Pixels *433*

Plane of the ecliptic *29*

Planes, Air *See* Airplanes

Planes, inclined (Simple machines) *359*

Planet X *36*

Planets
    Color *32*
    Day, Length of *34*
    Diameter *32*
    Distance from sun *31*
    Gravitational force *33*
    Inferior *34*
    Jovian *34*
    Moons *36*
    Naming of surface features *44*
    Revolution, Period of *32*
    Rings *33*
    Rotation *34*
    Superior *34*
    Terrestrial *34*
    Time to orbit sun *32*
    See also names of individual planets, i.e., Jupiter

Planktonic crustaceans (Krill) *244*

Plant patents *439*
    Color drawings *439*

Plant products
    Industrial applications of *440*
    See also Essential oils; Luffa sponges; specific plants, i.e. Pine (trees)

Plantae (Kingdom) *199*

Plantar warts *See* Warts

Plants
    Bulbs, tubers, corms, etc. *213*
    Butterfly attractants *220*
    Carnivorous *214*

    Chloroplasts *205*
    Classification *199, 200*
    Climbing *206*
    Cloning *193*
    Clover with most leaves *216*
    Endangered species (U.S.) *168*
    Energy sources *152*
    Founder of plant science *212*
    Fuel use *See* Biomass energy
    Herbs *See* Herbs
    Hummingbird attractants *220*
    Kingdom category *199*
    Living stones *217*
    Movement in response to stimuli *206*
    Music effect on growth *207*
    Photosynthesis *197*
    Pollination *205*
    Products *See* Plant products
    Stimuli *206*
    Symbolism *215*
    Tendrils *206*
    Threatened species (U.S.) *168*
    Transplanting comparison *219*
    Water-lilies *217*
    Wormwood *217*
    See also Flowers

Plasmas (State of matter) *8*

Plastic surgery, Most frequently performed *331*

Plastics
    Biodegradable *135*
    Glass-fiber-reinforced plastics *131*
    Invention *135*
    Recycling *184*
    Waste, Solid *180, 181, 183*

Plate glass *129*

Plate tectonics *71*

Platelets, Lifespans of *268*

Platinum (Element) *13*

Platypuses
    Breath-holding capacity *257*
    Lifespans *227*
    Reproduction *259*

Pleaching *222*

Pleistocene epoch *85*
    Biological events *187*
    Great Ice Age *74*

Plessors (Plexors) *325*

Plimsoll line or mark *389*

Plimsoll, Samuel *389*

Credits

p. 14 Illustration from *Organic Chemistry: The Name Game*. Copyright © 1986 by the American Chemical Society. Used with permission; pp. 18, 98, 107: Illustrations from *Aviation Weather for Pilots and Flight Operation Personnel*, prepared by U.S. Department of Transportation, Federal Aviation Administration, Flight Standards Service and Department of Commerce, National Oceanic and Atmospheric Administration, National Weather Service, Washington, DC, revised 1975; p. 30: Illustration from *A Field Guide to Stars and Planets*, 3rd edition, by Jay M. Pasachoff and Donald H. Menzel. Houghton Mifflin Company, 1992. Copyright © 1983 by the Estate of Donald H. Menzel and by Jay M. Pasachoff. Reprinted by permission of Jay M. Pasachoff; pp. 38, 92: Illustrations from *The New York Public Library Desk Reference*. Copyright © 1989 by The New York Public Library and The Stonesong Press, Inc.; p. 64: Illustration by Peter Gardiner from *New Scientist*, February 16, 1991. Reprinted by permission of *New Scientist Magazine*; p. 69: Illustration from *A Dictionary of Earth Sciences*, edited by Stella E. Stiegeler, BSc. Pan Books Ltd., 1978. Copyright © 1976 by Laurence Urdang Associates Limited.

pp. 72, 74, 112: Illustrations from *Our Changing Continent*, prepared by the U.S. Geological Survey, U.S. Department of the Interior; pp. 76, 97, 301, 224: Illustrations from *Survival*, prepared by U.S. Department of the Army, Washington, DC. March 26, 1986; pp. 77, 81: Illustrations from *Volcanoes*, by Robert I. Tilling. U.S. Department of the Interior/ U.S. Geological Survey; pp. 78, 84, 366: Illustrations from *What's What: A Visual Glossary of the Physical World*, by Reginald Bragonier, Jr. and David Fisher. Copyright © 1981 by Reginald Bragonier, Jr. and David Fisher. Reprinted by permission of Hammond Incorporated, Maplewood, NJ; p. 82: Illustration from *Eruptions of Mount St. Helens*, by Robert I. Tilling, Lyn Topinka and Donald A. Swanson, prepared by U.S. Department of the Interior/Geological Survey; p. 83 top: Illustration by M. Holford, Loughton from *The Genius of China: 3,000 Years of Science, Discovery, and Invention*, by Robert Temple. Text copyright © 1986 by Robert Temple. Compilation copyright © 1986 by Multimedia Publications (UK) Ltd.; p. 83 bottom: Illustration from *The Golden Guide to Geology*, by Frank H.T. Rhodes. Copyright © 1972 by Western Publishing Company. Illustration by Raymond Perlman. Used by permission; pp. 84, 101, 249: Illustrations from

Environmental Trends, prepared by Executive Office of the President, Council on Environmental Quality, Washington, DC, 1989; p. 128: Illustration from *The Almanac of Science and Technology: What's New and What's Known*, edited by Richard Golob and Eric Brus. Copyright © 1990 by Harcourt Brace Jovanovich, Inc. Reprinted by permission of Florante A. Quiocho, Rice University; p. 152: Illustration from *Annual Energy Review*, 1990, prepared by Energy Information Administration, Office of Energy Markets and End Use, U.S. Department of Energy, Washington, DC.

pp. 159, 213: Illustrations from *The Encyclopedia of Environmental Studies*, by William Ashworth. Copyright © 1991 by William Ashworth. Reprinted with the permission of Facts on File, Inc., New York; p. 169: Illustration from *Environmental Quality*, edited by Dale Curtis and Barry Walden Walsh. Council on Environmental Quality, March 1992; p. 181: National Solid Wastes Management Association, 1990; p. 245: Illustration from The *Science Teacher*, April 1992. Copyright © 1992 by the National Science Teachers Association, 1742 Connecticut Avenue, NW, Washington, DC, 20009. Reproduced by permission of National Science Teachers Association; p. 253: Illustration by Anne Senechal Faust; p. 258: Photograph by Ernest P. Walker; p. 260: Illustration by Betty Fraser from *Everglades Wildguide*, by Jean Craighead George. U.S. Department of the Interior, Washington, DC, 1988. Reproduced by the Division of Publications, U.S. National Park Service; p. 261: Illustration from *An Introduction to the World's Oceans*, 3rd edition, by Alyn C. Duxbury and Alison B. Duxbury. Copyright © 1991 by Wm. C. Brown Communications, Inc., Dubuque, Iowa. All Rights Reserved. Reprinted by permission; p. 262: K.Fink/ARDEA LONDON LTD.

pp. 273 top, 286: Illustrations by Fred Maschall from *First Aid Book*, prepared by U.S. Department of Labor, and Mine Safety and Health Administration, reprinted 1991; pp. 273 bottom, 322, 324, 380: Illustrations by Peter Bull from *The Random House Book of How Things Work*, by Steve Parker. Kingfisher Books, 1990. Copyright © 1990 by Grisewood & Dempsey Ltd. Reprinted by permission of Grisewood & Dempsey Ltd.; p. 280: Illustration from *The University Desk Encyclopedia*. Copyright © 1977 by Elsevier Publishing Projects S.A., Lausanne. Reprinted by permission of Elsevier Trading & Copyrights; pp. 290, 318 right: Illustrations from *Harver Ready Reference Encyclopedia*, International Edition. Illustrations copyright © 1969 by N.V. Uitgeversmaatschappij Elsevier, Amsterdam-Brussels. Reprinted by permission of Elsevier Science Publishers BV Academic Publishing Division; p. 294: Illustration from *The Injury Fact Book*, by Susan P. Baker, Brian O'Neill, Marvin J. Ginsburg and Guohua Li. Copyright © 1992 by Oxford University Press, Inc. Reprinted by permission of Oxford University Press, Inc.; p. 298: Illustration from *Eating to Lower Your High Blood Cholesterol*, prepared by U.S. Department of Health and Human Services, Public Health Services, National Institutes of Health, reprinted June 1989; p. 303: Illustration from *Van Nostrand's Scientific Encyclopedia*, edited by Douglas M. Considine, P.E. Copyright © 1983 by Van Nostrand Reinhold Company, Inc. Illustration by A.M. Winchester.

p. 309: Illustration from *Working Safely With Video Display Terminals*, prepared by U.S. Department of Labor, and Occupational Safety and Health Administration, revised 1991; p. 359: Cover illustration from *The Bulletin of the Atomic Scientists*, Volume 47, No. 10, December 1991. Reproduced by permission of *The Bulletin of the Atomic Scientists*, 6042 South Kimbark Avenue, Chicago, IL 60637, USA. A one-year subscription is $30; p. **522** 360: Illustration by Neil Bulpit from *The Random House Book of How Things Work*, by

Steve Parker. Kingfisher Books, 1990. Copyright © 1990 by Grisewood & Dempsey Ltd. Reprinted by permission of Grisewood & Dempsey Ltd.; pp. 370 top, 376: Illustrations from *The Macmillan Encyclopedia of Architecture and Technological Change*, by Pedro Guedes. Copyright © 1979 by Reference International Publishers Ltd.; pp. 370 bottom, 371: Illustrations from *Wood-Frame House Construction*, by Gerald E. Sherwood, PE, U.S. Department of Agriculture, Forest Service, Forest Products Laboratory, Madison, WI and Robert C. Stroh, PhD, National Association of Home Builders, National Research Center, Washington, D.C., United States Department of Agriculture, revised September 1989; p. 373: Illustration from *Technical Manual: Carpenter*, prepared by Headquarters, U.S. Department of the Army, Washington, D.C., July 19, 1971; p. 378: Courtesy of Federal Highway Administration; p. 420: Illustration from *The Home Satellite TV Installation and Troubleshooting Manual*, by Frank Baylin and Brent Gale. Copyright © 1985 by Frank Baylin and Brent Gale. Reprinted by permission of Baylin Publications, (303) 449-4551

p. 423: Illustration from *The New Book of Popular Science*, Volume 6. Grolier Inc., 1987. Copyright © 1987 by Grolier Incorporated. Reprinted by permission; p. 425: Illustration from *What's What: A Visual Glossary of the Physical World*, by Reginald Bragonier, Jr., and David Fisher. Copyright © 1981 by Reginald Bragonier, Jr., and David Fisher. Illustration copyright © 1981 by International Business Machines Corporation. Reprinted by permission of International Business Machines Corporation; p. 437: © Rube Goldberg from King Features Syndicates, Inc., 1978. Reprinted with special permission of King Features Syndicates; p. 440: Illustration from United States Patents, Patent Number 6,469, May 22, 1849.